Instrument Technology

Volume 1

Dedicated to
P.R.H.

Instrument Technology
Volume 1

MEASUREMENT OF PRESSURE,
LEVEL, FLOW AND TEMPERATURE

E. B. JONES

B.Sc., F.Inst.P., F.Inst.M.C.

BUTTERWORTHS

London–Boston–Durban–Singapore–Sydney–Toronto–Wellington

First published 1953
  Revised/reprinted 1960
  Reprinted 1961, 1964
Second edition 1965
  Reprinted 1970, 1972
Third edition 1974
  Reprinted 1976, 1979, 1981, 1983

© Butterworth & Co (Publishers) Ltd, 1974

ISBN 0 408 70535 3

Printed and bound in Great Britain by
Redwood Burn Limited, Trowbridge, Wiltshire

PREFACE TO THE THIRD EDITION

The enormous and rapid increase in the measurement, recording and control of critical variables in industrial processes has led to a corresponding expansion of the staffs responsible for the installation and maintenance of instruments which, incidentally, represent considerable investment of capital.

To gain the full advantages of instrumentation, the instruments should be installed and maintained by those who bring understanding as well as manual skill to their work. This book is written with the object of helping the reader to understand the 'why' as well as the 'how' of his work. It is presumed that he has some acquaintance with physics, but to help those whose knowledge of physics is small, some basic principles are stated at the beginning of each section.

The mathematics have been kept as simple as possible to avoid embarrassing students whose attainments in mathematical subjects are limited.

The selection of material has been a difficult problem as the subject is so wide, but the aim has been to give as complete a picture as possible while emphasising the more important and the more common types of instruments.

It is hoped that this volume together with its companion volumes will cover the requirements of students studying both craftsmen and technician courses of the City and Guilds of London, and instrument personnel studying the courses established by the various industrial training boards.

The usefulness of the book is not, however, limited to candidates for the examination courses mentioned above, but it is expected that instrument and chemical engineers and others, will find a great deal of useful information within its covers which will help them solve the instrument problems which may occur during their training and their daily work.

The introduction of SI units, development of new instruments, and the new standards required for the use of electrical instruments where an explosive hazard exists, has made it essential to produce this third edition. Many changes are taking place as we seek to align ourselves with the practices followed on the continent of Europe. Many of our existing instrument standards will change as the European influence increases, but although much of our practice may change, the fundamental principles described in this book will remain valid.

The author cannot adequately express his appreciation of the considerable help received from his colleagues past and present, and manufacturers and users of industrial instruments, in bringing this work into being.

The instrument documents of the British Standards Institution and of B.A.S.E.E.F.A. have been a valuable source of information and due acknowledgement is made to these bodies. The documents referred to are too numerous to mention here but extracts are noted in the text.

In conclusion, the writer acknowledges with gratitude the assistance of his wife in undertaking the typing and the multitude of other tasks associated with this work.

Holywell

E. B. JONES

CONTENTS

INTRODUCTION

Metric units

The system of units of measurement used in all three volumes is the internationally agreed 'Système International d'Unités' usually abbreviated to SI units. This system is closely associated with the metric system and the starting point is six base units, each already precisely defined, from which all other units are derived. As the SI system is a coherent system any derived unit is the result of the product or quotient of two or more base units, e.g. unit area results when unit length is multiplied by unit length, unit velocity when unit length is divided by unit time and unit force when unit mass is multiplied by unit acceleration. The base units are given in Table 1.

Table 1

Quantity	Unit	Symbol
Length	metre	m
Mass	kilogramme	kg
Time	second	s
Electric current	ampere	A
Thermodynamic temperature	kelvin	K
Luminous intensity	candela	cd

In addition there are supplementary units for plane angle, the radian (rad) and solid angle, the steradian (sr). The units Kelvin (K) and degree Celsius (°C) of temperature interval are identical and the International Practical Temperature Scale is used for all practical purposes. A temperature expressed in degrees Celsius is equal to the temperature expressed in Kelvin less 273·15.

Derived units

The derived unit of force is the kilogramme metre per second per second and is known as the newton (N). The newton is that force which when applied to a body having a mass of 1 kg gives it an acceleration of one metre per second per second and is therefore independent of the gravitational force.

The derived unit of pressure is the pressure produced when a force of one newton acts over an area of a square metre, i.e. N/m^2. Because this unit is very small, 100 000 N/m^2 being equal to approximately 1 atmosphere, the bar = 10^5 N/m^2 and m bar are used as the practical units of pressure. In order to avoid errors in calculations it is advisable to use the basic coherent SI units.

Owing to the convenience and ease of use of the simple manometer, pressures on industrial plant will still be measured where appropriate in terms of the height of a column of liquid. The column of liquid, however, can be calibrated in terms of SI units although it will be some time before the use of lengths of liquid column will cease to be used to express the values of pressures which can be conveniently measured by manometers.

The unit of energy in all forms is the joule, and of power is the watt. Thus the variously defined units such as the calorie, British Thermal Unit and horsepower are all superseded.

For some of the derived SI units special names and symbols exist, these are shown in Table 2.

Table 2

Quantity	Name of SI unit	Symbol	Expressed in terms of SI base units or derived units
Frequency	hertz	Hz	$1\ \mathrm{Hz} = 1/s$
Force	newton	N	$1\ \mathrm{N} = 1\ \mathrm{kg\ m/s^2}$
Work, energy, quantity of heat	joule	J	$1\ \mathrm{J} = 1\ \mathrm{N\ m}$
Power	watt	W	$1\ \mathrm{W} = 1\ \mathrm{J/s}$
Quantity of electricity	coulomb	C	$1\ \mathrm{C} = 1\ \mathrm{A\ s}$
Electric potential, potential difference, tension, electromotive force	volt	V	$1\ \mathrm{V} = 1\ \mathrm{W/A}$
Electric capacitance	farad	F	$1\ \mathrm{F} = 1\ \mathrm{A\ s/V}$
Electric resistance	ohm	Ω	$1 = 1\ \mathrm{V/A}$
Flux of magnetic induction, magnetic flux	weber	Wb	$1\ \mathrm{Wb} = 1\ \mathrm{V\ s}$
Magnetic flux density, magnetic induction	tesla	T	$1\ \mathrm{T} = 1\ \mathrm{Wb/m^2}$
Inductance	henry	H	$1\ \mathrm{H} = 1\ \mathrm{V\ s/a}$
Luminous flux	lumen	lm	$1\ \mathrm{lm} = 1\ \mathrm{cd\ sr}$
Illumination	lux	lx	$1\ \mathrm{lx} = 1\ \mathrm{lm/m^2}$

The SI definition of other quantities will be dealt with as they arise in the text but students wishing to study the system of units further are advised to study PD 5686, 'The use of SI units' available from British Standards Institution. All dimensions used in the Figures are in millimetres unless otherwise stated. Some instruments illustrated have Imperial Scales and these will be revised in subsequent editions when such instruments having SI scales are available from the manufacturers. Likewise the section on instruments for hazardous areas is based on the latest acceptable practice but when the new British Codes based on the International Electrotechnical Commission's recommendations become available this may be revised.

In describing the performance of an instrument four terms are frequently used, *viz.* 'accuracy', 'precision', 'sensitivity' and 'rangeability'.

Accuracy. The 'accuracy' of a reading made with an instrument may be defined as 'the closeness with which the reading approaches the true value'.

Precision. The 'precision' of the readings is the 'agreement of the readings among themselves'. If the same value of the measured variable is measured many times and all the results agree very closely then the instrument is said to have a high degree of precision, or reproducibility. A high degree of reproducibility means that the instrument has no 'drift'; i.e. the calibration of the instrument does not gradually shift over a period of time. Drift occurs in flowmeters because of wear of the differential-pressure-producing element, and may occur in a thermo-couple or a resistance-thermometer owing to changes in the metals brought about by contamination or other causes. As drift often occurs very slowly it is likely to be unnoticed and can only be detected by a periodic check of the instrument calibration.

A high degree of precision is, however, no indication that the value of the measured variable has been accurately determined. A manometer may give a reading of 1·000 bar for a certain pressure to within 0·001 bar, but it may be up to 0·2 bar in error, and the true value of the pressure will be unknown until the instrument has been calibrated. It is accurate calibration that makes accurate measurement possible.

All measurement is comparison. When a length is measured it is compared with a fixed length, or 'standard of length', and the number of times the unknown length is greater than the standard is found. The standard is chosen so that the 'number of times' or 'numeric' is not too large or too small.

To ensure that the length, or any other measured quantity, as measured by one person, shall agree with that as measured by a second person, the standards must be absolutely fixed and reproducible with precision. Once a standard of length has been established, all measuring instruments based on this standard are made to agree with the standard. In this way, the length of an object as measured by one instrument will agree with the length as measured by any other instrument.

Accuracy can be obtained only if measuring instruments are periodically compared with standards which are known to be constant; i.e. from time to time the instruments are calibrated.

In some cases the standard is reproducible on the spot so that comparison is easy, while other standards are not so easy to reproduce. In the latter case, it is sometimes necessary to send the instrument back to the maker for calibration. The maker compares the instrument with a 'standard' which has been compared with an absolute standard in the National Physical Laboratory. In this way the measured quantity is being indirectly compared with the absolute standard.

If all users of instruments do this then the results obtained will agree among themselves, for two measurements which are each equal to a third must be equal to each other. In this way precision is obtained.

The aim of all instrument users should be to obtain as high a degree of accuracy as possible without involving unreasonable labour and expense, but the relative sizes of errors should always be kept in mind. An error of 1 metre in the measurement of the length of a room is a serious error, but the same error in the measurement of a distance of several kilometres is negligible. To ensure that this idea of relative size of an error is kept in mind, the error is usually expressed in terms of the true value of the quantity being measured, as a fraction, or, more conveniently, as a percentage: e.g., if a pyrometer is used to measure a furnace temperature at say 1,650° C and it is accurate to $\pm 5°$ C, then the percentage error is

$$\pm \frac{5}{1650} \times 100 \text{ per cent } = \pm \frac{10}{33} \text{ per cent } = \pm 0\cdot 3 \text{ per cent}$$

If, however, an error of $\pm 5°$ C occurs in the measurement of the temperature of boiling water at 100° C, then the percentage error is

$$\pm \frac{5}{100} \times 100 = \pm 5 \text{ per cent}$$

a much more serious error.

The accuracy of an instrument is usually expressed in terms of its inaccuracy, i.e. it is expressed in terms of the percentage error.

The accuracy of an instrument may be expressed in a number of ways.

Makers usually state the 'intrinsic accuracy' of the instrument. This is the accuracy of the instrument when calibrated in the laboratory in the absence of vibration, ambient temperature changes etc. This is not necessarily the same as the practical accuracy obtained with the instrument under plant conditions. To ensure that the instrument shall give in use an accuracy which approaches its intrinsic accuracy it must be carefully sited, maintained, and standardised and readjusted at intervals.

The instrument must also be suited to the plant on which it is used. It must not be corroded by the atmosphere in which it is placed, and should as far as possible be protected from vibration and wide variations of ambient temperature. In some cases it is necessary to sacrifice intrinsic accuracy in order to obtain an instrument which has a more robust construction, and which is more likely to maintain a higher degree of practical accuracy under difficult plant conditions.

Point Accuracy. The 'limits of error' of an instrument may be expressed in a number of ways. In some cases the 'point accuracy' is given. This is the accuracy of the instrument at one point on its scale only, and does not give any information on the general accuracy of the instrument. The general accuracy may be given, however, by a table giving the point accuracy at a number of points throughout its range. For example, the accuracy of a pyrometer may be given at the fixed points on the International Scale of Temperature which happen to be within the range of the pyrometer.

Accuracy as 'Per cent of Scale Range'. When an instrument has a uniform scale its accuracy is often expressed in terms of the scale range, i.e. 'accurate to within $\pm 0\cdot 5$ per cent of the scale range'. Accuracy expressed in this way may be misleading, for an error of $0\cdot 5$ of the scale range at the upper end of the scale may be negligible, but at 10 per cent of the full scale this is an error of 5 per cent of the true value.

Accuracy as 'Per cent of the True Value'. Probably a better conception of the accuracy of an instrument is obtained when the error is expressed in terms of the true value, i.e. accurate to within $\pm 0\cdot 5$ per cent of the true value. Such a statement means that as the reading gets less so does the size of the error, so that at 10 per cent of the full scale the accuracy of the instrument would be ten times better than that of an instrument which is accurate to within $\pm 0\cdot 5$ per cent of the scale range. When an instrument consists of several units (such as, say, the orifice plate and the manometer in a flow meter), each unit will have its own limits of error. Suppose an instrument consists of three units, the limit of error for the units being $\pm a$, $\pm b$, $\pm c$ respectively. Then the maximum possible error will be $\pm (a+b+c)$. It is unlikely that all the units will have the maximum possible error at the same time, so the accuracy is often expressed in terms of the root square error $\pm \sqrt{(a^2+b^2+c^2)}$.

A Complete Statement of Accuracy. If it is required to give a complete picture of the accuracy of an instrument, a graph should be drawn showing the error at several points on the scale plotted against the true value as measured by some reliable standard. The instrument should be calibrated as the measured variable is increased in steps, and the process repeated as the measured variable is decreased by the same steps. The error at each reading can then be plotted against the true value at each point. In this way, two curves are obtained which give the error and the 'hysteresis' at each reading. The 'hysteresis' is the difference between the readings obtained when a given value of the measured variable is approached from below, and when the same value is approached from above, and is usually caused by friction or backlash in the instrument movement, or by changes in the controlling spring. Curves of this nature give a complete statement of the accuracy of the instrument, and may be used for correcting the instrument reading. Where the curves coincide there is no hysteresis error.

Sensitivity. The sensitivity of an instrument is usually taken to be 'the size of the deflection produced by the instrument for a given change in the measured variable'. It is, however, quite frequently used to denote 'the smallest change in the measured quantity to which the instrument responds'. The largest change in the measured variable to which the instrument does not respond is called the 'dead zone'.

Rangeability. The rangeability of a measuring instrument is usually taken to mean the ratio of the maximum meter reading to the minimum meter reading for which the error is less than a stated value. For example: In a positive displacement flow meter a certain quantity of liquid passes through the meter without being registered because of the leakage between the fixed and moving parts. The calibration curve is therefore of the form shown in *Figure 1*. The maximum quantity which can be measured by the meter is

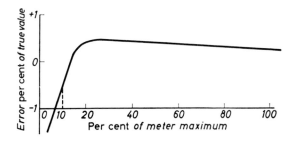

Figure 1   Graph relating error to meter reading

usually fixed by the meter size. Increasing the flow above the meter maximum will shorten the life of the meter owing to greatly increased wear and is therefore highly undesirable. It will be seen from the graph that the minimum flow for which the meter accuracy is plus or minus 0·5 per cent is 10 per cent of the maximum reading. The rangeability for an accuracy of plus or minus 0·5 per cent of true value is therefore 10:1. If an accuracy of plus or minus 1·0 per cent is acceptable, the rangeability is 100:7 or 14:1. It will be seen that for an instrument having a calibration curve of the form shown, increasing the tolerance on the acceptable accuracy does not greatly increase the rangeability.

In the case of meters having a pointer or pen the inability of the user to interpret accurately small movements of the pointer, or pen, limits rangeability. Even a perfect meter is no better than it can be read. Thus the reading error should be added to the other factors which limit accuracy in an actual meter, such as friction, lost motion etc., in determining the meter accuracy.

Where the instrument has a square root scale, as in the differential pressure method of measuring flow, the influence of the reading error will be quite different from that on an instrument with a linear scale. As the pen or pointer movement for a change of flow from 0 to 10 per cent of the measured maximum will only be 1 per cent of the pen movement for the meter maximum, the reading error will have a greater significance at the bottom end of the scale than in a meter having a linear scale. The rangeability can be relatively large provided the resulting error or uncertainty is acceptable. If this type of flow meter is being used to measure the total quantity which has flowed, and most of the flow is measured at a flow rate near the top of the scale with infrequent measurement at the low end of the scale then a low accuracy at the lower end of the scale is permissible. The effect of the error on the total quantity will be very small and a high rangeability will be acceptable. On the other hand, where the instrument is used for control, or for metering ingredients into a plant, high accuracies as percentage of the true value will be required and this will limit the acceptable rangeability.

Application Errors. An instrument which has a high degree of intrinsic accuracy may, in use, have a poor practical accuracy because of inproper use. It is essential to ensure that an instrument is measuring what it is intended to measure, and nothing else, if its reading is to be accurate. This may seem a very obvious statement, but a surprisingly large number of instruments are thought to be faulty when the real fault lies in the way in which they are used. Differential-pressure types of flow meters will read inaccurately if they are placed immediately after a valve or bend. Thermometers will not read correctly if the sensitive portion is insufficiently immersed, or is radiating heat to a colder portion of the installation, while conduction along a sheath to, or away from, a thermometer will cause an erroneous reading. Points to be noted in the method of use of various instruments will be dealt with in the sections describing each instrument.

It is very important to use an instrument under conditions which approximate as closely as possible to those under which the instrument was calibrated. Ideally the instrument should be calibrated under its working conditions, but this is not always possible.

The scale of the standard instrument used to calibrate the instrument should be more open than the scale of the instrument which is being checked. It is then possible to read the check instrument more accurately than that which is being checked.

Calibration. In order that the calibration of an instrument may be corrected when it is in use two fundamental adjustments are essential. It is necessary to be able to adjust the instrument zero and it is essential to be able to adjust the scale span. The adjustment of the scale span should not be confused with range changing.

As an example, consider a simple Bourdon tube type pressure indicator having the mechanism shown in *Figure 1.35*. Owing to continued use, it is possible that after some time, the tube will suffer from a permanent set causing a zero error. In addition, the elastic properties of the tube may change causing it to straighten a greater amount for a given change of pressure resulting in the pointer deflecting through a greater angle. The zero error may be corrected by adjusting the position of the pointer relative to the spindle. The increased angle of deflection may be corrected by reducing the magnification of the quadrant and pinion arrangement by moving the point of attachment of the link connecting the tube to the quadrant further away from the pivot of the quadrant.

In general, in any type of instrument it is essential that zero adjustment and span adjustment should be possible. It will greatly increase the student's understanding of instruments if this fundamental principle is realised, and every time he meets a new instrument he discovers how these adjustments are made.

Methods of display

The measured value of a process variable may be displayed in one of two ways, analogue or digital.

Analogue. In this form of display, the value of the variable is indicated by a pointer moving over a scale which may be horizontal, vertical or circular. In order to increase the scale length for a given panel space for vertical or horizontal scaled indicators, the scale may be marked on the edge of a drum which is rotated by the measuring system, and a portion of the scale is visible through a window having at its mid point a fixed pointer parallel to the scale markings.

Likewise the effective length of a circular scale may be increased by permitting the pointer to make more than one complete rotation of the scale while a second pointer indicates the number of complete revolutions. For example a level indicator may have a scale indicating depth in mm up to one metre while a second pointer indicates the number of complete metres. Such a display is described as 'multiple speed' or multiple shaft analogue display.

The accuracy with which a value may be read is increased by increasing the scale length and reducing the width of the pointer but high reading accuracy has little advantage unless it is combined with high accuracy of measurement. Scale markings should be plain straight lines and on a circular scale should be radial. The major unit markings should be longer than the sub-divisions of the major units. Very closely spaced marks can result in greater reading errors than interpolation by eye particularly where the numbering system of the scale is sparse. In addition, a misleading impression of the accuracy may be obtained if the divisions on the scale are 0·1 per cent of the scale length when the measuring accuracy is only to ±1·0 per cent of the scale length.

Digital. In this form of display the measured value is indicated as a series of numbers as in the mileage indicator on a car or the quantity indication of a quantity meter described in Section 3. Usually a decimal system of numbers is used so the indication is the total number of units that has passed through the meter together with an indication of the decimal fraction. As the quantity which has passed through the meter increases the digits increase until the number indicated is 9 when that digit will revert to zero and the neighbouring digit to the left will increase by one.

Use of electrical instruments in an area where an explosive hazard may exist

This is a complex subject and it is not the purpose of this section to discuss it in detail. Sufficient information only will be given to enable the reader to understand the subject in as far as it affects him as an instrument technician, and to indicate where further information may be found should he desire to study the subject further.

In certain installations on chemical plants and in the oil industry an explosive hazard may occur where there is a possibility of flammable dust, gases or vapour existing in the atmosphere at concentrations capable of ignition by an electrical spark or arc, or hot spot. Electrical equipment can, however, be used under certain circumstances in such an area provided the apparatus is designed so that the risk of ignition is reduced to an absolute minimum.

The Factories Act 1937 and other electrical Regulations, while giving no specific rules regarding electrical apparatus for use in situations where explosive hazards exist, place the onus on industry to recognise and assess an explosion hazard and to take all necessary precautions to meet it. Thus, in addition to complying with the statutory Regulations, Orders or Acts of Parliament currently governing the use of electrical apparatus, the apparatus must be such that it is safe to use in its own particular location. Instruments suitable for detecting flammable atmospheres will be described in Volume 2.

Area classification (IEC79–10 1972)

For the purpose of assessing the risk of ignition, plants or locations are graded according to the probability of the existence of dangerous concentrations of flammable substances in the atmosphere.

Areas are graded as follows:

Zone 0 (Division 0) Continuously dangerous area

This is an area in which flammable substances are continually present in dangerous concentrations in the atmosphere under normal operating conditions.

Zone 1 (Division 1) Intermittently dangerous areas

This is an area in which flammable substances may from time to time be present in dangerous concentrations in the atmosphere under normal operating conditions.

Zone 2 (Division 2) Remotely dangerous area

This is an area in which any flammable or explosive vapour, dust or volatile liquid, although processed, handled or stored, is so well under control that it is likely only under abnormal conditions to produce an ignitable concentration in a sufficient quantity to constitute a hazard, and then only for periods of short duration. In such situations it is considered that, provided electrical equipment does not spark or have any hot spots in normal operation, additional protection can be reduced to a minimum.

As a general guide, if a year is regarded as 10 000 hours and a hazardous atmosphere exists for 1000 hours or more, the area would be classified as Zone 0. If the hazard existed for between 1 and 1000 hours the area would be Zone 1. If less than 1 hour the area would be Zone 2.

Safe areas

This is an area where no flammable or explosive gas, vapour or dust is processed, handled or stored.

Whilst areas of hazard and types of protection can be classified, no two plants are identical so that it is not feasible to apply rigid rules in the selection of suitable equipment and many local factors have to be taken into consideration.

The accepted practice as it exists in 1973 for the use of electrical equipment in hazardous areas is summarised below.

The Department of Trade and Industry have set up a certifying body known as the British Approvals Service for Electrical Equipment in Flammable Atmospheres (BASEEFA) who test equipment to British Standards, or other Standards where they exist, or produce their own standard where necessary. Many new British Standards are being produced in order to bring British practice into line with the recommendations of the International Electrotechnical Commission (IEC).

Equipment which has been certified by BASEEFA will carry a certification mark comprising the letters Ex within the outline of a crown. In addition, the user will require to know the method of protection, the apparatus group and the temperature class so that the equipment will bear a label indicating these features, e.g. Ex d 11B T6 which identifies flameproof equipment (d) suitable for use in gases requiring Group 11B apparatus having a minimum ignition temperature equal to or greater than that of Temperature Class T6, i.e. greater than 85° C.

Grouping of apparatus

The grouping of apparatus is based on flamepath dimensions or levels of igniting current. The IEC grouping of apparatus has been adopted and this applies to both flameproof enclosures and intrinsically safe equipment. The equipment is grouped according to the dimensions of the flamepath relative to the volume of the enclosure in accordance with B.S. 4683 (IEC 79–1).

> Group 1. For use in methane (firedamp)
> Group 11A. For use in propane, etc.

Group 11B. For use in ethylene, etc.

Group 11C. For use in hydrogen.

Apparatus marked 11C includes Groups 11B and 11A and apparatus marked Group 11B includes Group 11A. A list of gases with appropriate apparatus grouping is given in B.S. 4683, Part 2 (1971) and in BASEEFA Certification Standard SFA 3012 (1972).

Table 3 shows the new grouping of apparatus in relationship to the earlier gas grouping for Flameproof Enclosures B.S. 229 and classification of intrinsically safe apparatus to B.S. 1259.

Table 3   GROUPING OF EQUIPMENT AND GASES

B.S. 4683 (1971) (IEC 79–1) *Group of Apparatus*	B.S. 229 (1957) *Group of Gases and Vapours*	B.S. 1259 (1958) *Class of Apparatus*
1	1	1
11A	11	2c
11B	111	2d
11C	(1V)	2e

Temperature classification of equipment

Apart from ignition from arcs and sparks, it is possible to ignite gas mixtures by hot surfaces and B.S. 4683, part 1, which is in line with IEC publication 79.8 specifies six maximum surface temperatures for electrical equipment for use in explosive atmospheres, and classifies the system of marking. In the case of small components in intrinsically safe equipment the temperature may exceed the limiting temperature of the Class provided the heated areas are sufficiently small that they do not constitute an ignition risk. When selecting equipment the user must ensure that the surface temperature does not exceed the ignition temperature of the gas or vapour which may be present in the atmosphere.

The classification is based on a peak ambient temperature of 40° C (but for special cases other temperatures may be used and are stated on the equipment) and is given in Table 4.

Table 4   ENCLOSURE SURFACE TEMPERATURE CLASSES

Class	Maximum Surface Temperature °C
T1	450
T2	300
T3	200
T4	135
T5	100
T6	85

TYPES OF PROTECTION

Flameproof enclosure Ex d

Electrical equipment may be housed in an enclosure, tested to a pressure at least 1·5 times the pressure produced by burning the appropriate gas mixture or 3·5 bar whichever is the greater, designed to withstand an internal explosion and prevent the ignition of external gases. Such enclosures are designed to B.S. 4683, Part 2 (1971), which replaces B.S. 229 (1957) and specifies the maximum permissible dimensions for any gaps which may exist between joint surfaces and the minimum length of flamepath between the inside and outside of the enclosure appropriate to the group of enclosure. Such equipment is considered suitable for Zones 1 and 2 areas.

Increased safety Ex e

Such protection relies on additional measures applied to equipment of normal construction to give increased security against excess temperature and the occurrence of arcs or sparks during the service life of the equipment. Such equipment is certified to BASEEFA standard 3008 (1970) based on IEC publication 79–7, and a new British Standard (B.S. 4683, Parts 4 and 5) covering such equipment will be available by the end of 1973. Such equipment is suitable for Zones 1 and 2.

Pressurised or purged equipment Ex p

In this form of protection, clean air or an inert gas is maintained at a positive pressure within the housing of the equipment to prevent the ingress of flammable gas. A flameproof pressure switch is arranged to sound an alarm or cut off the electrical supply to the equipment should the purging supply fail. The British Standards Institution intends to produce a standard specification which it is hoped will line up with an amended IEC recommendation. This will be available in a few years time. Such equipment is suitable for Zone 1 and possibly Zone 0.

Special protection Ex s

Certain equipment, such as hermetically sealed relays and switches and encapsulated solenoids, do not meet the requirements of other categories but may be tested to BASEEFA standard SFA 3009 (1972). This specifies certain basic requirements of design and construction, and the degree of safety will not be less than that achieved by other means for equipment suitable for Zone 1 locations.

Zone 2 equipment Ex N

Special equipment recognised in the U.K. as being suitable solely for Zone 2 is being developed. This equipment is covered by B.S. 4683, Part 3 and BASEEFA is producing a standard for control gear for Ex N equipment.

Intrinsic safety Ex ia, Ex ib

Intrinsic safety provides a high degree of security and Ex ia is the only safety concept which is generally accepted in this and many other countries for Zone

0 areas. In this method of protection, applicable to a circuit or part of a circuit, the energy available to produce any spark or temperature rise normally, e.g. by breaking or closing the circuit, or accidentally, e.g. by a short circuit or earth fault, is incapable under the prescribed test conditions of causing the ignition of the prescribed gas or vapour.

The IEC has agreed that there should be two categories of intrinsic safety, the higher level of security suitable for Zone 0 to be known as ia, and a lower level suitable for Zone 1 to be known as ib. The British Standard applicable to Intrinsically Safe Equipment is B.S. 1259 but BSI is currently engaged in producing a new standard intended to line up with the requirements of the IEC. Until this standard is published, BASEEFA certifies equipment to BASEEFA standard SFA 3012.

The grouping of intrinsically safe apparatus is related to the maximum energy available to produce a spark under fault conditions. This energy must always be considerably less than the minimum energy required to ignite a fuel-air mixture which may be present. Thus, the group of apparatus which may be used in a given area will be related to the minimum energy required to ignite the fuel or fuels responsible for the hazard.

The minimum energy required to ignite a methane-air mixture under ideal conditions is of the order of 0·29 millijoules, while the minimum energy required for a whole range of fuels such as Propane, n Butane, n Hexane and n Heptane is of the order of 0·25 mJ. Ethylene-air mixture on the other hand requires only 0·12 mJ, while a Hydrogen-air mixture is ignited by 0·019 mJ. Thus Methane requires apparatus in Group 1, Propane etc. Group 11A, Ethylene Group 11B, and Hydrogen Group 11C.

Use of shunt diode safety barriers

The purpose of the safety barrier is to limit to a safe level the energy fed to equipment in hazardous areas, which may readily be designed to be intrinsically safe, by associated equipment which is not intrinsically safe but is located in a non-hazardous area such as a pressurised control room. All power sources and as much of the electronic equipment as possible is located in the safe area. The equipment in the hazardous area is designed to operate at low energy levels and the signal level to the control room may be of the order of 20 milliamps and 20 volts.

The Shunt Diode Safety Barrier provides a coupling device between equipment in the safe area and the intrinsically safe equipment in the hazardous area, and is located at the interface between the two areas, usually at the cable terminations of the control panel. The barrier consists of a network of Zener diodes and resistors designed to have a negligible effect on the normal instrument signals. Should the voltage in the safe area rise above a certain value the diodes conduct, and provided the barrier is suitably earthed, limit the power which can pass into the hazardous area. They are designed for the worst possible fault conditions such as the application of the full mains voltage. Thus, as far as the equipment in the hazardous area is concerned the output of the barrier device is a fixed d.c. voltage from a fixed resistance source so that, by limiting the values of the circuit capacitance and inductance of the external circuit, an intrinsically safe circuit for use in an area containing a known gas or vapour may be designed.

A great advantage of such a system is that provided supply and the earthing of the system follow sound practice, the choice of equipment in the control room, and any modifications which are subsequently made, should not affect the safety of the system. Further any inadvertent application of excess voltage for example during instrument maintenance cannot cause a dangerous amount of energy to be applied to the equipment in the hazardous area.

Shunt Zener Diode Barrier—Description and mode of operation

A typical shunt diode barrier consists of a simple combination of a fuse, resistors and zener diodes as shown in *Figure 2*.

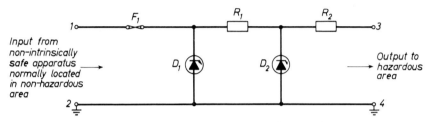

Figure 2   Theoretical circuit of typical shunt Zener diode barrier

The action of the shunt zener diode barrier is to pass a signal with negligible distortion up to the avalanche voltage of the zener diodes. Above this voltage the diodes become conducting and hold the voltage at terminals 3 and 4 at a prescribed level provided that terminals 2 and 4 are properly earthed to ensure fuse operation under fault conditions, and to guard against overriding of the current limiting resistor R_2 by earth faults. (A maximum permissible total earth loop path impedance of 1·0 ohm is specified.)

Resistor R_1 enables diode D_2 to hold the voltage at terminals 3 and 4 to its zener voltage during any transient pulse conditions which may occur while F_1 is blowing when elevation of the zener voltage of D_1 may occur. R_1 also enables each diode to be tested independently after assembly. Resistor R_2 is the safety resistor which limits the current in the hazardous area to a prescribed limit.

In some designs fuse F_1 is replaced by a resistor R_3 which must be so designed that it is not liable to become faulty in service. Such an assembly is a resistor protected shunt diode safety barrier.

Shunt diode barrier

These barriers are basically similar to the zener barrier except that two shunt diodes connected in opposite directions are used in place of each zener diode and therefore the voltage at the terminals 3 and 4 is held to the forward voltage of the diodes (see *Figure 3*).

In order to prevent interference with the barrier the whole is encapsulated, and positive barriers are coloured red and negative barriers black. Diode (or alternating current barriers) are coloured grey. The design of the barrier is

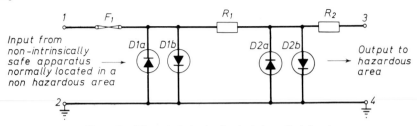

Figure 3   Theoretical circuit of typical shunt diode barrier

asymmetrical about the mounting studs so that it can readily be seen to be mounted correctly to its earth bus bar.

The earthing bus bar should be insulated from all other equipment, such as the control panel and earthed independently. Otherwise the inadvertent application of a high voltage to the panel or other equipment could raise the voltage of the earth connections of the barriers momentarily with the consequent risk of applying excess energy to the circuit in the hazardous area.

The non-earthed terminal for the hazardous area must be separated from the non-earthed terminal to the safe area by at least 50 mm and be protected to prevent contact by any other leads. Further details of the design and testing of safety barrier devices will be found in the **BASEEFA** Certification Standard SFA 3004 (1971), 'Shunt diode safety barriers'.

TYPICAL INSTALLATION DIAGRAMS AND CABLE PARAMETERS FOR SOME SIMPLE, WIDELY USED APPLICATIONS

Installation diagrams

Configuration A. 2-wire system with single barrier.

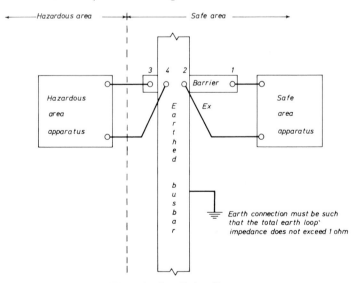

Figure 4   Installation diagram
Barrier can be either positive or negative. *(By courtesy, BASEEFA)*

Configuration B. 2-wire system with two barriers of like polarity.

Configuration C. 2-wire system with two barriers of opposite polarity.

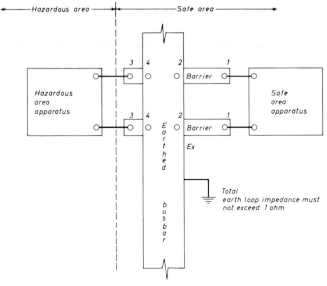

Figure 5   Installation diagram
Configuration B note. Barriers must be either both positive or both negative.
Configuration C note. One barrier must be positive, the other negative.
(By courtesy, BASEEFA)

Table 5

Barrier type	Permissible installation configuration	Max. permissible cable capacitance (microfarads)	Max. permissible cable inductance (millihenries)	Max. permissible L/R ratio of cable (microhenries/ohm)
27 Volt 270 Ohm	A	0·15	3·7	55
22 Volt 150 Ohm	A	0·2	1·5	40
15 Volt 100 Ohm	A	0·8	1·5	60
	B	0·8	1·5	60
10 Volt 47 Ohm	A	3·0	1·0	80
	B	3·0	1·0	80
	C	0·2	1·0	40
4·7 Volt 10 Ohm	A	>1 000	0·16	100
	B	>1 000	0·16	100
	C	3·0	0·16	50
1 Volt 2 Ohm	A	>1 000	0·16	320
	B	>1 000	0·16	320
	C	>1 000	0·16	160

The maximum values for Class IIC (Hydrogen) are given. For most practical purposes, the values for gases of Class IIB are four times these values and for gases of Class IIA are eight times these values.

The L/R ratio of the cable is defined as follows: L/R ratio $= \dfrac{\text{Inductance per unit length (Microhenries)}}{\text{Resistance per unit length (Ohms)}}$

(By courtesy, BASEEFA)

Cable parameters

In the circuit located in the hazardous area connected to the terminal (3) in *Figures 4 and 5*, the total capacitance of the cables and either (a) the total inductance of the cables or (b) the L/R ratio of any of the cables must not exceed the values in Table 5.

Maintenance of flameproof and intrinsically safe equipment

All electrical installations in hazardous areas must comply with the usual regulations governing installations in non-hazardous areas but in addition must satisfy further requirements. No modifications whatsoever may be permitted to a certified flameproof, intrinsically safe or approved apparatus which may invalidate its certificate.

Adequate precautions must be taken to prevent contact between the conductors of an intrinsically safe circuit and those of any other circuit, and to prevent currents being induced in intrinsically safe circuits by the proximity of other circuits. Apparatus which is not intrinsically safe should on no account be connected in the same circuit with intrinsically safe apparatus. Intrinsic safety is often obtained by the insertion of additional components to limit the spark energy. When the apparatus fails owing to these components becoming defective, the equipment can be made to function by shorting out or removing these components. This is a most dangerous practice and is strictly forbidden as the equipment is no longer intrinsically safe.

Conduit systems should be avoided. Mineral-insulated metal-sheathed cables may be used and terminated in flameproof glands. All sheathing and screening armour should be effectively earthed and where necessary protected from corrosion. All cables, joints and terminations should be effectively sealed to prevent the ingress of hazardous material.

It is essential that all apparatus used in hazardous areas should be installed and maintained in accordance with the design details accepted by the testing authority. The whole of a flameproof enclosure should be examined on completion of its installation to ensure that all joints are bolted up tightly, that no bolts are missing, that all cover plates are in position and that the flange gaps do not exceed the permissible maximum for the group for which the enclosure is certified. When a flameproof enclosure is opened, all flange faces exposed should be thoroughly cleaned, lightly smeared with petroleum jelly, and care is necessary to avoid scoring the faces. In use, the apparatus must be examined regularly to ensure it remains in the state it was in when installed. This is particularly important in areas where corrosion may occur.

Classification of instruments

In this book, instruments are classified according to the physical principle upon which they are based. The early appearance of an instrument in the classification does not indicate that it is any better or any worse than an instrument which appears later, neither does it indicate that the instrument is more common.

In each section the introduction deals largely with the mathematics and physics of the instrument, and the instruments are then classified according to the decimal system.

The book is divided into four main divisions:

1. Instruments for measuring pressure
2. Instruments for measuring level
3. Instruments for measuring flow
4. Instruments for measuring temperature.

Instruments in each main division are then classified, according to the physical principle upon which they are based, into classes, which in turn are divided into sections, and then into subsections. Where necessary the subsections may be further subdivided. The numbers before each type of instrument indicate the subsection, section, class and division to which the instrument described belongs. For example, in the division on the measurement of flow: 3.1.3.3 Rotating impeller type, is the heading of the portion describing the third type of instrument in subsection 3.1.3 Volumetric meters for gases. This subsection, 3.1.3, is in the second section, 3.1, of the first class of instruments. It is therefore a quantity meter of the volumetric class, used for gases and is of the rotating impeller type.

Where possible, figures are given which indicate the range for which the instrument is suitable. These figures, however, are merely a guide and must not be regarded as being absolute limits. The instrument industry is developing at a very rapid rate and new materials and techniques are constantly being developed. In many cases, the instrument with a range which is exceptional today will be the commonplace instrument of tomorrow. Even during the short time taken to write this book a great deal of progress has been made.

I

MEASUREMENT OF PRESSURE

As an introduction to this section the principles upon which the measurement of pressure and the transmission of force by a fluid depend are briefly stated.

1. Density may be defined as the mass of a unit volume of a substance. In SI units it is the mass in kilogrammes of 1 cubic metre of the substance.

2. Specific gravity is the number of times the mass of any volume of a substance is greater than the mass of the same volume of water under identical conditions, i.e.

$$\frac{\text{mass of any volume of substance}}{\text{mass of the same volume of water}}$$

both masses being measured in the same units.

3. The free surface of a liquid at rest is horizontal; or, liquids find their own level.

4. When a fluid is in contact with a boundary it produces a force at right angles to that boundary. The force produced per unit area is called the pressure due to the fluid. In SI units, pressure may therefore be defined as force in newtons acting on a square metre. Pressure may also be expressed in terms of heights of columns of water or mercury.

5. The pressure due to the liquid at any point in a liquid is proportional to the depth of the point below the surface of the liquid, and is therefore the same at all points on the same horizontal level. At a depth of h m in a liquid of density ρ kg/m³ the pressure will be the product of the depth and the weight per unit volume of the liquid.

The weight of a body is the force acting on the body due to gravity. From Newton's second law of motion we know that the force acting on a body is the product of its mass and its acceleration. At any location a body will be subject to an acceleration g m/s² owing to gravity. Therefore, the force acting on a body, i.e. its weight, if it has a mass of 1 kg will be g kgm/s² or g N.

Thus the pressure at a depth of h m in a liquid density ρ kg/m³ at a point where the acceleration due to gravity is g m/s² will be $h \rho g$ N/m².

For most purposes the International Standard value of gravity $g = 9.80665$ m/s² defined as the value of g at sea level at latitude 45°, may be used, but for work of very high precision the actual value of gravity g at the location must be used. This is related to the latitude and the height above sea level by the following equation

$$g_1 = g_e \left(1 + \beta \sin^2 \theta - 5.9 \times 10^{-6} \sin^2 2\theta \right) - 0.0003086 \, H \qquad (1.1)$$

where $g_e = 9.78049$ m/s² the value of the acceleration due to gravity at the equator

$\beta = 0.005\,288$.

θ = geographical latitude.

H = height above sea level (m).

The pressure due to a column of water 1 mm high will be $10^{-3}\rho_w\, g$ N/m². As pressures are expressed in terms of liquid column, the pressure due to a standard mm water gauge has been defined as the pressure due to a column of water of maximum density 1 mm high, i.e. at temperature of 3·98° C where the density $\rho_s = 1000$ kg/m³, and the acceleration due to gravity $g_s = 9.806\,65$ m/s².

Thus the standard mm water gauge $= 10^{-3} \times 1000 \times 9.806\,65$ N/m². At 20° C the density of water is 998·203 kg/m³ thus the pressure due to a 1 mm column $= 0.998\,203 \times 9.806\,65$

$\quad = 9.789\,03$ N/m².

Likewise the pressure due to a 1 mm column of mercury at 0° C will be

$\quad 13.595\,5 \times 9.806\,65$

$\quad = 133.326$ N/m².

At 20° C the pressure due to a 1 mm column of mercury will be

$\quad 13.546\,2 \times 9.806\,65$

$\quad = 132.843$ N/m².

Where the temperature is other than 0° C or 20° C the actual density of the water or mercury may be obtained from tables or calculated as explained in section 1.3.1.1.

If a body is immersed in a liquid, the underside will be at a greater depth than the upper side, so that there will be a greater pressure on the lower side than on the upper side. Therefore there will be a net upward force on the body. The size of this force is given by Archimedes' Principle which states: 'When a body is immersed in a fluid it appears to lose weight and the apparent loss of weight is equal to the weight of fluid displaced by the body.' A body having a volume of 1 m³ will, when immersed in water, displace 1 m³ of water, which weigh approximately 9.81×10^3 N. Thus the body will appear to weigh 9.81×10^3 N less when in water than when it is in air.

A body less dense than water will sink until the upward force on it is equal to the downward force, i.e. the weight of fluid it displaces is equal to its own weight. Thus, the depth to which a standard body sinks in a liquid is a measure of the density of the liquid. This principle is used in the hydrometer.

6. The pressure at any point in a liquid is the same in all directions. If it were not, then the liquid at the point considered would move owing to the unbalanced force in some direction.

1.1   ABSOLUTE AND DIFFERENTIAL PRESSURE MEASUREMENT

Before considering the actual methods of measuring pressure it is essential to make clear the difference between absolute and differential pressure.

The 'absolute pressure of a fluid' is the difference between the pressure of the fluid and the 'absolute zero of pressure'. It is, therefore, the difference between the pressure of the fluid and the pressue in a complete vacuum. A barometer is an example of a gauge for measuring 'absolute' pressure. The height of the mercury column is a measure of the difference between the atmospheric pressure and the pressure in the Torricellian vacuum over the top of the mercury in the tube. It is, therefore, a measure of the 'absolute' pressure of the atmosphere. If, therefore, a gauge of any form is required to measure the absolute pressure of a fluid it must do so by comparing the pressure of the fluid with the pressure in a complete vacuum.

It is very difficult to obtain a complete vacuum in some forms of instruments, but with modern techniques it is possible to get pressures which are very near the absolute zero of pressure, so that for all practical purposes the instruments indicate the absolute pressure.

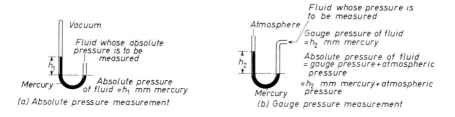

(a) Absolute pressure measurement   *(b) Gauge pressure measurement*

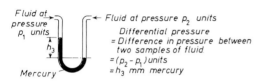

(c) Differential pressure measurement

Figure 1.1   Comparison of types of pressure measurement

Most pressure gauges measure the difference between the absolute pressure of fluid and the atmospheric pressure, and the pressure measurement so obtained is called the 'gauge pressure'. This is, of course, a differential pressure measurement, the 'gauge pressure' being the difference between the absolute pressure of the fluid and the atmospheric pressure. Therefore:

Gauge pressure = absolute pressure of fluid − pressure of the atmosphere,

or

Absolute pressure of fluid = Gauge pressure + pressure of the atmosphere

$$(1.1a)$$

In gauges which are sometimes spoken of as indicating vacuum or suction, they are really indicating the amount the absolute pressure of the fluid is less than atmospheric pressure, and

Gauge pressure = pressure of atmosphere − pressure of fluid

therefore

Absolute pressure of fluid = pressure of atmosphere − Gauge pressure

$$(1.1b)$$

In other forms of differential pressure measurement the gauge is used to measure the difference between the absolute pressures of two samples of the same fluid, or, the difference between the absolute pressures of two fluids.

Figure 1.1 shows a U tube used for the various types of pressure measurement. The same principle applies when other forms of pressure measuring elements are used.

1.2   METHODS OF MEASUREMENT OF PRESSURE

Pressure may be measured directly in two ways. In the first place, the pressure due to a fluid may be balanced by the pressure produced by a column of liquid of known density. In the second place, the pressure may be permitted to act over a known area, and since

$$\text{Pressure} = \frac{\text{Force}}{\text{Area}}$$

or, Force = Pressure × Area, a force will be produced whose magnitude depends upon the pressure.

This force may be measured by balancing it against a known weight, or, by the strain or deformation it produces in an elastic medium. Other methods of measuring pressure depend upon indirect means.

Methods of measuring pressure may, therefore, be classified as follows:

1. Pressure measurement by balancing against a column of liquid of known density,
   Simple U tube with vertical or inclined limb,
      The simple U tube in practice.
      Absolute pressure measurement.
      Differential pressure measurement.
      Vernier reading manometer.
      Precision calibrating standard manometer.
      Sonar manometer.
   Industrial types of manometer.

2. Pressure measurement by balancing against a known force,
      Piston type pressure gauge.
      Ring balance type pressure gauge.
      Bell type pressure gauge.

3. Pressure measurement by balancing the force produced on a known area against the stress in an elastic medium
      Bourdon tubes,
         The 'C' type Bourdon tube.
         The spiral Bourdon tube.
         The helical Bourdon tube.
      Diaphragm types,
         Stiff metallic diaphragm or bellows.
         Slack diaphragm and drive plate.

4. Other methods.

1.3 PRESSURE MEASUREMENT BY BALANCING AGAINST A COLUMN OF LIQUID OF KNOWN DENSITY

1.3.1 Simple U Tube

1.3.1.1 SIMPLE U TUBE IN PRACTICE

Consider a simple U tube containing a liquid of specific gravity ρ_m (*Figure 1.2*) A and B are at the same horizontal level in the liquid and the liquid at C stands at h mm above B.

Therefore,

Pressure at A = pressure at B

Fluid pressure at A = atmospheric pressure + pressure due to column of liquid BC

= atmospheric pressure + $h\rho_m$ mm H_2O

$h\rho_m$ mm H_2O is the 'gauge pressure' and is written $h\rho_m$ mm H_2O gauge.

Wet leg correction

If the fluid in the left-hand limb has a density which cannot be neglected in comparison with the density of the liquid in the gauge, then an allowance must be made for the pressure due to the fluid in the gauge and connecting pipes. For example, suppose the simple gauge, *Figure 1.3* was being used to measure steam pressure, and the pipes between A and the main were filled with water whose level stands h_1 mm above A. Suppose the liquid in the gauge is mercury specific gravity 13·55 at 20° C and let the gauge indicate h mm as before.

Pressure at A = Pressure at B

h_1 mm H_2O + steam pressure = 13·55 h mm H_2O + atmospheric pressure

Steam pressure = $(13\cdot55h - h_1)$ mm H_2O + atmospheric pressure

i.e. Steam pressure = $(13\cdot55h - h_1)$ mm H_2O gauge

= 9·79 $(13\cdot55h - h_1)$ N/m² gauge   (1.2)

When the gauge is being used to measure the pressure differential produced by a throttling device as in steam flow measurement, both limbs of the gauge above the mercury and the pipes to the main are filled with water up to the

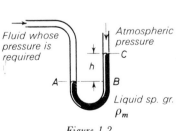

Figure 1.2
Simple U tube

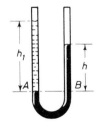

Figure 1.3
Simple U tube with
wet leg connection

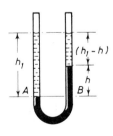

Figure 1.4
Simple U tube with
wet leg connection

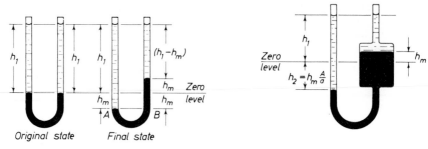

Figure 1.5
Simple U tube with wet leg connection

Figure 1.6
Well-type U tube

same horizontal level, say h_1 mm above A. If A is connected to the high pressure side then:

$$\text{Pressure at A} = \text{Pressure at B } (\textit{Figure 1.4})$$
$$h_1 \text{ mm H}_2\text{O} + \text{high pressure} = (h_1 - h) \text{ mm H}_2\text{O} + 13{\cdot}55\, h \text{ mm H}_2\text{O} + \text{low pressure}$$
$$\text{Differential pressure} = \text{High pressure} - \text{low pressure}$$
$$= \{(h_1 - h) + 13{\cdot}55\, h - h_1\} \text{ mm H}_2\text{O}$$
$$= (13{\cdot}55 - 1)\, h \text{ mm H}_2\text{O}$$
$$= 12{\cdot}55\, h \text{ mm H}_2\text{O} \tag{1.3}$$

or in terms of mm mercury

$$\text{Differential pressure} = \frac{12{\cdot}55}{13{\cdot}55} h \text{ mm Hg} \tag{1.4}$$

Or in the general case when the specific gravity of the gauge liquid is ρ_m and of the covering liquid is ρ_c.

$$\text{Differential pressure} = h \frac{\rho_m - \rho_c}{\rho_m} \text{ mm gauge liquid} \tag{1.5}$$

If the differential pressure is being measured by the amount the mercury in the right-hand limb rises above a fixed level, then this correction must be treated differently.

Suppose the change in level in the right-hand limb is h_m mm. Then provided the tube is of uniform cross-section, rise in level in the right-hand limb equals fall in level in left-hand limb.

$$\text{Pressure at A} = \text{Pressure at B } (\textit{Figure 1.5})$$
$$(h_1 + h_m)\, \rho_c \text{ mm H}_2\text{O} + \text{high pressure} = \{(h_1 - h_m)\, \rho_c + 2\, h_m\, \rho_m\} \text{ mm H}_2\text{O} + \text{low pressure}$$
$$\text{Differential pressure} = \{(h_1 - h_m - h_1 - h_m)\, \rho_c + 2\, h_m\, \rho_m\} \text{ mm H}_2\text{O}$$
$$= (2h_m\, \rho_m - 2h_m\, \rho_c) \text{ mm H}_2\text{O}$$
$$= 2\, h_m\, (\rho_m - \rho_c) \text{ mm H}_2\text{O} \tag{1.6}$$

Another way of looking at this is as follows:
The greater pressure on the liquid in the left-hand limb is being balanced because a column of liquid of length $2h_m$ mm has a specific gravity of ρ_m instead of ρ_c.

$$\text{Differential pressure} = 2h_m\, (\rho_m - \rho_c) \text{ mm H}_2\text{O}$$

If the limbs have a different diameter, as in the case of the well type of mano-meter (*Figure 1.6*), then the rise in one limb will not equal the fall in the other. If the well has an area A mm, while the tube has an area a mm, and the rise in the right-hand limb is h_m mm as before, then since:

loss of liquid to the left-hand limb = gain of liquid in the right-hand limb,

$$h_m A = h_2 a$$

$$\text{or} \quad h_2 = h_m \frac{A}{a}$$

$$\text{Differential pressure} = h_m \left(1 + \frac{A}{a}\right) (\rho_m - \rho_c) \text{ mm } H_2O \qquad (1.7)$$

Effects of Temperature

The densities to be used in the foregoing equation are the densities of the fluids at the working temperature. If there is any variation in temperature, there will be a variation in the reading owing to the reduction of density of the fluid in the gauge as its temperature rises and the fluid expands.

$$\text{Density} = \frac{\text{Mass}}{\text{Volume}}$$

Consider any mass of liquid, and let its volume at $t_0°$ C be v_0 and its density ρ_0. If its temperature rises to $t_1°$ C, then its new volume will be v_1 and its density ρ_1, where $v_1 = v_0 [1 + \beta (t_1 - t_0)]$ where β is the average co-efficient of cubical expansion of the liquid for the range considered.

$$\therefore \rho_1 = \frac{\text{mass}}{v_1} = \frac{\text{mass}}{v_0 [1 + \beta (t_1 - t_0)]} = \frac{\rho_0}{1 + \beta (t_1 - t_0)} \qquad (1.8)$$

As the pressure in a liquid depends upon the depth and density only, the fact that the gauge tube expands does not affect the reading. In accurate work, however, the expansion of the scale should be taken into account. The complete correction factor is worked out in the section on temperature measurement.

Liquids Used in the Gauge

The type of liquid used in the gauge will depend upon the pressure and the nature of the fluid whose pressure is being measured. If a small pressure differential, or a pressure which is very little different from that of the atmos-phere, is being measured, then the maximum reading will be obtained by using a liquid of the lowest density; and a convenient and cheap liquid which can often be used in these circumstances is water. Water, however, evaporates readily, and the gauge needs frequent topping up. Water is also difficult to see unless it is coloured.

Other liquids which may be used are indicated in Table 1.1 overleaf.

Table 1.1.

Liquid	Specific Gravity at 15.5° C	Applications	Advantages	Disadvantages
Transformer Oil	0·864	This is a suitable indicating liquid for ammonia gas flow-meters and measurement of small pressure differences.	Low density gives a large difference of level for small pressure difference. Unaffected by ammonia. Easily seen. Does not readily evaporate.	Tends to cling to the inside of the gauge. Specific gravity is subject to variation so it is advisable to buy a drum which will last the lifetime of the instrument.
Aniline	1·025	Suitable for low pressure air or gas flowmeters with the exception of ammonia and chlorine.	Low density. Evaporates slowly. Does not mix with water. Easily seen.	Attacks paintwork. It is a poison which penetrates the skin and causes blood poisoning. Container must be kept well stoppered as aniline darkens on contact with air.
Dibutylphthalate	1·047	Suitable indicating liquid for ammonia gas installations.	Does not mix with water.	
Carbon Tetrachloride	1·605	Suitable as indicating liquid in chlorine gas flowmeters and as a temporary measure when it is required to measure a higher pressure differential than can be conveniently measured with above liquids.	Not attacked by chlorine.	Not easily seen. Evaporates readily in hot places, so that the gauge needs frequent topping up.
Tetrabromoethane	2.964	Suitable indicating liquid for ammonia gas installations.	Evaporates slowly. Useful when liquid of high specific gravity is required.	
Bromoform	—	Supplies an indicating liquid of density between that of carbon tetrachloride and mercury.	Measures pressures intermediate between those conveniently measured with water and mercury.	Density uncertain. It is poisonous, freezes easily, and is subject to attack. It also attacks rubber if this is used as a gland or in tube form.
Mercury	13·56	Used in water, steam, liquid and compressed gas flowmeters.	High density. Easily seen. Does not mix with other liquids. Does not evaporate or wet the sides of the tube.	High cost. Mobility and density are affected by contamination. Cannot be used in industries such as the photographic industry as tiny quantities escaping into the product will cause a lot of damage.

1.3.1.2   MEASUREMENTS OF ABSOLUTE PRESSURE

Pressures Greater than two Atmospheres

In order to measure pressures of several atmospheres, one limb of the manometer would have to be inconveniently long. This difficulty may be overcome
by closing one limb of the manometer, the closed limb containing dry air
(*Figure 1.7*). The pressure is then measured by the degree to which the air is
compressed. If the original pressure of the air is atmospheric—say, p_0 mm of
mercury—when the mercury stands at the same level in both limbs, then
when a pressure of 2 atmospheres is applied the volume of the air will be
reduced to a little more than half its original volume. The gas in the closed
limb will obey Boyle's Law. If the bore of the closed limb is uniform, volumes
of enclosed gas can be regarded as being proportional to length. If the original

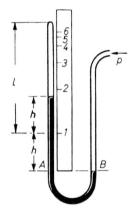

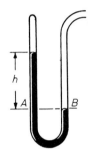

Figure 1.7
Absolute pressures greater than two atmospheres

Figure 1.8
Absolute pressures (small)

length of the column of gas is l mm, a pressure of p mm of mercury is applied
to the right-hand limb, and the levels of the mercury in the limbs change by
h mm.

Then pressure of the air $= (p-2h)$ mm of mercury.

Since the product of pressure and volume is constant,

$$p_0 l = (p-2h)(l-h)$$

or $\qquad p_0 \dfrac{l}{l-h} = p-2h$ (dividing both sides by $l-h$)   (1.10)

or $\qquad\qquad p = 2h + \dfrac{p_0 l}{l-h}$

l and p_0 are fixed for any instrument, so p will depend upon h, and the scale
can be graduated directly in bar.
    This instrument will be satisfactory for a pressure of a few atmospheres,

but at high pressures a small change of pressure will not produce an appreciable change in the volume of the air, so the instrument becomes insensitive.

Small Absolute Pressures

A mercury vacuum gauge is used when it is required to measure a small absolute pressure. This may be done by having a U tube sealed at one end as shown (*Figure 1.8*). The sealed limb is completely filled with mercury. When the source at low pressure is connected to the open limb, the mercury falls, leaving a vacuum in the closed limb. The pressure at A will then be h mm Hg or $10^{-3} p_m$ hg N/m² absolute, where h mm is the difference in height of the mercury columns, and p_m is the density of mercury in kg/m³.

When the pressure to be measured is extremely low, the change in level produced in the closed limb by the very small change of pressure is measured by means of an optical lever (*Figure 1.9*).

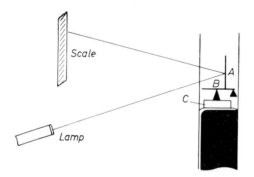

Figure 1.9   Optical lever

This consists of a mirror (A) carried by a holder B pivoted on two knife edges, one carried by a steel float C which rises and falls with the mercury level, the other fixed to the wall of the tube. A beam of light falls on the mirror and is reflected on to a scale. A small change in the position of the float changes the angle at which the light is reflected from the mirror, and causes a large movement of the spot of light on the scale.

Very Small Absolute Pressures

The McLeod type of gauge is used for measuring very low pressures down to $2 \cdot 5 \times 10^{-4}$ mm of mercury. A very convenient form shown in *Figure 1.10* is the modification due to Gaede. Gas from the system whose pressure is required enters the gauge through B and fills the tubes down to the level of the mercury reservoir. The reservoir G is then raised, cutting off the gas present in the bulb H and compressing it into the capillary extension which lies along the scale S. The mercury rises faster in the left-hand limb, and may be made to stand at any height in the tube A, above that in the closed limb. The diameter of the tube A and of the closed capillary must be the same to avoid differences of pressure arising from capillary depression of the mercury in the narrow tube.

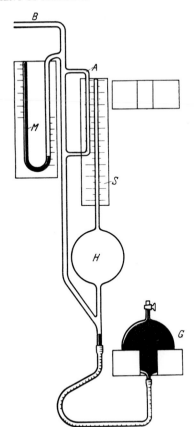

Figure 1.10   McLeod gauge

Suppose the initial pressure of the gas is p mm of mercury, the final difference in level between the mercury in A and in the capillary is h mm, the initial volume of gas in the bulb v and the final volume v_1, then:

By Boyle's law,

$$pv = (p+h)\, v_1$$
$$= pv_1 + hv_1$$
$$p(v - v_1) = hv_1$$
$$p = \frac{hv_1}{v - v_1} \tag{1.11}$$

Now v_1 is usually very small in comparison with v so this equation may be written

$$p = \frac{hv_1}{v} \tag{1.12}$$

It is often arranged so that when the reservoir is raised the mercury at A always rises to the level of the top of the closed capillary, which is the zero of the scale S. Then $v_1 = hv_0$, where v_0 is the volume represented by one division of the scale, so that substituting in equation 1.11 we have

$$p = \frac{h^2 v_0}{v} \tag{1.13}$$

The scale is graduated directly according to this law. A large range of pressures is thus obtained on a relatively short scale. Pressures from 100 mm of mercury down to 12 mm of mercury can be read directly on the manometer M, and from 12 mm of mercury down to $2\cdot5 \times 10^{-4}$ mm on scale S. At this extremely low end of the scale, allowance must be made for the vapour pressure of mercury.

The common Bourdon brass spiral gauge described later has also been used for low pressure measurement. Other gauges take the form of a spiral of thin glass tubing. The movement of the end of the glass spiral is measured by an optical lever.

Another method is that devised by Pirani and Hall. At low pressure the quantity of heat conducted through a gas is proportional to the pressure of the gas. The amount of electrical energy required to keep a heated tungsten wire at a constant temperature, and therefore at a constant resistance, is a measure of the pressure of the gas. Thus a convenient gauge for very low pressures can be based on this principle.

1.3.1.3 MEASUREMENT OF DIFFERENTIAL PRESSURES

Small Differential Pressures

U *tube with inclined limb*. The pressure at any depth in a liquid depends only upon the density of the liquid and the depth below the surface. It is not affected by the cross-section of the vessel which contains the liquid. When it is required to measure a small difference of level this can be done by using a U tube having an inclined limb, while the other limb is reduced to a bulb as shown in *Figure 1.11*.

Suppose a tube is inclined at a slope of 1 : 20 to the horizontal, the 20 units being measured along the tube as shown in *Figure 1.12*. A rise of h mm in the level of the liquid in the tube will mean that the movement of the liquid along the tube will be $20h$ mm. Thus, the movement for a small change in level is more easily detected than in a vertical limbed manometer. Great care must be taken, however, to keep the tube clean if the readings are to be accurate; for errors due to the changes in the force of adhesion between the

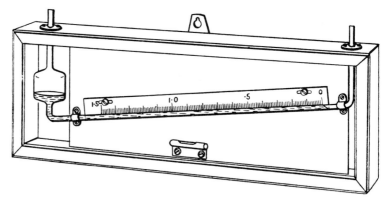

Figure 1.11   Inclined limb manometer

liquid and the tube will be magnified in the same ratio as the movement due to change of pressure. For accurate work, allowance must also be made for the change in level in the bulb B. This can be corrected in the graduation of the scale if the sides of the bulb are parallel. If the diameter of the bulb is large in comparison with the diameter of the tube, the change of level in the bulb will be very small.

Frequently the tube is made of a plastic material, and the liquid used is a light oil so chosen that the line of the meniscus and the lines of the scale form a straight line as shown in *Figure 1.13*. This enables the position of the liquid to be read more easily. Gauges of this kind usually have a range up to 40 mm water gauge, and can be read to 0·25 mm.

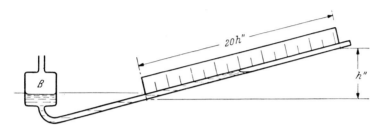

Figure 1.12   Inclined limb manometer

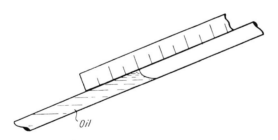

Figure 1.13   Shape of meniscus

In flow measurement of low-pressure gas, this type of gauge is used, and the tube is curved so that in spite of the square root relationship between flow and pressure differential it is possible to obtain approximately equal spacing of the divisions on the flow scale from about 1/15th of flow up to full flow. Such a gauge is shown in *Figure 1.14*

Vernier Reading Manometer

Where smaller differential pressures in terms of mm w.g. are required, for example when measuring air flow in ducts by means of a pilot static tube, the

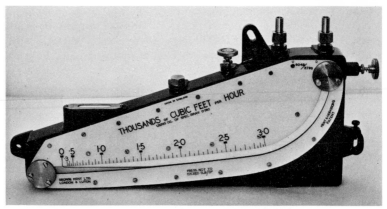

Figure 1.14　Curved limb manometer. (By courtesy, George Kent, Ltd.)

vernier reading manometer shown in *Figure 1.15* may be used. The instrument consists of a flexible U tube manometer with one end connected to a reservoir and the other to a measuring chamber. Both chamber and reservoir may be moved vertically and the movement accurately measured.

The instrument contains sufficient water to half fill chamber and reservoir when they are at the same height. In the base of the measuring chamber a stainless steel needle is fixed point upwards and the needle and its reflection, produced by internal reflection, can be viewed at 45° from a window in the bottom. When the tip of the reflection just touches the tip of the needle, the needle is just in contact with the underside of the water surface. Thus, the position of the water surface is accurately determined without the interfering effects of surface tension. The measuring chamber   made of aluminium alloy and is attached to a vernier which is moved over the engraved column by means of a lead screw. The chamber is connected to the brass reservoir by means of plastic tube with a Terylene braided insert which reduces change of the internal shape of the tube owing to movement or pressure change. A circular spirit level is provided and two levelling screws fitted to the base for setting up.

The instrument has a range of 0 − 200 mm w.g. and the vernier can be read to 0·02 mm w.g. and the instrument will stand a static pressure of 7 bar. The types of manometer described above are extremely useful as workshop standards. Pressures can be easily reproduced because the calibration is in terms of lengths and densities which can be measured directly. For this reason many other forms of pressure measuring devices are calibrated by means of columns of water or mercury.

Precision Calibrating Standard Manometer

When used as a calibrating standard, a manometer is usually provided with features which enable readings to be made with a very high degree of precision and accuracy. In order to enable readings to be made with precision, the user is provided with aids which make reading the instrument easy and eliminate as far as possible the need for the observer's judgement to enter

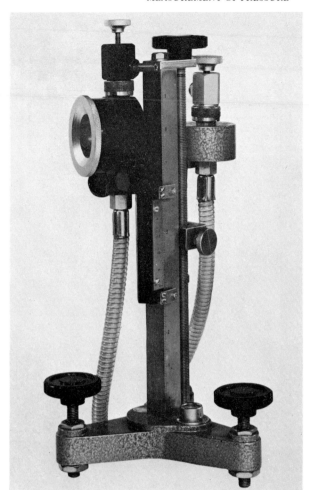

Figure 1.15   Vernier reading manometer showing zero point and its reflected image as seen through the eyepiece. (By courtesy, C. F. Cassella & Co.)

into the reading. For example, a vernier is provided to eliminate the necessity to estimate subdivisions of the scale, and a device introduced to eliminate the necessity to estimate the correct position in relationship to the meniscus of the liquid. In addition, provision must be made to ensure that the liquid column is vertical and that allowance can be made for temperature change. An example of a manometer provided with these necessary features is shown in *Figure 1.16*. This manometer has a range of $0-1080$ mbar and may be read to $0 \cdot 1$ mbar by means of the vernier. The makers guarantee an accuracy of 1 part in 5000 and a sensitivity of 1 part in 25 000.

The base (*Figure 1.16a*) is a sturdy metal tripod which supports the manometer tube and the four supporting columns. The manometer tube is gasket-sealed to a mercury well within the base. This well is lined with stainless steel to prevent contamination of the mercury and is filled through the pressure connection into the well. Each leg of the base has a levelling screw

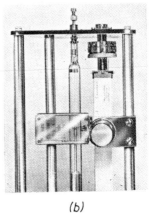

(b)

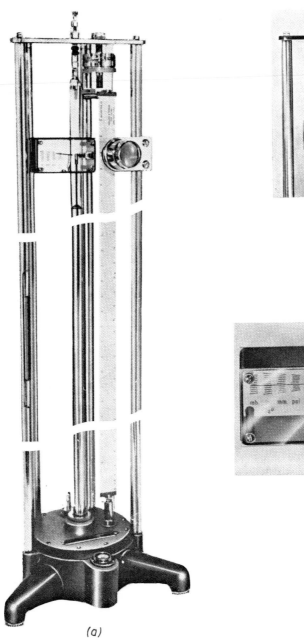

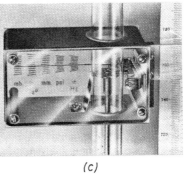

(c)

(a)

*Figure 1.16   Precision calibrating
standard manometer. (By courtesy,
Wallace & Tiernan Ltd.)*

and the built-in bubble type level indicator shows when the column is vertical.

The manometer tube (*Figure 1.16 a* and *b*) is a large-diameter precision-bore glass tube treated to prevent electrostatic charge accumulating on the inside, and extends from the well in the base to a place at the top supported by the four columns. It has a mercury-sealed valve at the top enabling the tube to be evacuated when used as an absolute pressure indicator. This valve consists of a stainless steel stem lapped to a seat in the tube, with a small quantity of mercury trapped above the seat to provide a vacuum-tight seal. When the instrument is used for differential or gauge pressure measurement the stem of the valve is withdrawn from its seat. When required for absolute pressure measurement the mercury level is raised to about 5 mm above the valve seat by evacuating the tube and applying a small pressure to the well. The valve is then reseated to seal off the tube by applying a reduced pressure to the well. To ensure complete evacuation of the tube this procedure is repeated with the valve seated.

In order to eliminate errors owing to uncertainty of the position of the meniscus, a float arrangement is used. This consists of a small aluminium float with a rod fixed to its upper surface carrying a horizontal disc at its upper end. This disc indicates directly on the scale the height of the mercury column above the well level. To permit greater resolution, a vernier is provided which enables readings to be taken to one tenth of the main scale division. This vernier (*Figure 1.16c*) consists of a magnetic float follower fixed on a horizontal pointer shaft and rotating in jewel bearings. Owing to the length of the pointer, movements of the float are amplified ten times so that readings on the auxiliary scale indicate the mercury level to 0·1 mbar. Accurate reading is facilitated by the magnifying eyepiece which may be focused anywhere on the scale. The scale is a durable white plastic with black graduations, graduated under controlled temperature conditions by a machine which has a guaranteed traceable accuracy of 5×10^{-3} mm in 1 m. The size of the graduations are such as to take into account the relative areas of cross-section of well and tube. Compensation for temperature changes and correction for the value of the acceleration due to gravity at the place of use (between 9·77 and 9·816 m/s²) are accomplished by setting an adjusting knob at the top of the scale (*Figure 1.16b*). This puts the scale under the correct tension to compensate for the changes, and the scale reading is then the true pressure reading at standard temperature and gravity. A thermometer is built into one of the support columns to enable the temperature correction to be easily made.

A wide range of scales within the range of the instrument are available, and the instrument is designed for a static pressure of 5 bar.

Sonar Manometer

A development of this instrument is the Sonar Manometer in which the difference in mercury level in the two columns of the manometer is measured digitally and displayed electronically as a pressure reading with a range of 1080 m mbar in units of 0·01 mbar.

This is done by a sonic-electronic system that is capable of reading out the pressure at any rate between 10 readings per second and 1 reading per 5 seconds depending on the setting of a manually adjustable sample-rate knob. At the beginning of each measuring cycle, an electrical pulse is sent simultaneously to an individual piezoelectric transducer located at the bottom of each mercury column. Each transducer converts the electronic pulse to an ultrasonic sound pulse that travels upwards through the column of mercury, is reflected from the meniscus, and returns to the transducer where it is converted to an electric pulse. The pulse involved in the shorter column returns to its transducer first and is used to start the counting so that pulses are fed from a crystal-controlled master clock oscillator to digital counter boards. The pulse involved in the longer column returns to its transducer later, and stops the pulses to the counter boards. The number of pulses is, therefore, directly proportional to the difference in column heights and is translated directly into pressure units. The pressure reading is stored in binary coded decimal form for actuation of the digital display unit and other data-processing equipment until the next measuring cycle occurs.

1.3.2. Industrial types of manometer

The simple U tube type of manometer is usually made of glass and contains the appropriate liquid. For industrial use, it is often more convenient to have the U tube made of some tougher material which is less likely to break in the course of use. The glass is therefore often replaced by steel. When this is done, however, it is no longer possible to read the level of the liquid directly. Thus some device must be adopted to move a pointer outside the tube to indicate the level of the mercury inside.

If one side of the tube is open to the atmosphere this is reasonably simple, for a float may be placed on the liquid in the open tube, and this float can actuate the mechanism which moves the pointer. When differential pressures are being measured, as in flow metering, both sides must be sealed and other methods used to transmit the position of the liquid level.

In such cases it is often more convenient to use a well type of manometer rather than a simple U tube. One limb of the U tube is greatly enlarged in comparision with the other limb. When a differential pressure is now applied to the manometer, the rise of the liquid on one side will not equal the fall on the other side, as it does in the simple U tube. The relative sizes of the rise and fall will depend upon the diameters of the tube and well. This principle is also used in single column gauges (*Figure 1.17*). Suppose a differential pressure of h mm of mercury is established between the two sides of the manometer as in *Figure 1.18*, and suppose the diameters of tube and well are d mm and D mm respectively. Suppose the level in the well falls x mm then:

Volume of mercury leaving the well = volume entering the tube

$$x \times \frac{\pi D^2}{4} \text{ mm}^3 = (h-x) \frac{\pi d^2}{4} \text{ mm}^3$$
$$xD^2 = hd^2 - xd^2$$
$$x(D^2 + d^2) = hd^2$$
$$x = h \frac{d^2}{D^2 + d^2} \text{ mm} \tag{1.13}$$

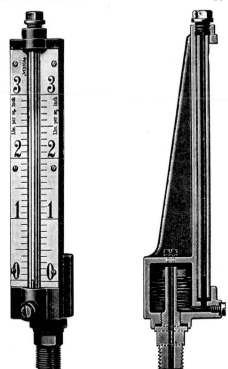

Figure 1.17
Well type of manometer.
(By courtesy, Budenberg Gauge Co.)

Thus, by adjusting the diameters of tube and well, we can make x any desired fraction of h. This has several distinct advantages.

1. The travel of a float placed on the liquid in the well can be limited to that which it is convenient to transmit to the pointer-moving mechanism.
2. By changing the diameter of the well or tube it is possible to change the range of the instrument without altering anything else.
3. In flow measurement the rate of flow depends upon the square root of the differential pressure produced. By shaping the well so that its section has a profile of a certain shape it is possible to make the motion of the float bear

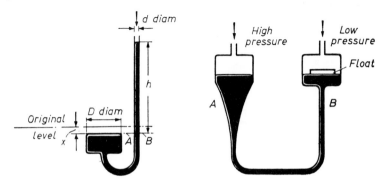

Figure 1.18   (left) Well type of manometer; (right) Well type of manometer with shaped well

a linear relationship to the flow; i.e., by making one limb of the shape shown in *Figure 1.18*, it is possible to make the float move equal amounts for equal changes in flow. In order to do this the volume of mercury moving from Chamber *A* to Chamber *B* and hence the movement of the float for equal changes of head, must get steadily less. If the instrument is designed for a differential head of 100 units, and the scale is divided into 10 units, then the pointer will move 1, 2, 3, 4, 5, 6, 7, 8, 9, 10 divisions for the corresponding values 1, 4, 9, 16, 25, 36, 49, 64, 81, and 100 units of differential head. It is not possible, however, to have a scale which is uniformly divided over the whole range of the instrument; the beginning of the scale will be unevenly divided. This portion of the scale in a flowmeter, is, however, rarely reliable, so this factor is not so important in flow indicators; but it has an important bearing on the design of mechanical integrators. Another disadvantage of the shaped-limb manometer is that the quantity of filling liquid is critical, so that it can be used only with mercury; and then care must be taken to ensure that the correct quantity of mercury is introduced and none lost while the instrument is in use.

4. Because of the large diameter of the well, a large float may be used. Thus the available force for moving the pointer mechanism can be made large, and the effects of friction easily overcome.

Methods of Transmitting the Position of the Float to the outside of the Tube

Several methods of transmitting the position of the float to the indicator are in common use.

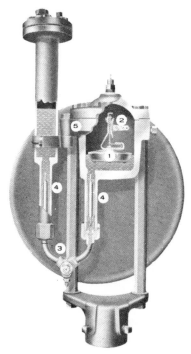

Figure 1.19
Well type of manometer
showing complete instrument.
(By courtesy, Foxboro-Yoxall Ltd.)

In one method (*Figure 1.19*) the rise and fall of the float is communicated to a segmental lever, which is fixed to a shaft carried in a pressure-tight bearing. This shaft transmits movement of the float to the pen or pointer mechanism. Great care must be taken in the design and production of this bearing for it must be pressure-tight yet free from friction.

Details of the bearing are shown in *Figure 1.20(a)*. This bearing is supplied as a complete unit and must be replaced as such because the shaft is hand-lapped into the bearing, and tested for pressure-tightness before being packed with grease. The correct grade of lubricant must be used for the service conditions which will be experienced, in which case the bearing is positively sealed against leakage and at the same time is, from all practical considerations, frictionless. The segmental link is connected to the float by means of a chain assembly (*Figure 1.20(b)*). By this means, the float is able to travel in a

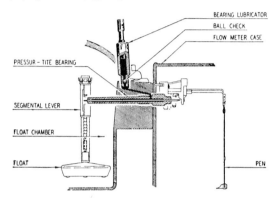

Figure 1.20(a)   Pressure-tight bearing for well type of manometer.
(By courtesy, Foxboro-Yoxall Ltd.)

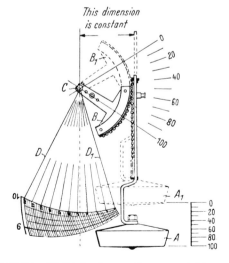

Figure 1.20(b)   Well type of manometer: segmental lever and float.
(By courtesy, Foxboro-Yoxall Ltd.)

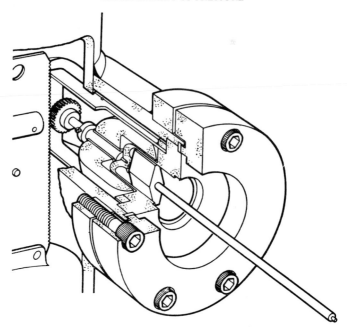

Figure 1.21   Well type of manometer: magnetic coupling

straight line, and its action is always transmitted at right angles to the pen-
arm shaft, so that the relationship between float travel and pen travel is linear.

Another method of obtaining a linear relationship between the vertical
motion of the float and the angular movement of the shaft is to attach a
vertical rack to the float. This rack engages with and rotates a pinion, and
the rotation of the pinion may be transmitted through the meter wall by
means of a magnetic coupling. *Figure 1.21* shows a photograph of a part-
sectioned magnetic coupling-system. The vertical rack, shown on the left-
hand side of the illustration, is attached to the float. This rack rotates a pinion
which is mounted on the same spindle as a cup-shaped two-pole magnet-
system. The magnet-system rotates on the outside of a cylindrical partition,
which must be of non-magnetic material and of sufficient strength to withstand the static pressure within the chamber. The armature which is a
lozenge-shaped magnet is mounted on an outer spindle, and rotates inside the
partition, and follows the rotation of the magnet-system. The recording and
indicating movement is driven from a pinion mounted on the outer spindle.
The great advantage of this system of transmission is that no pressure-tight
bearing is required as there is no mechanical connection between the float
mechanism and the external meter movement.

Yet another method of transmitting the level of the mercury within the
well to the outside is shown in *Figure 1.22*. In this case, as the mercury rises
it shorts out resistances so that the level of the mercury is represented by the
electrical resistance between the insulated rods and the casting of the well.
A rise in the mercury level reduces the resistance in the circuit.

A device used to transmit the position of the liquid in a glass U tube to a
recorder or indicator, consists of a ring around the bottom of the U tube

and another ring around the limb of the tube and carried by a lead screw driven by a servo-motor. The liquid in the U tube acts as one plate of a condenser, with the movable ring as the other plate. Any change in the capacity of the condenser affects an electronic capacity-sensitive circuit which operates the servo-motor so that the ring carried by the lead screw follows

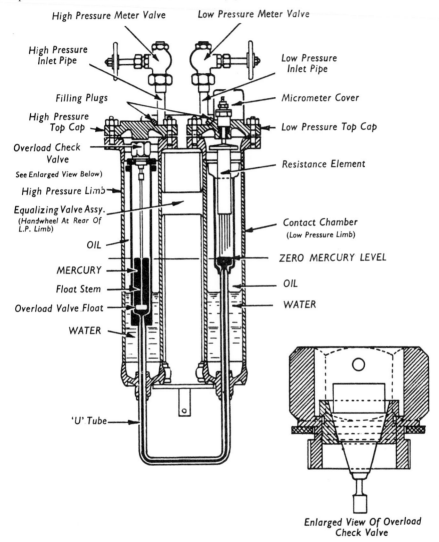

Figure 1.22   Well type of manometer: electrical transmitter. (By courtesy, G.E.C. Elliot Process Instruments Ltd.)

any movement of the surface of the liquid. The movement of this ring is then transmitted to the pen arm.

Another method of detecting the level of the mercury in a manometer is to use a floating cylindrical magnet which moves up and down in a cage inside one limb of the manometer according to the level of the mercury.

Outside the manometer is a pivoted strip magnet, which is repelled by the floating magnet. The tendency for this magnet to be deflected is balanced by the magnetic force produced by an electromagnet. The value of the current through the electromagnet is therefore a measure of the differential pressure applied to the manometer. The device will be dealt with at greater length in the section on transmitters. One of its great advantages is that it can be arranged to transmit a current proportional to the differential head, or a current proportional to the square root of the differential head; which in flow measurement would mean a current proportional to the flow.

Overload and damping devices

If for some reason the differential pressure between the two sides of the pressure gauge greatly exceeds the normal value, the mercury will be blown out of the instrument into the pipes and plant, unless some form of check valve is provided. The most common form is shown in the diagram (*Figure 1.23*). It consists of a heavy float *A* with a conical valve *B* at the lower end,

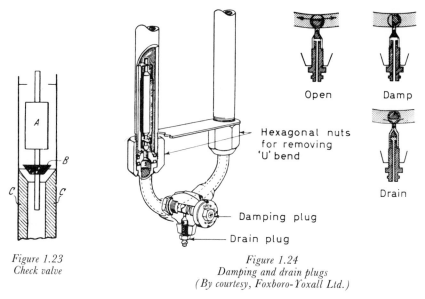

Figure 1.23
Check valve

Figure 1.24
Damping and drain plugs
(By courtesy, Foxboro-Yoxall Ltd.)

and a seating *C*. Normally the upthrust due to the mercury keeps the valve *B* clear of the seating. If, however, there is a sudden flow downwards the float is carried along with the stream and the valve beds down on the seating, sealing off that limb of the manometer and preventing further flow. It also comes into operation when the level of the mercury has fallen to such an extent that it no longer supports the float, for, in that case, the weight of the float will cause the valve to bed down on its seat and cut off further flow. In this instance, therefore, if each chamber of the gauge is large enough to hold all the mercury, the instrument will return to normal operation as soon as the overload is removed. A check valve may be fitted in each limb, as shown in *Figures 1.19* and *1.24*.

In installations where the differential pressure to be measured is pulsating, a restricting plug, which can be adjusted while the meter is in operation, is fitted in the tube connecting the two limbs of the manometer. This, by damping down the fluctuations due to the pulsations, smooths out the record produced by the instrument; the plug can be seen at the bottom of the U bend in *Figures 1.20* and *1.24*.

1.4 PRESSURE MEASUREMENT BY BALANCING THE FORCE PRODUCED ON A KNOWN AREA BY A MEASURED FORCE

1.4.1. Piston type

Even when mercury is used as the fluid in a manometer, the pressure which may be measured by the simple type soon reaches a limit. This limit is set by the height of a column which is practicable. A convenient type of gauge which can be used for higher pressures, and, in particular, for checking the elastic diaphragm or Bourdon type of gauge, is the free piston type of gauge.

In this type of instrument, the force produced on a piston of known area is measured directly by the weight it will support. In *Figure 1.25*, if the pressure

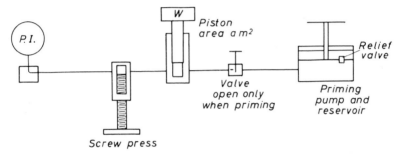

Figure 1.25 Diagrammatic arrangement of the dead-weight tester

acting on the piston is $\dfrac{p \text{ N}}{\text{m}^2}$ g and the area over which it acts is a m², then the force produced will be pa N. If there is no friction between the piston and the walls, then the piston will support a weight W, where $W = pa$ N.

The Dead-weight Pressure Tester, shown diagramatically in *Figure 1.25* and in the practical form in *Figure 1.26* is often used as a standard of pressure measurement. The accuracy largely depends upon the accuracy of the manufacture of the piston which must be finished to very narrow limits of error in diameter, roundness and straightness. The pistons are therefore made of hardened and tempered steel accurately ground and lapped to size, and for the high pressure models a National Physical Laboratory report on the accuracy is supplied with each tester. The piston is then fitted into the cylinder with the minimum of clearance and the effective diameter is presumed to be the mean of the piston and cylinder diameter. The cylinder is filled with a light acid-free and resin-free mineral oil for pressure up to 550 bar and with castor oil for higher pressures.

In order to eliminate the effects of friction, the piston is rotated while a reading is being taken. In the low pressure types (i.e. up to 550 bar) the weights are placed directly on the top of the piston, but for higher pressures this method is not suitable, since the stack of weights becomes unwieldy, and excessive frictional errors may be introduced if the weights are piled out of centre. In the high pressure models (up to 8000 bar) an overhang design is used, in which the piston, which may have an area of as little as 32 mm², is fitted with a head-and-ball socket to support the weight carrier. The weight carrier consists of a platform around the bottom of a long tube. The top of the

Figure 1.26   Dead-weight tester with dual range piston unit. (By courtesy, Budenberg Gauge Co. Ltd.)

tube is domed and rests on the ball, and is therefore free to set itself so that the weight acts vertically downwards through the centre of the piston, avoiding all side stresses.

The dead-weight tester is fitted with a pump for priming and a screw press for producing the pressure, the size and design of the press depending upon the highest pressure involved. The general arrangement is shown in *Figure 1.25*.

No oil must enter the gauge when testing oxygen pressure gauges, because the oil vapour together with the oxygen will produce an explosive mixture. For pressures up to 20 bar oxygen gauges may be tested with water or air. For pressures between 20 bar and 550 bar a seal of the form shown in *Figure 1.27* may be used.

As the load applied to the piston, to maintain equilibrium, divided by the area of the piston, which is a fixed quantity, is a direct measure of the pressure within the cylinder, the weights are supplied in convenient multiples of units of pressure. These units may be N/m², bar etc.

In the dual range instrument shown in *Figure 1.26* suitable for pressures from 1–1100 bar g, a patented piston unit is used having areas of 80·64 and 4·032 mm². This feature cuts down the number of weights required to cover the range and the work involved in lifting them on to the piston. Thus, to test a 140 bar g gauge at the beginning and at the end of the scale, weights corresponding to 7 bar g, on the larger piston are put on, and the pressure raised to 7 bar g. Further turning of the screw press locks the larger piston out of use and at 140 bar g, the smaller piston comes into action without any further weights being added. A colour indication on the side of the unit shows whether the larger or smaller piston is in use. Certificates of accuracy are provided with each instrument and the makers guarantee the error does not

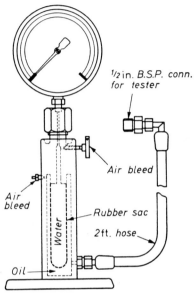

Figure 1.27   Seal unit for oxygen pressure gauges. (By courtesy, Budenberg Gauge Co. Ltd.)

exceed 0·03 per cent of the pressure being measured. Low-pressure air-operated Dead-weight Testers are available for calibrating pneumatic equipment such as telemetering and control equipment. These are used in a similar manner to the liquid filled ones but clean dry air or nitrogen from a main or cylinder is used as the operating medium. Typical ranges for such testers are 0·1 to 7 bar g or 1 to 35 bar g.

As the weight of a mass m kg placed on the platform of the piston is mg N it is dependent on the local gravity and a correction will have to be made for any departure of the local gravity from the value for which the dead weight tester was designed (see equation 1.1).

1.4.1.1   COMBINED PISTON AND LIQUID COLUMN GAUGE

When gas such as North Sea Gas is delivered to a consumer who uses large volumes, the pressure at which the volume of the gas is measured becomes important and the industrial pressure gauge is not sufficiently accurate. A device which may be used to measure pressures in gas mains in the range 4–15 bar with a resolution of 0·5 m bar is the liquid column dead weight gauge illustrated in *Figure 1.28*.

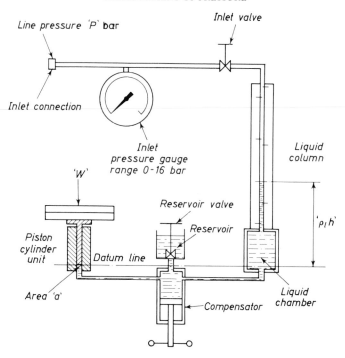

Figure 1.28   Layout of liquid column dead-weight gauge.
(By courtesy, Budenberg Gauge Co. Ltd.)

The pressure to be measured is applied to the column of liquid. The datum line of the system is the mid float position of the piston and if the piston face is at this line and the system in equilibrium

$$\frac{W}{a} = P + g p_i h$$

or
$$P = \frac{W}{a} - g p_i h \qquad (1.14)$$

where W is the weight applied to the piston, p_i the density of the liquid and h the height of the liquid column in self-consistent units. The weights are marked in bar and are arranged in steps of 0·02 bar while the column is calibrated in units of 1 m bar. The liquid is a red coloured hydraulic mineral oil and if the reading is taken with the piston floating slightly above or below the datum line the variation is small and may be ignored.

The instrument is housed in a cabinet which is levelled by means of two screws at the front of the cabinet before readings are taken.

1.4.2   Ring balance type

This type of instrument is frequently used for the measurement of low differential pressures of the order of 100 mm of water gauge. The essential portion of this instrument (shown in *Figure 1.29*) consists of a hollow ring of circular section, partitioned at its upper part and partially filled with a liquid in order to form two pressure-measuring chambers. The body of the ring is supported at its centre by a knife edge resting on a bearing surface, or by roller-bearings or ball-bearings. The ring may be of metal, such as aluminium alloy, or of a plastic moulding, the material depending upon the nature of the gas whose pressure is being measured. The nature of the gas will also determine the nature of the filling medium. The quantity and the nature of this medium has no influence upon the calibration of the instrument, as it acts only as a seal. The

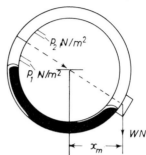

Figure 1.29   Ring balance

force which operates the instrument is due to the difference between the pressures on the two sides of the partition. The cross-sectional area of the ring is therefore made large when the differential pressure to be measured is low, and less when the differential pressure is higher.

The fluids whose pressure difference is required are led into the ring by flexible connections. These are placed so that their length and movement are at the minimum. The ring is balanced by a control weight which is at its lowest point when the pressure is the same on both sides of the partition.

Suppose the cross-sectional area of the ring is A m², the mean radius of the ring is r m, and the balance weight is r_1 m from the pivot. If a pressure p_1 N/m² is applied to one side and a pressure p_2 N/m² $(p_2 > p_1)$ to the other side then:

$$\text{Force acting on the partition} = (p_2 - p_1)\,A\ \text{N}$$

This will have a turning effect or moment of $(p_2 - p_1)\,r\,A$ Nm about the centre.

The ring will therefore rotate in a counter-clockwise direction until this moment is balanced by the moment of the counterweight.

If x m is the horizontal distance from the weight to a vertical line through the pivot, then:

$$(p_2 - p_1)\,Ar = Wx \tag{1.15}$$

i.e. $$x = (p_2 - p_1)\,\frac{Ar}{W} \quad \text{or} \quad (p_2 - p_1) = \frac{xW}{Ar} \tag{1.15a}$$

Since A, r and W are fixed, x is a measure of the differential pressure $(p_2 - p_1)$.

But $x/r_1 = \sin a$, where a is the angle through which the ring has turned, or $x = r_1 \sin a$.

Substituting in equation 1.15,

$$(p_2 - p_1) = \frac{r_1 W}{Ar}\sin a \tag{1.16}$$

By changing the counterweight *W*, the calibration of the instrument may be altered.

In the case of an overload, the excessive pressure will equalise by depressing the filling liquid on one side down to the bottom of the ring so that the gas on the high pressure side will bubble through to the low pressure side. The quantity of filling liquid should be such that bubbling takes place before the level of the filling liquid reaches the level of the low pressure connection. No special overload device is therefore necessary. When required to give a linear relationship between the deflection of a pointer and rate of flow, the ring balance is used to drive a cam, and the follower rotates the spindle of the pointer or the recording mechanism. Another method of obtaining a linear scale is to suspend the control weight by means of a flexible metal strip which passes over a cam mounted on the ring. By shaping the cam the control torque can be varied over the entire range of the instrument so as to obtain a linear scale.

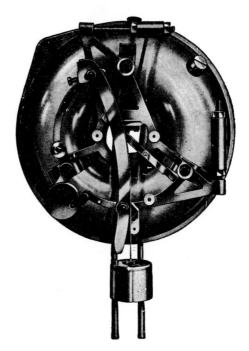

Figure 1.30   Ring balance

Yet another method is to attach a shaped displacer to one side of the ring. This displacer dips into a vessel of mercury or other liquid. It can be so arranged that increasing amounts of the displacer leave the liquid as the ring rotates so that the control force increases at an increasing rate giving a linear scale.

This type of instrument can be used for measuring at a large range of static pressures. The range is fixed by the nature and thickness of the material from which the ring is made and the pressure which the flexible connections will stand. For very high static pressures a half ring of steel is used with flexible connections consisting of hypodermic tubing. The fact that there is only half a ring does not alter the principle of the instrument.

The range of differential pressure for which the instrument can be used is fixed by the size of the ring and the nature and the quantity of the sealing liquid.

Although the calibration of the instrument does not depend upon the sealing liquid, the range is fixed by the maximum differential pressure which can be applied before the level of the sealing liquid falls to the bottom of the ring and allows the gas to bubble through from the high pressure to the low pressure side. This will not depend upon the size of the counter-weight, but only upon the product of the density of the liquid and the maximum permissible difference of levels. The mass of the counterweight and its position should be such that the ring rotates a convenient amount for the range of the instrument.

An actual ring balance portion of an instrument for low static pressure is shown in *Figure 1.30*.

1.4.3 Bell type

The Bell type of pressure gauge is really a bridge between sections 1.4 and 1.5, for in this type of gauge, the force produced by the difference of pressures on the inside and outside of a bell is balanced against a weight, or against the force produced by the compression of a spring.

The instrument consists of a bell suspended with the open end downwards in a sealed chamber, usually made of cast iron, containing a liquid such as oil or mercury. The liquid covers the open end of the bell and acts as a seal, so forming two chambers. In this type of instrument (shown in *Figure 1.31*), where gravity provides the controlling force, the higher pressure is led into the inside and the lower pressure acts on the outside of the bell. The resulting force causes the bell to rise until equilibrium is reached between the upward force and the apparent weight of the bell. As the bell rises, there is less of it immersed in the sealing liquid, so that the upthrust on it due to buoyancy is reduced. Its apparent weight will therefore increase. The thickness and density of the material from which the bell is made, its cross-sectional area, and the

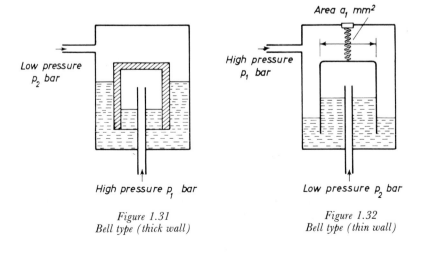

Figure 1.31
Bell type (thick wall)

Figure 1.32
Bell type (thin wall)

density of the sealing liquid, are determined by the range for which the instrument is to be used.

Since the pressure within the bell is greater than that outside, it will cause the level of the liquid on the outside of the bell to be greater than the level on the inside, as well as causing the bell to rise. The mathematical analysis of the relationship between differential pressure and the amount the bell rises is therefore complicated, and will not be dealt with here, but is given in the appendix.

In the second type (*Figure 1.32*), however, where the bell is made of thin material and the controlling force is obtained by means of a spring, the

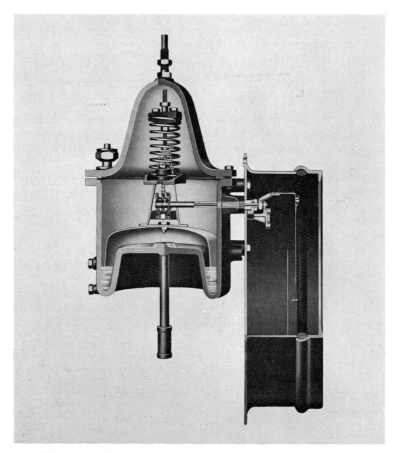

Figure 1.33   Bell type (thin wall). (By courtesy, Foxboro-Yoxall Ltd.)

mathematics are simple, as in this case the effects due to displacement of liquid are small, and if the sealing liquid is not very dense they may be neglected.

If the bell is made of thin material, the areas over which the pressures act will be equal on the inside and the outside. In this type of instrument it is usual to apply the high pressure to the outside of the bell and the low pressure to the inside. Then: Difference between force due to pressure on inside and force due to pressure on the outside $= (p_1 - p_2)\, a_1 \times 10^{-1}$ N. If the spring obeys Hooke's law;

$$\frac{\text{change in length}}{\text{original length}} = \frac{\text{applied force}}{\text{modulus of elasticity}}$$

$$\therefore \text{Change in position of bell} = \frac{\text{original length of spring} \times (p_1 - p_2) a_1}{\text{modulus of elasticity of spring}}$$

$$\therefore \text{Travel of bell} = \text{constant} \times (p_1 - p_2) \tag{1.17}$$

i.e. Travel of bell is proportional to the differential pressure to be measured.

The same methods may be adopted to transmit the motion of the bell to the outside of the chamber as were used to indicate the position of the float in instruments of the type described in section 1.3.2.

No overload device is necessary in the bell type of instrument, for, when the differential pressure becomes great enough to force the liquid down to the level of the edge of the bell, the gas will bubble through the sealing liquid and tend to equalize the pressure until the differential returns to a measurable quantity (*Figure 1.33*).

The range of the instrument will be determined by the modulus of elasticity of the spring and by the density of the sealing liquid. For low ranges up to a few mm Hg an organic liquid is used as a seal. For higher ranges mercury is used.

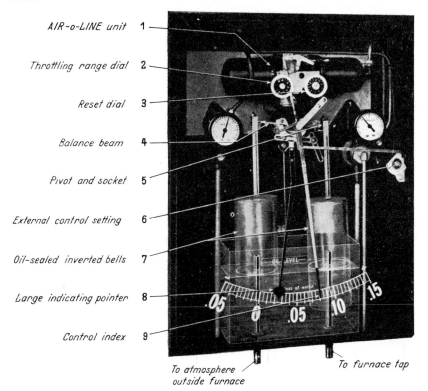

AIR-o-LINE unit   1

Throttling range dial   2

Reset dial   3

Balance beam   4

Pivot and socket   5

External control setting   6

Oil-sealed inverted bells   7

Large indicating pointer   8

Control index   9

To atmosphere outside furnace

To furnace tap

Figure 1.34 Brown furnace pressure controller. (By courtesy, Honeywell Controls Ltd.)

Owing to the large area of the bell this type of instrument is very useful for measuring the difference between two low static pressures.

In the pressure-measuring element shown in *Figure 1.34*, two bells inverted in a bath of oil are used. These bells are suspended from the balance beam (4), which is carried on pivot-and-socket type bearings having a very small surface contact giving small friction. The two pressures to be compared are led into the inside of the bells, and the pressure differential is indicated by the pointer (8) which moves with the balance beam. Since the restoring force is small (being produced by the change in position of the Centre of Gravity of the balance beam as it rotates) the instrument is sensitive to very small changes of pressure. Both bells are subject to the same changes of ambient temperatures, so that the instrument is unaffected by temperature changes. Changes as small as $2 \cdot 5 \times 10^{-2}$ mm of water gauge may be detected, and the instrument is used for controlling furnace pressure.

Models are available having differential ranges between 0–5 mm w.g. and 0–300 mm w.g. at static pressures up to 10 bar g.

1.5 PRESSURE MEASUREMENT BY BALANCING THE FORCE PRODUCED ON A KNOWN AREA AGAINST THE STRESS PRODUCED IN AN ELASTIC MEDIUM

1.5.1 The Bourdon tube

Because of its simplicity and versatility probably the most commonly used type of pressure gauge is that using a Bourdon tube and its modifications.

1.5.1.1 THE 'C' TYPE BOURDON TUBE

In its simplest form the Bourdon tube consists of a tube of oval section, bent in a circular arc. One end of the tube is sealed and attached by a light link-work to the mechanism which operates the pointer. The other end of the tube is fixed, and is open for the application of the pressure which is to be measured. The internal pressure tends to change the section of the tube from oval to circular, and this tends to straighten out the tube. The resulting movement of the free end of the tube causes the pointer to move over the scale. In gauges measuring pressures less than atmospheric, the free end of the tube tends to move towards the boss, so that the pointer-operating mechanism must be reversed if the difference between the applied pressure and the atmospheric pressure is to be indicated in a clockwise direction on the scale.

The tube is made from a variety of materials in a variety of thicknesses. The material chosen depends upon the nature of the fluid whose pressure is being measured, and the thickness of material upon the range of measurements required. The actual dimensions of the tube used will determine the force available to drive the pointer mechanism, and this should be large enough to make any frictional force negligible.

Where no corrosion problems exist, solid drawn phosphor-bronze tubes with soft soldered or brazed joints are used for ranges of 1–70 bar g; solid drawn heat-treated beryllium-copper tubes with brazed joints for up ranges to 350 bar g; and solid drawn alloy steel tubes with screwed and welded joints for ranges 70 to 6000 bar g. When required to resist corrosion by the measured fluid, solid drawn carbon steel tubes with soft soldered or welded joints are used for ranges 1 to 35 bar g; drawn stainless steel tubes with welded joints for ranges 2 to 70 bar g and solid drawn 'K' monel tubes with screwed and welded joints are used for ranges 70 to 1400 bar g.

The Bourdon gauge may also be used for measuring pressures less than atmospheric, and can be used to measure steam pressures on boilers of stationary or locomotive engines. It is used extensively to measure water pressure, air pressure, carbon dioxide pressure and for measuring the pressure of a very large variety of other liquids and gases.

The construction of the standard concentric pressure gauge is shown in *Figure 1.35*. The motion of the free end of the tube is communicated by means

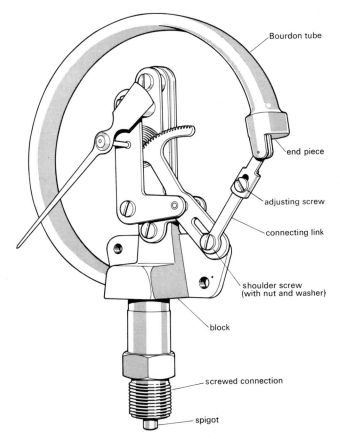

Figure 1.35   Chassis of Bourdon gauge. (By courtesy, Budenberg Gauge Co. Ltd.)

of a connecting link to the lower end of the pivoted quadrant. The upper part of this quadrant consists of a toothed segment which engages with the teeth of the central pinion which rotates the pointer. The play between quadrant and pinion is taken up by a fine phosphor-bronze hair-spring.

In the eccentric-pattern pressure gauge the motion of the end of the tube is translated into pointer rotation by means of a connecting link which applies the motion directly on to an eccentric carried by the pointer spindle.

In applications where the gauge is subject to vibration, Bakelite quadrants carried by Bakelite plates are found to withstand the rough treatment better than the standard ones made of brass or phosphor-bronze.

In applications such as the measurement of the pressure of a liquid delivered by a high-speed gear pump, the vibrations of the pointer may be so small that they are scarcely visible. They are, nevertheless, just as destructive to the mechanism, and are therefore damped out by completely filling the case of the instrument with glycerine, which damps out these vibrations.

Accuracy and temperature errors

Provided the power from the tube is sufficient to overcome friction at all points of relative movement, the tube has no hysteresis, the gauge is used at or near the temperature at which it was calibrated, and the standard of comparison an error less than $\pm 0 \cdot 1$ per cent; an industrial pressure gauge can be made with an accuracy of $\pm 0 \cdot 5$ per cent of the scale range. However, as the conditions of use often fall short of ideal, in practice gauges for plant use are required to have an error of less than ± 1 per cent of the scale range for the middle 80 per cent of the scale, and $\pm 1\frac{1}{2}$ per cent scale range over the first and last 10 per cent of the scale range. These figures are for tests conducted under both rising and falling pressure at a temperature of $20°$ C.

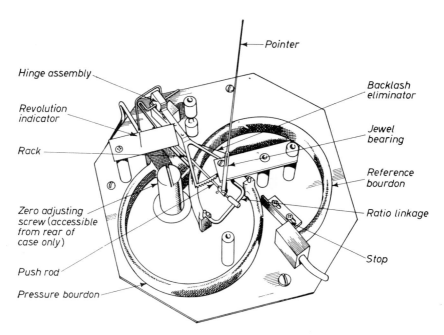

Figure 1.36   High precision type of Bourdon gauge. (By courtesy, Wallace & Tiernan)

Deviation of the temperature of use from the temperature at which the gauge is calibrated causes both a zero and ranging error. The magnitude of the zero error depends upon the coefficient of expansion of the materials of the Bourdon tube, link and movement. For a gauge with a brass block and link and a phosphor-bronze tube the error is $-0\cdot1$ per cent of full scale deflection for a temperature rise of $20°$ C. A similar gauge with a steel tube will have an error of $-0\cdot5$ per cent full scale for the same temperature. As the gauge movement magnifies the tip movement of the Bourdon tube, increase in the movement magnification will cause a corresponding increase in the error.

Increase in temperature causes a decrease in Young's modulus for the tube material. This causes the deflection of the tip of the Bourdon tube for a given applied pressure to increase with increasing temperature. The error caused by this effect over normal temperature ranges is about $0\cdot7$ per cent of the applied pressure over $20°$ C rise for phosphor-bronze tubes and about $0\cdot5$ per cent for alloy steel tubes. As Ni-Span-C has a constant Young's modulus over the range of ambient temperature usually met, a gauge using a tube of this material will have a negligible error from this effect.

The high precision type of Bourdon tank gauge

Figure 1.36 illustrates a high precision type of Bourdon tube pressure gauge. The type shown is for absolute pressure measurement, and the reference Bourdon is completely evacuated and sealed. The pressure to be measured is applied to the second Bourdon. The free ends of Bourdon tube are connected by a ratio linkage which through the push rod transmits the difference in Bourdon movements to a rack assembly which in turn rotates the pinion and pointer. Jewel bearings are used to reduce friction to an absolute minimum. Backlash is eliminated by a nylon thread which connects a spring on the rack with a grooved pulley on the pinion shaft thus maintaining a uniform tension for all positions of the rack and pinion.

The Bourdon tubes are made of Ni-Span-C which has a very high temperature stability and resistance to corrosion. As both Bourdons are subject to the same changes of atmospheric pressure, the error in gauge reading does not exceed the instrument tolerance for barometric pressure changes of ±130 mm Hg. The dial diameter is 216 mm and the full range of the instrument is covered by two revolutions of the pointer giving a scale length of $1\cdot14$ m with high scale readability. A revolution indicator shows which scale should be read. Instruments are available for ranges from 0–7 bar a, to 0–35 bar a. In the gauge pressure indicator one Bourdon tube only is used, and instruments are available for ranges from 0–7 bar g to 0–35 bar g.

The instrument is normally calibrated with the dial vertical and has a sensitivity of $\frac{1}{80}$ per cent of full scale, an accuracy of $\pm0\cdot1$ per cent of full scale, and a total temperature effect $0\cdot1$ per cent of the range per $10°$ C change. A bleed system is provided enabling the pressure system to be purged when it is required to measure the pressure of a variety of gases.

When Bourdon types of pressure gauges are used, the normal working pressure should not be more than 60 per cent of the maximum pressure indicated by the gauge. If this rule is adhered to, the instrument is much less liable to become inaccurate owing to changes in the material of the tube brought about by elastic fatigue.

The nature of the liquid or gas whose pressure is being measured must also be considered when choosing the suitable material for the tube, as corrosion of the tube to the fluid it contains will soon alter its elastic properties and so render the instrument unreliable. If a gauge has been overloaded so that its pointer has gone off the scale, it should no longer be regarded as being accurate until recalibration has shown it to be so. Calibration of pressure gauges is dealt with at the end of this chapter.

1.5.1.2   THE SPIRAL BOURDON TYPE

The amount of movement of the free end of a Bourdon tube varies inversely as the wall-thickness and depends upon the cross-sectional form of the tube. It also varies directly with the angle subtended by the arc through which the tube is bent. In a tube having an arc of 180°, the movement of its end will be twice that of a similar tube having an arc of 90°. The movement of the free end of the tube may be increased without changing its wall-thickness, by increasing the length of the arc of the tube. When the arc through which the tube is bent reaches 360°, its length can be further increased in two ways: the tube can be made in the form of a spiral, or it can be made in the form of a helix. By increasing the number of turns in the spiral or helix an enlarged

Figure 1.37   Spiral pressure element. (By courtesy, Foxboro-Yoxall Ltd.)

movement of the free end of the tube is obtained and the need for a further magnifying movement avoided. In this way the need for the quadrant and pinion is eliminated, and with them the backlash which tends to occur when they become worn owing to continued use or to the presence of vibration.

The spiral form of the pressure element illustrated in *Figure 1.37* is used for lower pressure measurement, while the helical form is used for higher pressures.

The movement of the free end of the tube is transmitted to the pen arm or pointer through a flexible metal connecting strip which joins the free end to the pointer shaft. This enables the end of the spiral to move freely in a radial direction as the spiral expands. In the element shown, the spiral is made from chrome-molybdenum steel tubing, all joints and closures are welded and the element heat treated to remove any stress which may have been set up in the material. This ensures uniform elastic properties in the tube. The junction between the spiral and the connecting tube is made by means of a special compression fitting.

1.5.1.3 THE HELICAL BOURDON TUBE

For higher pressures the tube is wound in the form of a helix, and this is illustrated in *Figure 1.38.*

Figure 1.38   Beryllium copper helical pressure element. (By courtesy, Foxboro-Yoxall Ltd.)

The material used for the seamless tube from which the helix is wound is determined by the nature of the fluid being metered and the range of the instrument, and follows the same practice as for 'C' type Bourdon tubes. Some indication of the present practice is given here but this is subject to wide variation. Where the nature of the fluid allows, a special bronze alloy is used for ranges 1–40 bar g; beryllium-copper for ranges between 40 and 700 bar g; and chrome-molybdenum steel for ranges between 7 and 300 bar g, particularly in the presence of ammonia. Stainless steel elements are being increasingly used. They are used in the petroleum industry where bronze is affected by corrosive compounds in the oils. In general, bronze elements are used for measuring the pressure of steam, water, air, nitrogen and similar gases.

Beryllium-copper is tough and has reliable elastic properties which make it very suitable for high pressure elements. The motion of the end of the helix is communicated to the pen in the same way as for the spiral element. The helical element is used in applications where multiple records are made on the same chart; for example, in the measurement of gas flow, where the measurement involves the knowledge of the static gas pressure and the pressure differential produced across an orifice.

1.5.2 Diaphragm pressure elements

Diaphragm elements may be of two forms: Stiff metallic diaphragms or bellows; and slack diaphragms and drive plate.

1.5.2.1   STIFF METALLIC DIAPHRAGM ELEMENTS

The simplest form of diaphragm gauge is the Schaffer diaphragm gauge shown in *Figure 1.39*. It consists of a hardened and tempered stainless-steel corrugated diaphragm of about 65 mm diameter held between the two flanges. Pressure is applied to the underside in the chamber shown, and movement of the centre of the diaphragm is transmitted through the ball-and-socket joint and high magnification link to the pointer as in the Bourdon Gauge.

The upper flange is flat to prevent further movement of the diaphragm when the pointer has reached the end of the scale, so that the diaphragm is not damaged by pressures considerably larger than the maximum measurable pressure.

Figure 1.39   Schaffer diaphragm gauge. (By courtesy, Budenberg Gauge Co. Ltd.)

When the gauge is to be used on corrosive fluids the chamber is made from a corrosion-resistant material such as Meehanite cast iron, manganese bronze or stainless steel, or it may be mild steel lined with a corrosion-resistant material such as lead. In addition, the diaphragm is protected by coating the underside with P.T.F.E. or by covering it with a thin disc of silver. This type of gauge may be used for measuring pressures which are either greater or less than atmospheric, but it is more difficult to protect the diaphragm of the gauges for pressures less than atmospheric because of the tendency for diaphragm

Figure 1.40   Aneroid barometer. (By courtesy, Short and Mason, Ltd.)

and protective coating to part under the influence of the reduced pressure. This type of gauge gives better and more positive indication than the Bourdon type of gauge for low-pressure ranges, particularly for gauges graduated below 1 bar. It is also suitable for measuring fluctuating pressures.

The type of stiff diaphragm element commonly used for low-pressure measurement, between 25 mm water gauge and 0·3 bar, is found in the aneroid barometer and the altimeter. When used in the barometer or altimeter this element consists of a flat circular box or capsule, having a corrugated lid and base as shown in *Figure 1.40*. Most of the air is pumped out of the capsule before it is sealed. It is prevented from collapsing by a strong spring carried by a bridge and attached to the centre of the top of the capsule. Attached to the spring is a lever which, acting through a bell crank and lever, tightens or slackens the chain which rotates the pointer. The chain is kept in tension by a hair-spring. When the atmospheric pressure increases, the capsule is pressed in and the lever moves downwards, moving the bell crank and slackening the chain and so allowing the pointer to turn in one direction. A decrease in the air pressure allows the capsule to expand, moving the lever and bell crank in the opposite direction and so tightening the chain and rotating the pointer in the opposite direction.

As the air pressure decreases in a definite manner with height above sea level it is possible to graduate the scale in height in m instead of mm of mercury in which case the instrument becomes an altimeter.

Figure 1.41 illustrates a high precision type of pressure indicator using a capsule type of pressure element. This instrument may be designed for differential pressure or gauge pressure measurement. For differential pressure measurement, the higher absolute pressure is applied to the inside of the

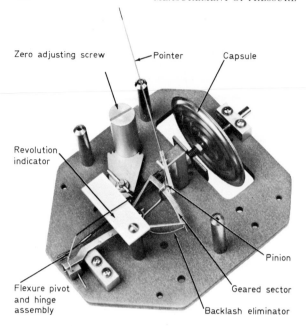

Zero adjusting screw   Pointer   Capsule

Revolution
indicator

Flexure pivot
and hinge
assembly

Pinion

Geared sector

Backlash eliminator

*Figure 1.41 High
precision capsule type
pressure indicators.
(By courtesy, Wallace
and Tiernan)*

capsule and the lower absolute pressure applied to the pressure-tight case. For gauge pressure measurement the measured pressure is applied to the capsule and the case is open to the atmosphere. Absolute pressure is measured by evacuating the case. Vacuum is measured by applying the measured pressure to the case and atmospheric pressure to the capsule.

The meter movement is similar to that of the precision Bourdon type gauge described earlier. The instrument is available in ranges from 0–300 m bar to 0–10 bar, has a sensitivity of 0·01 per cent of full range, an accuracy of 0·1 per cent of full scale and a temperature effect of 0·01 per cent per °C indicated reading about the calibration temperature of 20° C. The case is normally suitable for a pressure of 2·5 bar, but a special case capable of withstanding 7 bar is available.

To increase the sensitivity of the barometer, several capsules may be placed one on top of the other. This increases the flexibility of the system, and is the device used in the barograph or recording barometer shown in *Figure 1.42.*

*Figure 1.42   Recording barometer.
(By courtesy, Short and Mason, Ltd.)*

A bimetallic strip is introduced into the lever connecting the pressure-sensitive element with the bell crank in all these instruments to compensate for changes in temperature, which will affect the strength of the spring, flexibility of the capsule, and the lengths of the levers. The size of the bimetallic strip is chosen so that the amount it curls with temperature change exactly counteracts the movement of the pointer due to the other effects of temperature change.

The type of capsule element described above is found in many types of pressure instruments, and in automatic boost control and mixture control in aircraft. If the pressure to be measured is introduced into the inside of the capsule, then the element will measure the difference between the pressure so introduced and the atmospheric pressure, i.e. the normal gauge pressure. If, however, the capsule is pumped out and sealed and placed in a sealed chamber into which the pressure to be measured is introduced, then the element will measure the difference between the pressure inside the capsule, which is extremely small and can be considered constant, and the applied pressure. It will therefore measure the absolute pressure.

For light pressures the diaphragms are generally made of non-ferrous alloys, but for higher pressures they are made of steel. The beryllium-copper diaphragms are hardened by heat treatment, and those of steel also are hardened and tempered.

Elements are built up of various numbers of diaphragms of various diameters, usually less than 100 mm, and material of thickness from about 0·05 mm upwards. They are suitable for measuring pressures from 25 mm water gauge up to 3·5 bar.

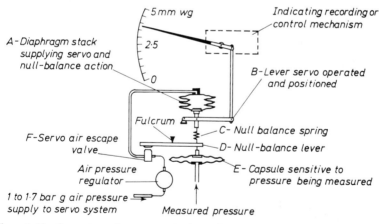

Figure 1.43   Servo-operated pressure mechanism. (By courtesy, Negretti & Zambra, Ltd.)

Figure 1.43 shows a servo-operated positioning mechanism, in which such capsules are used to measure low pressures and to amplify the working force sufficiently to operate the pointer mechanism satisfactorily.

Pressure being measured extends capsule *E* to move lever *D* and close air escape valve *F*. Air supplied to the servo-system increases the pressure extending diaphragm *A* to move lever *B*. Simultaneously the null balance spring *C* is increasingly compressed until it balances the upward pressure reaction of diaphragm E to move lever D and open the valve F. The action is such that while the pressure being measured remains constant, the valve will be held slightly open to allow additional air supplied to the servo-system to escape and thus maintain the pressure and consequently the position of lever B to correspond exactly with the pressure being measured. It follows, therefore, that the measured pressure is only employed to react upon the null balance lever D; whilst the relatively large effort of positioning lever B is the function of a powerful servo-mechanism.

The service air pressure supply passes through an air filter to a constriction and pressure regulator to regulate the pressure to the escape valve at 0·8 bar g, and it will be found that a service change of 0·7 bar will not affect the positioning accuracy.

For low-pressure measurement a built-up diaphragm element (*Figure 1.44*) is often used with small coils of non-ferrous spring metal attached to the edges of the diaphragms on one side to restrain the motion on that side. In this way the movement of the opposite side is almost doubled, and is transmitted to the pointer, as before, by a link mechanism.

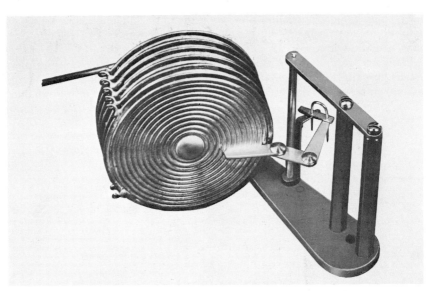

Figure 1.44   Low-pressure diaphragm unit. (By courtesy, Foxboro-Yoxall Ltd.)

Another instrument employing stiff diaphragms is the diaphragm meter illustrated in *Figure 1.45*. This type of instrument is now used very frequently in place of the well-type mercury manometer for measuring differential pressure. *Figure 1.45a* shows the principle of the operation while *Figure 1.45b* shows a cut-away view of the differential unit.

The compensating diaphragm element A and the range diaphragm element B with the range spring C, the drive unit D and the damping unit E, are assembled on the body. The pressure-tight covers F form a high pressure and a low pressure chamber. The diaphragms A and B are connected, internally, through the damping valve E, and are completely filled with a stable, non-freezing liquid.

An increase in pressure in the high pressure chamber compresses diaphragm A, displaces liquid into, and expands diaphragm B until the force of the range spring C equals the difference between the forces on diaphragms A and B. The linear motion of diaphragm B moves the inner end of the drive unit D and the outer end moves correspondingly through the bellows-sealed flexure. The pen arm is driven from the motion of the outer end of the drive unit.

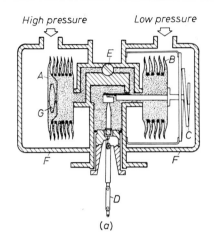

(a)

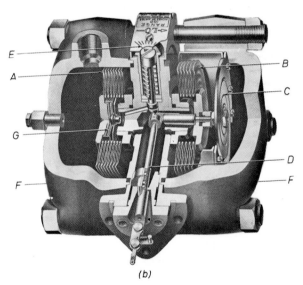

(b)

Figure 1.45   Diaphragm meter. (By courtesy, Foxboro-Yoxall Ltd.)

A bimetallic temperature compensator G, inside the compensating diaphragm A, adjusts the capacity of the diaphragm assembly to the changing volume of the filling liquid resulting from any change in the ambient temperature. The temperature compensator is correct for all differential ranges and no zero adjustment is necessary.

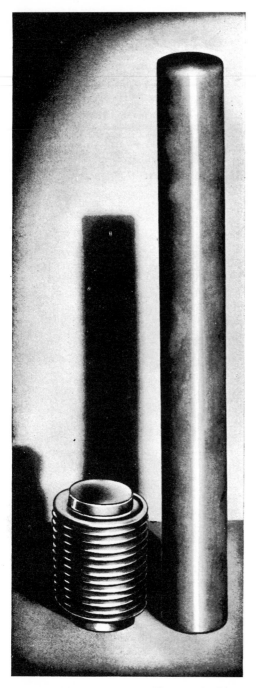

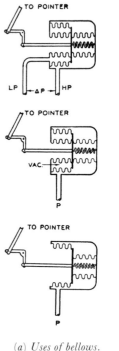

(a) Uses of bellows.
(By courtesy,
Taylor Instrument Companies Ltd.)

(b) Hydroflex bellows.
(By courtesy, Drayton Hydroflex Ltd.)

Figure 1.46   Bellows elements

The diaphragm elements are made from pre-formed disc welded by a cycle-stitch-seam process. Each disc is shaped to nest into the adjacent disc and each is welded to a spacing ring. Excessive pressure in either chamber will compress the diaphragm assembly until each pair of discs is nested and the spacing rings have contacted to form a metal-to-metal stop.

The diaphragm assembly built up in this way has a greater flexibility and lower spring rate than equivalent spun or hydraulically formed bellows.

Drive unit D, consisting of solid rods joined by an Elgiloy metal flexure, forms a direct connection between the range diaphragm and pen lever. The fulcrum is formed by two similar flexures; the seal is a flexible bellows capable of withstanding pressures 50 per cent above the meter rating.

The body of the meter is made from forged Type 316 stainless steel and the diaphragms of Type 316 stainless steel.

In order to make the range change spring independent of temperature it is manufactured from Ni-Span-C alloy. The instrument is suitable for a maximum operating pressure of 150 bar and differential ranges from 0–50 to 0–500 m bar are available.

Bellows element

The built-up pile of capsules is replaced in many instruments by a bellows. For many years such bellows were produced by a series of spinning or rolling operations, interspersed with the necessary annealing, and subsequent pickling and washing. The great improvement in the production of alloys having continuity of composition and freedom from impurities has made it possible to produce a bellows by a hydraulic process which involves one step only.

The bellows is made from a tube sealed at one end, and is produced by forming the bellows under high internal fluid-pressure in a collapsible die. The die consists of a series of plates equal in number to the number of convolutions in the bellows and spaced equidistantly on, and surrounding the tube. The internal pressure causes the tube to flow between the plates as the tube collapses endwise to form the bellows in one continuous operation.

Since the internal pressure is several hundred thousand bar the bellows is destroyed in the making if the tube alloy has any imperfections. *Figure 1.46b* shows a tube and the formed bellows. These bellows are produced in a variety of forms in brass and in 'Alumbro', an alloy similar to brass with the addition of about 2 per cent aluminium, which increases its resistance to corrosion and makes it suitable for use in salt air or sea water.

The 'flexibility of a bellows', which is defined as the change in length when a pressure of 1 N/m^2 is applied, is proportional to the number of convolutions and inversely proportional to the wall-thickness and modulus of elasticity of the bellows material.

The 'spring rate' or 'compression modulus' of a bellows is the load in newtons which, when applied to the bellows at its free length, will compress it by 1 mm. It varies directly as the modulus of elasticity of the material of the bellows and the cube of the wall-thickness. It is inversely proportional to the number of convolutions and to the square of the outside diameter of the bellows. In order to increase the spring rate of the bellows in certain uses, the force tending to compress the bellows may be opposed by a spring contained in the bellows. By varying the spring rate of the internal spring the calibration of the instrument may be varied. *Figure 1.46a* shows the ways in which bellows units may be used to measure differential, absolute, or gauge pressure.

Bellows type

Figure 1.47 shows a sectional view of the Barton differential pressure unit. Bellows units of this type are widely used for measuring differential pressures particularly in flow measurement replacing the liquid U tube type of instrument. They are capable of withstanding static pressures as high as 420 bar while measuring a wide range of differential pressures from 250 mm wg to 210 bar. They have a calibrated accuracy of $\frac{1}{2}$ per cent full scale range for ranges up to 0·75 bar, and $\frac{3}{4}$ per cent full scale for ranges above 0·75 bar. Changes of static pressure as large as 175 bar produce less than $\frac{1}{4}$ per cent error in scale reading while a temperature change of 55° C produces less than 0·4 per cent error.

By leaving the low pressure side open to the atmosphere and applying pressure to the high pressure side, gauge readings of pressure will be obtained. By applying process pressure to the low pressure side and leaving the high pressure side open to the atmosphere measurement of vacuum may be obtained. Models suitable for measuring absolute pressure in which the low pressure side is evacuated and sealed are also available.

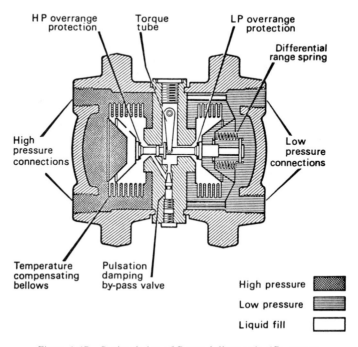

Figure 1.47   Sectional view of Barton bellows unit. (By courtesy, G.E.C.-Elliott Process Instruments Ltd.)

The range of the instrument is determined by the bellows diameter and the spring range of the spring. Zero suppression may also be achieved by use of suitable springs.

The bellows are manufactured from En 58J stainless steel. To protect the bellows from overpressure the bellows are liquid filled. The liquid used is ethylene glycol and water for normal service, ethylene glycol for service at pressures below atmospheric, and Fluorolube an inert fluorocarbon for use on oxygen.

Two bellows are secured to opposite sides of the centre plate and their free ends attached to a rigid rod passing through a hole in the centre plate. Each bellows is contained in a separate housing and the differential pressure to be measured is applied across the two housings. The bellows move in the direction of the low pressure housing the amount of movement being determined by the range spring opposing the movement. Movement of the bellows rod is transmitted by a drive arm which rotates the free end of a torque tube. The torque tube consists of a beryllium copper tube sealed to the meter body at one end and carrying a stainless steel rod which is sealed to the tube at its free end. As the drive arm rotates the free end of the torque tube, the tube will be twisted but the inner rod will transmit the rotation to the remainder of the mechanism providing a completely leak-free method of transmission.

Figure 1.48   Absolute pressure element. (By courtesy, Foxboro-Yoxall Ltd.)

As the bellows move, liquid will be transferred from one to the other, but a valve forming part of the bellows rod closes when the bellows have reached their full travel preventing further transfer of liquid. This sealed-in liquid being incompressible protects the bellows from distortion by over-pressure.

In order to provide adjustable pulsation damping the flow of liquid through its normal channel is restricted and a by-pass controlled by an adjustable needle valve is provided. The range springs may be housed within the bellows or outside depending upon the type of unit and its service requirements. In order to allow for expansion and contraction of the bellows filling liquid, an expansion bellows is connected to the main high pressure bellows but not to the bellows rod.

Absolute pressure element

The bellows described above can be used in pairs to produce an absolute pressure element. The mechanism shown in *Figure 1.48* consists of two carefully matched bellows. One bellows is pumped down to an absolute pressure of less than 0·05 mm mercury, and sealed. The other bellows is connected to the source of pressure to be measured. The effect of changes in the barometric pressure on the pumped-out bellows opposes the effect on the measuring bellows so counterbalancing it, and giving a measure of the absolute pressure. An element of this kind is shown in *Figure 1.48*. The movement of the pressure bellows is transmitted through a yoke and link to the pen arm or pointer.

1.5.2.2   SLACK DIAPHRAGM AND DRIVE PLATES

This type of pressure measuring element can be used to measure either 'gauge pressure' or 'differential pressure' and the construction is shown in *Figure 1.49* and in the sectional view shown in *Figure 1.50*.

The essential portion of the instrument consists of a flexible diaphragm of specially selected and prepared leather for the lower ranges (up to 200 mm

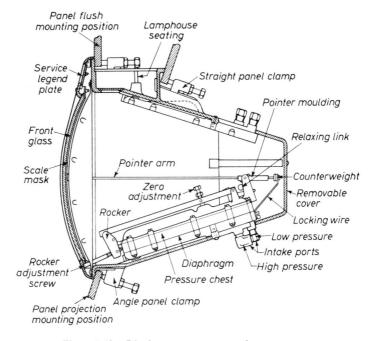

Figure 1.49   Diaphragm pressure gauge for gauge pressure.
(By courtesy, Bailey Meters & Controls Ltd.)

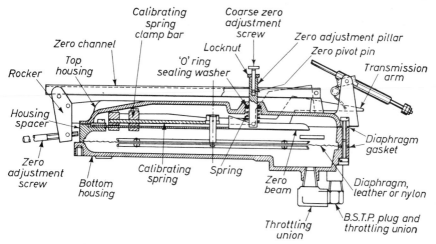

Figure 1.50 Diaphragm pressure gauge for differential pressure.
(By courtesy, Bailey Meters & Controls Ltd.)

w.g.) or impregnated nylon fabric for the higher ranges. The diaphragm is supported on either side by light metal plates. The diaphragm is arranged so that there is considerable slack allowing full movement over ample travel. The travel of the diaphragm is limited by the supporting plates coming to rest against the diaphragm limit stop so that the plates are supported in the overload condition. The principle of operation is the same as that of the piston type of manometer, the diaphragm acting as a seal. Thus the force on the diaphragm will be the product of the effective area of the supporting plates and the difference between the pressures acting on either side. The effective area of the plates is directly affected by the degree of slackness of the diaphragm and never exceeds the area of the plates plus one-half of the annular space between the plates and the case. Movement of the diaphragm is opposed by a flat beryllium-copper calibrating spring of suitable spring rate for the scale range, and stressed well below its elastic limit. Final adjustment of the calibration is obtained by varying the position of the pointer actuating link relative to the pointer pivot (i.e. radius adjustment).

When designed for measuring gauge pressure, the movement of the diaphragm is transmitted to the pointer directly by means of a transmission arm and pointer linkage. For measuring differential pressure both sides of the chamber containing the diaphragm are sealed and the motion of the transmission arm is transmitted through the sealed chamber by means of a magnet assembly. A magnet is attached to the end of the transmission arm within the sealed chamber. On the outside of the chamber, which is made of a non-magnetic material, opposite the inner magnet is a follower magnet which actuates the pointer mechanism. Thus, any movement of the diaphragm results in a corresponding pointer movement. The measurement is independent of the static pressure and the transmission is virtually frictionless.

If when the instrument indication is compared with that of a standard low-range manometer it is found necessary to make any adjustments they may be carried out as follows:

Zero adjustment. The fine zero adjustment which alters the position of the zero channel and the pivot pin is accessible through a small hole in the front of the instrument case. The course zero adjustment is accessible from the back of the case after the rear cover has been removed. Adjustment of these screws causes movement of the zero beam, its associated mechanism and of the pointer.

Range-span adjustment. Fine adjustment is made by means of the radius adjustment on the pointer assembly. An adjusting sleeve which carries a screwdriver slot is located at the pivot end of the pointer assembly. Rotation of the sleeve moves the radius assembly along the line of the pointer, thus varying the point of application of the pull of the operating link, resulting in a change in the calibration range of the instrument.

For a range adjustment greater than that obtainable by means of the fine range adjustment it is necessary to adjust the length of the calibrating spring by resetting the calibrating spring clamp. To do this the unit must be removed from its case, and the top cover of the diaphragm unit taken off, revealing the calibrating spring beam. By slackening off the holding screws the clamp bar may be moved along the calibrating spring until the desired range is obtained. Movement of the bar towards the free end of the spring will increase the range, and vice versa. After reassembly, the precise adjustment of the range is obtained by means of the radius adjustment as described above.

If the unit has sufficient volume it is insensitive to rapid pulsations of pressure while remaining sensitive to the changes in mean pressure. As the volume of the chambers increases, however, the speed of response of the instrument will decrease.

1.6   OTHER METHODS OF MEASUREMENT

1.6.1   Force balance type

In the transmitting types of pressure measuring elements, a force balance device is used. Such devices are described in the section on transmitters. The principle of the system is that the pressure on a diaphragm is arranged to control the flow of air into, or out of, the chamber on the opposite side of the diaphragm, until a balance is obtained. This balancing pressure is then transmitted to a receiver which indicates or records the pressure measured.

Another development in the force balance type of pressure measuring instrument is the electrical pressure gauge for roof pressure measurement in open hearth furnaces. In this type of instrument, the force produced on a diaphragm due to the difference in pressure on the inside and outside of the roof of the furnace, is balanced against the force on the core of a solenoid, by controlling the current electronically. As the value of the current is a measure of the force produced, the instrument can be used to indicate the difference between the pressure just under the roof of the furnace and that outside.

1.6.2   Katharometer type

This type of instrument is described in greater detail in the section on gas analysis instruments but it can be used as a method of measuring gas pressures.

As the pressure of a gas is reduced, its density and also its thermal conductivity are reduced. If then the conductivity of a gas is compared with that of a standard sample it is possible to calibrate the indication in terms of gas pressure. The advantage of this type of instrument for low-pressure measurement is that the change of e.m.f. in the Katharometer unit with change in pressure is not a linear relationship, but, up to a point, increases as pressure decreases. Thus the lower the pressure measured, the more open the scale becomes.

1.6.3 Piezo-electric type

If a quartz crystal is cut in a special way and placed between two plates, then the e.m.f. set up between the plates will be a measure of the pressure applied to the crystal. This property of the crystal, called the piezo-electric effect, forms the basis of a method of measuring high pressure. By measuring the e.m.f. set up, the applied pressure can be found. This type of device is extremely useful in research on engines for, with its aid and a cathode-ray oscillograph, it is possible to record the pressure cycle while the engine is actually running.

1.6.4 Strain gauge method

The pressure acting on a diaphragm may be measured in terms of a change of electrical signal by means of the pressure transducer illustrated in *Figure 1.51*. The pressure to be measured is applied to a stainless steel diaphragm which moves until the force acting on the diaphragm is balanced by the force pro-

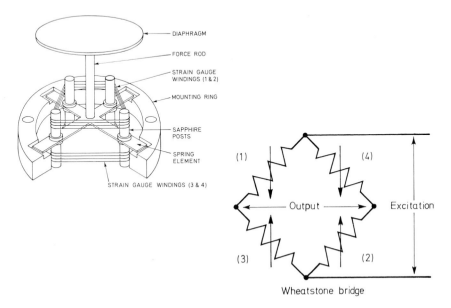

Figure 1.51 · Strain gauge pressure transducer. (By courtesy, Bell and Howell Consolidated Electrodynamics Division)

duced by the deflected spring element. The resulting flexure of the spring element tilts the sapphire posts on which the strain gauge windings are mounted increasing the strain, and hence the resistance in the two windings at one end of the posts, and decreasing the strain and hence the resistance in the windings on the other end. These changes in resistance produce an unbalance in the Wheatstone bridge. The thermal and mechanical summetry of the unit and the precise construction methods used ensure that the output of the bridge is a linear function of the pressure applied to the diaphragm. For pressures of 7 bar and above the diaphragm is machined integral with the instrument case. For lower ranges the diaphragm is stitch welded to the case. By venting the back of the diaphragm to atmosphere the gauge may be used to measure gauge pressure. By sealing the housing for the strain gauges and filling with helium at a low pressure an absolute version is produced. In the vented version heat sinks having a large thermal mass relative to the windings surround the element to attain thermal stability. Before calibration each transducer is mechanically and thermally aged by subjecting it to pressure cycles while energised so that the sensitivity, linearity and hysteresis characteristics are stable.

Transducers are available having ranges from 0–0·75 bar to 0–700 bar, absolute or gauge. Special versions for use on corrosive fluids are available. The gauge can be used at temperatures from $-54°$ C to $+150°$ C with a thermal zero shift of not more than 0·05 per cent of full scale range per °C and has a very high resistance to the effects of acceleration and shock. Intrinsically safe versions of the instrument are available.

1.7   CALIBRATION OF PRESSURE MEASURING INSTRUMENTS

Instruments in section 1.5, which depend for their calibration upon the stress produced in an elastic medium when it is strained, must from time to time be checked against a standard, if their accuracy is to be guaranteed. This is necessary because elastic media when stressed do not return exactly to their original state when the stress is removed, and this causes what is known as 'zero set' or 'zero shift'. If an elastic member such as a Bourdon tube, bellows, diaphragm, spring or combination of these is stressed repeatedly it will be found that the first zero shift is greatest and then the shift gets less after each stressing. For this reason elements are stressed through their maximum range several times before they are calibrated. The amount of 'permanent set' or 'zero shift' is then very small unless the 'elastic limit' of stress for the material of the element is exceeded.

To illustrate this, consider a rod of metal in tension and let the stress applied, i.e. the

$$\frac{\text{load}}{\text{area of cross section}}$$

be plotted on a graph against the strain produced;

where $\text{strain} = \dfrac{\text{Increase in length}}{\text{Original length}}$

A graph is obtained of the form shown in *Figure 1.52*. This shows that for small stresses the strain produced is proportional to the stress, i.e.

$$\frac{\text{Stress}}{\text{Strain}} = \text{Constant}$$

(This constant is known as Young's modulus).

This relationship is Hooke's law. When the stress is increased beyond the 'limit of proportionality', it is found that this linear relationship no longer holds, and at the 'elastic limit' the strain increases a great deal with very little change in the stress. When the stress is further increased, the strain again increases until the 'yield point' is reached. At this point the rod continues to increase in length (i.e. strain increases), even if some stress is removed, until finally the specimen fractures.

If the stress is removed the specimen will return practically to its original length as long as the stress has not exceeded the 'limit of proportionality', or, what is for many materials the same stress, the elastic limit.

If, however, the stress is increased beyond the elastic limit to a stress represented by the point A, then, when the stress is removed, the graph will follow the dotted line AB, and the specimen will have a 'permanent set', or strain, represented by OB.

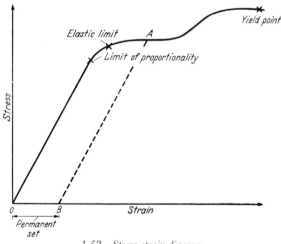

1.52   *Stress-strain diagram*

The difference between the original length and the length after a stressing cycle is known as 'hysteresis'.

In industrial instruments the hysteresis is often measured along with friction and lost motion, since the three effects produce the same form of undesirable errors. The total effect of these errors at any point on the scale of the instrument, say at half scale, is determined as follows: Increase the pressure applied to the instrument until it is reading at half scale. Measure the applied pressure on a standard manometer free from the effects being measured. Increase the pressure to the full-scale reading, and then reduce the pressure to original half-scale value. The difference between the original and final position of the pointer is a measure of the sum of hysteresis, friction, and lost motion at half-scale. This error is usually expressed as a percentage of the full-scale reading and is sometimes spoken of as the instrument 'hysteresis', although it includes the other errors.

In good quality instruments, the error rarely exceeds $\frac{1}{2}$–1 per cent of the full-scale reading, but where great accuracy is not required, much less costly instruments are used, when these errors may amount to 4 or 5 per cent of the full-scale reading.

If an instrument has been subjected to an overload, the elastic measuring element may have been stressed beyond its elastic limit, in which case there will be a considerable 'zero shift'. It is not sufficient to reset the zero, for the elastic properties of the measuring element or its calibrating spring may have been changed. In addition the instrument must be checked throughout its range against a standard instrument.

1.7.1   Actual calibration

The calibration of the instrument is carried out by applying to it an air or liquid pressure whose value is accurately known. The applied pressure should preferably be measured by a method which does not depend upon elastic media for its accuracy. The pressure must therefore be balanced against a column of liquid of known density, or, if the pressure is too large, a 'dead-weight' piston type of instrument must be used. The method used will depend upon the range of the instrument. *Figure 1.53* shows a low pressure instrument under test. For very low pressures, up to 200 mm w.g. or a Vernier reading manometer, or an inclined water gauge may be used. The actual instrument used depends upon the accuracy required, and the range to be covered, but it should be such that it can be read more accurately than the instrument

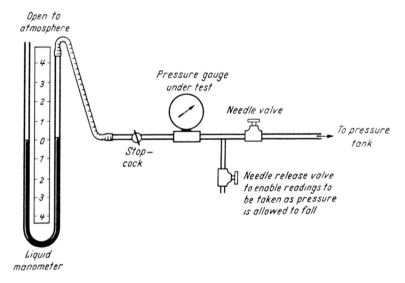

Figure 1.53   Calibration of pressure gauge

under test. For pressures up to 4 m w.g. or 300 mm Hg or 0·4 bar, a water manometer may be used, although at the upper part of this range a mercury manometer becomes more manageable. For higher pressures up to 2 m Hg or 2·7 bar a mercury manometer may be used. Where high accuracy is required a precision calibrating standard manometer would be used up to 1·35 bar or 1 m Hg. For still larger pressures a 'dead-weight' piston type of pressure measurer must be used.

If the reading on the calibrated instrument is noted as the applied pressure is increased in steps, and the procedure repeated as the applied pressure is decreased by the same steps, both the percentage error at each reading and the 'hysteresis' at each point on the instrument scale may be determined.

Periodically, depending upon the nature of the service and the importance of the reading, gauges should be returned to the instrument workshop for servicing.

If the gauge is on special service this must be carefully noted. For example a gauge may be on oxygen service, so that oil may not be used for calibration. On the other hand, a gauge may have a suppressed zero to allow for the head of liquid between the gauge and the point at which the pressure measurement is required.

If the gauge is obsolete, badly damaged or corroded it may be more economic to scrap the gauge than to effect a repair.

The gauge is first checked on the calibrating rig in the manner described below.

Raise the pressure to about 25 per cent above the scale maximum and hold there for ten seconds as a test for leakage.

Release the pressure slowly to zero when the pointer should return to zero. Tap the gauge if necessary. If the pointer fails to return to zero it is a sign that the tube has deteriorated and requires replacing.

Raise the pressure slowly to gauge maximum and release slowly. If the action of the gauge is jerky it usually indicates wear or stickiness in the movement, or leakage, or a damaged hair spring.

Raise the pressure in steps and note the readings for rising and falling pressures. If these readings show an error less than $+10$ per cent and the hysteresis does not exceed $\frac{1}{4}$ per cent it can be assumed that the tube is satisfactory.

Dismantle the gauge, removing the pointer by means of a special jack. Renovate the case, glass, bezel and dial. Most dials can be cleaned by wiping gently with a sponge moistened in soap and water.

Remove the movement and if the teeth of the gearing or the holes in the bearing plates are badly worn, the movement should be replaced as a whole.

Reassemble the gauge in its case omitting bezel and glass. Subject the gauge for a few seconds to a pressure about 20 per cent in excess of the dial reading. The gauge is now ready to be adjusted so that the pointer follows the scale. This can only be done by trial and error. Two points at say 10 and 90 per cent of the dial range are chosen as test points and the magnification and reading at the low point are checked and adjusted alternately, the former by adjusting the slider in the quadrant slot and the latter by removing and resetting the pointer until the gauge reads correctly at both points. The reading at the midpoint of the dial is then checked and corrected if necessary by adjusting the length of the connecting link between quadrant and tube. All adjusting screws are then tightened and the gauge finally checked at all points on the scale. The pressure is then brought to the lower point and the pointer driven on to the tapered pinion shaft by a gentle blow on a hollow drift. The assembly of the gauge is then completed. Time may be saved if a special testing dial is used having the same calibration as the gauge dial but having a cut-out to give access to the slider without the need to remove the dial.

Generally, it is sufficient to obtain agreement within 1 per cent of the full range of the gauge throughout the range. Where greater accuracy is required the whole installation must be carefully checked. If the gauge cannot be moved, it may be checked *in situ* by comparison with a 'master gauge' of similar type, mounted on the side of the gauge being tested. On some installations special adaptors are fitted, so that a 'master gauge' can be fitted from time to time, and the working gauge checked.

In liquid-filled manometers, the instrument must be level, the liquid and the chambers must be kept at a reasonably constant temperature, and errors must not be introduced in the connecting pipes. Proper allowance must be made for the pressure due to liquid in the pipes where liquid enters the mano-meter, or the instrument must be calibrated under conditions exactly similar to those under which it is used.

In setting up connecting lines for boiler draught gauges, the piping leaving the boiler should rise vertically for at least 0·5 m, to minimize the possibility of collecting soot, ash or condensed vapour. All changes in direction should be made with crosses so that the plugs can be removed to facilitate inspection and cleaning of the line.

Owing to the very large variation in design and construction of instruments, it is always necessary to adhere to the manufacturer's recommendations for installation, servicing and maintenance.

In general, gauges in section 1·5 are normally used for pressures of about 60 per cent of their maximum range, and must never be subject to pressures higher than the gauge maximum.

All pressure gauges should be mounted correctly, generally in a vertical position, protected from heat, corrosion and vibration.

In high-pressure installations, protective guards are usually fitted to minimize the dangers arising from a burst tube. In addition blow-out vents are provided at the back of the case. Restriction rings and chokes are often fitted in the joint between gauge and pipe to reduce the leakage of fluid due to a burst pipe.

The zero of the instrument should be checked daily, or weekly, depending upon the variation found, and the instrument corrected if this is found necessary. The calibration should be checked every 3, 6 or 12 months, depending upon the use and the accuracy expected.

If the instrument contains mercury, the mercury chamber should be checked every six months and the mercury cleaned or replaced if necessary.

1.8   SEALS AND PURGE SYSTEMS

The use of seals and purge systems has been avoided in recent years in certain applications by using instruments made from materials such as stainless steel which are not attacked by the fluid being metered.

In other applications, where metals or mercury are attacked by the fluid being metered a U tube or well type of manometer may be used and the liquid in it and the material of the tube chosen to resist attack. Great care must be used in choosing the liquid for it must give a suitable range, not mix with the metered fluid, evaporate very slowly at the working temperature in addition to resisting attack by the metered fluid.

In many applications, however, it is not possible to do this. In these cases, seals or purge systems must be used to prevent the metered fluid from entering the measuring instrument.

Seals may be of two kinds; they may be solids or liquids. The solid type consists of a diaphragm (*Figure 1.54*) or bellows (*Figure 1.55*) made of a suitable material, introduced between the tank and line, or line and the instrument. By making the diaphragm or bellows large and flexible, the pressure can be transmitted with very little loss. The space between the diaphragm and the instrument is generally completely filled with liquid, although in some installations it is filled with compressed air.

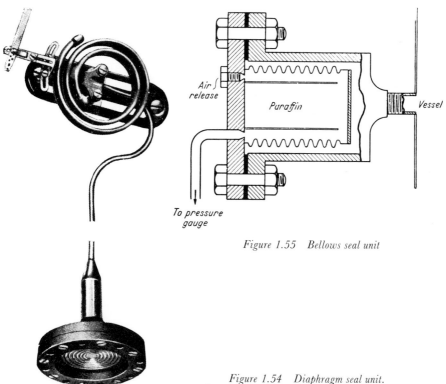

Figure 1.55   Bellows seal unit

Figure 1.54   Diaphragm seal unit.
(By courtesy, Taylor Instrument Companies Ltd.)

In Bourdon gauges and similar types of instruments, the system should be evacuated before filling, in order to remove all air, which, being compressible, would decrease in volume as the pressure increases. The volumetric change in the system is then no longer negligible and the stage is reached where the spring characteristic of the diaphragm affects the calibration of the instrument. For this reason, units consisting of instrument, line and seal are supplied by the makers completely filled and sealed ready for use.

The liquid used in the tube system is chosen to meet the temperature conditions, as vaporization must be avoided in high temperature installations.

The material chosen for the diaphragm must resist the corrosive action of the fluid being metered and may be chosen from nickel, Monel, silver, aluminium, platinum, tantalum or stainless steel, while in other applications, P.T.F.E., Neoprene or synthetic rubber may be used.

1.8.1   Liquid seals

1.8.1.1   STEAM AND VAPOUR

Many Bourdon gauges are not designed for use at high temperatures, and in order to keep live steam out of a simple pressure measuring installation 'pig tails' are often used (*Figure 1.56*). The loop in the line helps to condense the steam, and retains the water when the apparatus is shut down. In this way, the water from the condensed steam is used as a seal, and the Bourdon tube

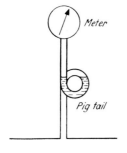

Figure 1.56   Pig tail

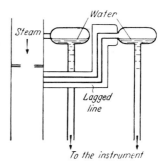

Figure 1.57   Condensing chambers

is not subjected to the high temperature. In more complex applications such as differential pressure measurement in steam flow measurement, condensing chambers (*Figure 1.57*) are used, and these serve two purposes. They keep live steam out of the instrument, and maintain the water above both sides of the instrument at a constant level. Similar devices can be used when metering other vapours.

1.8.1.2   CORROSIVE LIQUIDS OR LIQUIDS WHICH WOULD SOLIDIFY IN THE PIPES

When it is desired to keep the fluid out of pressure pipes and instrument, a chamber is fitted near the tapping into the main or tank, and this chamber is connected to the instrument. Part of the chamber, the pressure piping, and the instrument, are then filled with the sealing liquid, while the metered fluids fills the remainder of the sealing chamber. If the metered fluid is less dense than the sealing fluid, the line from the pipe is brought into the upper part of the seal chamber as shown in *Figure 1.58*, while if the metered fluid

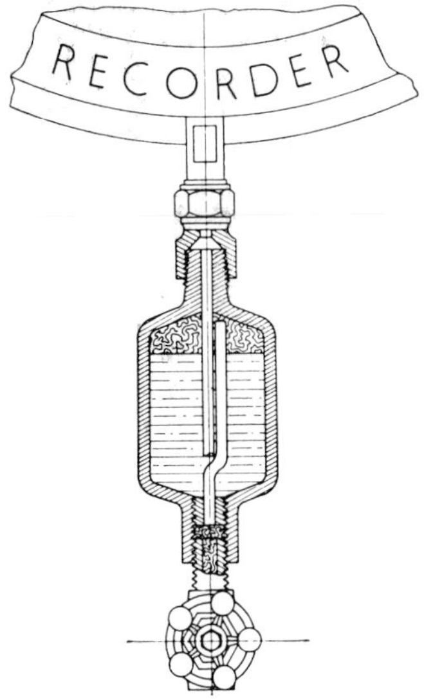

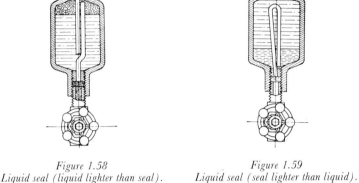

Figure 1.58
Liquid seal (liquid lighter than seal).
(By courtesy, Foxboro-Yoxall Ltd.)

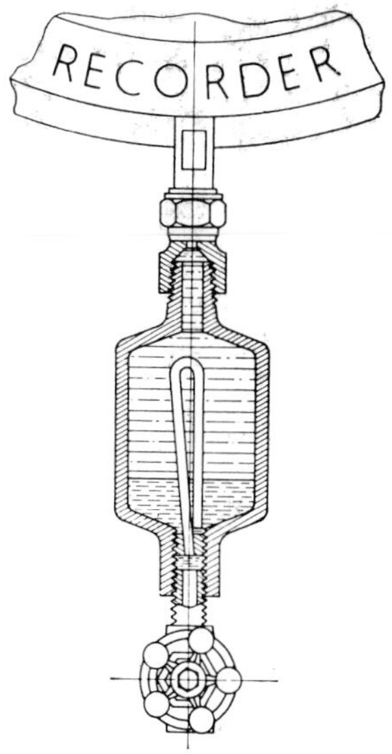

Figure 1.59
Liquid seal (seal lighter than liquid).
(By courtesy, Foxboro-Yoxall Ltd.)

is the more dense the line is brought in near the bottom as shown in *Figure 1.59*. In many cases it is also necessary to heat the lines by lagging in with them a steam tracing, while in more difficult cases steam purges are necessary.

1.8.2   Effect of size seal chamber on the meter calibration

When seal chambers are used in conjunction with pressure transmitters or bellows or diaphragm types of meters in which the displacement of the measuring element is small, the error owing to changes of level in the seal chambers is negligibly small. When used in conjunction with well-type mercury manometers, however, where a significant volume of mercury is displaced from the high-pressure side of the instrument to the low-pressure side, the size of the seal chamber will have a marked influence on the meter calibration, particularly if the density of the sealing liquid is very different from the density of the process fluid. One method of reducing the error is to make the seal chambers sufficiently large in area that the change of level in the seal chambers can be neglected. In this case, the effective density of the mercury will be the difference between the density of the mercury and the density of the sealing liquid. As the quantity of mercury displaced is usually of the order 16×10^4 mm³, an area of sealing chamber of $6 \cdot 4 \times 10^4$ mm² would be required to reduce the change of level to $2 \cdot 5$ mm.

A method of eliminating the error is to make the seal chamber intermediate in area between that of the float chamber and the range chamber of the manometer as shown in *Figure 1.60*.

Suppose a volume V mm³ flows from the high-pressure side to the low-pressure side.

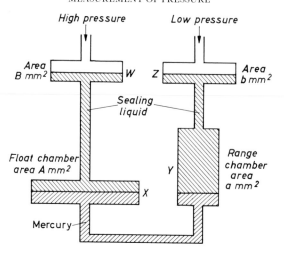

Figure 1.60   Seal chambers for mercury manometers

Fall in level in the float chamber $X = V/A$ mm.

Fall in level in the seal chamber $W = V/B$ mm.

If $B < A$, reduction in the height of sealing liquid on H.P. side $= (V/B) - (V/A)$ mm.

Similarly, reduction in the height of the sealing liquid on L.P. side $= (V/a) - (V/b)$ mm.

For calibration to be independent of the sealing liquid, these must be equal.

i.e.
$$\frac{V}{B} - \frac{V}{A} = \frac{V}{a} - \frac{V}{b}$$

or
$$\frac{1}{B} - \frac{1}{A} = \frac{1}{a} - \frac{1}{b}$$

If $B = b$, this reduces to

$$\frac{1}{B} - \frac{1}{A} = \frac{1}{a} - \frac{1}{B}$$

or
$$\frac{2}{B} = \frac{1}{a} + \frac{1}{A} = \frac{A+a}{A \cdot a}$$

or
$$B = \frac{2 \cdot A \cdot a}{A+a}$$

For example, in an actual well-type flow indicator of dry calibration 2·5 m w.g.

Area of the float chamber A	$= 6\cdot36 \times 10^3$ mm²
Float travel	$= 24\cdot4$ mm
Volume of mercury displaced	$= 1\cdot552 \times 10^5$ mm³
Difference in level at 2·5 m w.g.	$= \dfrac{2500}{13\cdot5} = 185\cdot2$ mm

Change in level in the range chamber $= 160.8$ mm

Area of range chamber $= \dfrac{1.552 \times 10^5}{160.8} = 9.651 \times 10^2$ mm²

Required area of the seal chamber $= \dfrac{2 \times 6.36 \times 9.651 \times 10^5}{7.325 \times 10^3}$

$= 1676$ mm²

i.e. a chamber of diameter 46·2 mm would be satisfactory.

1.8.3 Purge systems

Purges may be of two kinds, either gas or liquid.

1.8.3.1 GAS PURGES

In this system (*Figure 1.61*) a supply of gas, usually air, at a pressure higher than that in the main is fed, at a convenient rate, into the pressure pipe through a needle valve. In order that the rate at which air is fed in, can be easily seen, the air is bubbled through a small quantity of oil or water contained in a sealed glass tube. It is usually arranged so that 60 bubbles a minute pass through the tube. In order to maintain a steady bubble rate under all pressure

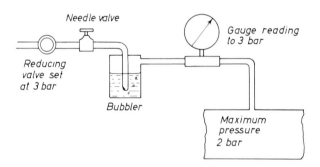

Figure 1.61   Air purge systems

conditions, a differential reducing valve is sometimes used. This is so connected that it maintains a constant differential pressure across the bubbler under all plant pressure conditions. The air pressure in the pressure pipe will build up until it balances the pressure in the main, and will expel all the metered fluid from the pipe. As long as the flow of air into the main is kept small, the gauge will register the pressure in the main. In more recent installations the bubbler is replaced by a flow indicator of the rotameter type. This type of indicator is described in the flow section.

In applications such as metering of fuel oil for furnaces, where the fluid is liable to solidify in the pipes if it becomes cold, a purge of steam is used. If the pressure in the main is liable to fluctuate through a large range, a vessel having a large capacity can be placed near the tapping in the main. This will prevent the metered fluid from entering the gauge when the pressure in the main suddenly increases.

1.8.3.2 LIQUID PURGES

Liquid purges are used in liquid metering only, and the principle is the same as for gas purges. In order to measure the rate of flow of the purge liquid, the liquid passes through a sealed glass vessel fitted with a flap or rotor, which indicates the rate of flow, or a flow indicator of the rotameter type is used.

Generally, water is used as the purge liquid, but in some applications other liquids are used; for example, paraffin in the oil industry.

In purge systems it is advisable to arrange for the capacity change of the measuring system which occurs over its whole range of measurement to be a small fraction of the total capacity of the system, so that when the pressure changes from the lower end of the range to the upper end there is no danger of fluid from the plant entering the measuring system.

1.9 PROTECTION OF PRESSURE MEASURING INSTRUMENTS FROM VIOLENT PRESSURE SURGES AND PULSATIONS

In addition to the methods used within an instrument for reducing the effects of pulsating pressure, damping devices may be used in the pipe-lines leading to the instrument. These deadeners, or snubbers, result in the instrument indicating or recording an average pressure, instead of recording each individual surge or pulse.

*Figure 1.62   Capillary deadener.
(By courtesy, Foxboro-Yoxall Ltd.)*

In general, these snubbers reduce the velocity of fluid flow to the instrument, and thus prevent sudden extreme changes in pressure from reaching the measuring element too rapidly. This reduction in velocity may be achieved by introducing a restricting device into the pipe-line as in the capillary deadener shown in *Figure 1.62*. This consists of a length of capillary tube and is screwed into a union, cock, or hydraulic connection.

Another form of snubber is shown in *Figure 1.63*. It consists of a piston attached to a pin which rises and falls with the pressure impulses, and so absorbs the effect of shock and surge, while giving an accurate pressure reading. Owing to the rise and fall of the piston the snubber is self-cleaning.

Still another type of pulsation damping unit is shown in *Figure 1.64*. It

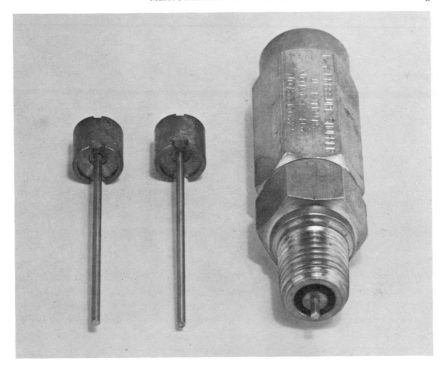

Figure 1.63   Ray pressure snubber. (By courtesy, Foxboro-Yoxall Ltd.)

consists of a series of small capacities built in tubular form and separated by orifices. A rod passes axially through the orifices, and by choosing a rod of suitable diameter it may be used for liquids or gases. The rod is easily removed to facilitate cleaning. Fine-meshed screens are fitted at each end to prevent solid particles entering. In operation, an 'above average' pressure enters the damping unit and in passing through the several orifices and capacities is reduced in magnitude. The following 'below average' pressure then enters the device causing a reversal of the flow process. This action is reduced as the wave proceeds, and finally the pressure is levelled out at the average value.

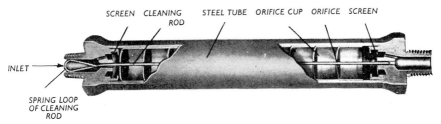

Figure 1.64   Pulsation damping unit (capacity type). (By courtesy, Taylor Instrument Companies Ltd.)

MEASUREMENT OF LEVEL

In many industrial processes it is very important to know the level of liquid in a tank or vessel. The need for an instrument to indicate this probably arose with the invention of the steam engine. It is essential to know the level of the water in the boiler while it is in use and under pressure, but it is impossible to view it directly.

Industrial methods of determining level can be divided into two groups. In the first group, the level of the liquid is measured directly by means of a hook type level indicator, a sight glass, or float-actuated mechanism. In the second group, use is made of the fact that the pressure due to a column of liquid does not depend upon the cross-sectional area of the column, but only upon the depth and density. Thus, if the pressure due to a column of liquid of known density is measured, then its depth may be calculated. The pressure may be measured directly, by balancing it against a mercury column, it may be measured by means of a gauge using some form of diaphragm or elastic pressure element, or some form of gas or liquid purge system may be used, when a fluid pressure equal to that of the column of liquid has to be measured. In this second group may be included weighing tubes and buoyancy types.

Level measurement is therefore described under the following headings.
1. Direct methods
   Hook type of level indicator
   Sight glass
   Float-actuated mechanisms and using other forms of interface detection
2. Pressure-operated types
   Simple pressure-actuated mechanisms indicating one or more levels
   Static-pressure-actuated mechanisms including purge systems
   Buoyancy types

2.1 DIRECT METHODS

Probably the simplest method of telling when the liquid in a tank has reached the required level is to have an overflow pipe at that level. At some convenient point in the overflow pipe, which can be easily seen by the operator, there is fitted a glass chamber with a flapper which is deflected, or a small rotor which is rotated, by the liquid flowing through the overflow pipe. Thus the fact that the liquid has reached the required level is indicated immediately to the operator. In other installations, where a constant level of liquid is required in a tank, this may be maintained by allowing liquid to flow continuously through the overflow pipe.

2.1.1 Hook type

When the level of liquid in an open tank is measured directly on a scale, which may be in the liquid or alongside it, it is sometimes difficult to read the level accurately. This is because it is not convenient to get one's eye on the same level as the liquid. In this case a hook type of level indicator shown in *Figure 2.1* is a great help. It consists simply of a wire of corrosion resisting alloy, such

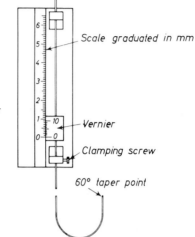

Figure 2.1   Hook type level indicator

as gun-metal or stainless steel, about 5 mm diameter, bent into a U with one arm longer than the other. The shorter arm is pointed with a 60° taper, while the longer one is attached to a slider, having a vernier scale, which moves over the main scale and indicates the level. The hook is then pushed below the surface of the liquid and gradually raised until the point is just about to break through the surface. It can then be clamped and the level read off from the scale.

2.1.2 Sight glass

In steam boilers and similar applications, a simple device for enabling the level within the boiler to be determined is the sight glass shown in *Figure 2.2*. It consists of a tube of toughened glass connected through unions and valves into the boiler, in which the water level is required. The top arm of the tube is asbestos packed and the bottom arm has rubber packing. If the diameter of the bore of the tube is not small enough to introduce errors due to capillarity, the liquid will stand at the same level in the boiler and tube. The two valves are provided so that the steam may be shut off in case of breakage of the sight glass. The smaller valve at the bottom is provided for blowing out the gauge for cleaning purposes.

  Figure 2.3 shows a sight glass with automatic shut off. If the gauge glass should be broken, the steam pressure would force both upper and lower ball valves against their seats and so shut off the flow of steam and hot water, and

2

86 MEASUREMENT OF LEVEL

eliminate the danger of anyone being scalded. The upper valve is sometimes arranged to allow a slight escape of steam which will attract attention to the damaged glass. Another interesting feature of the gauge shown is the use of a main and check valve in the tryvalve. The main valve always opens before, and closes after, the check valve, so that the main valve face is not subjected to scoring action.

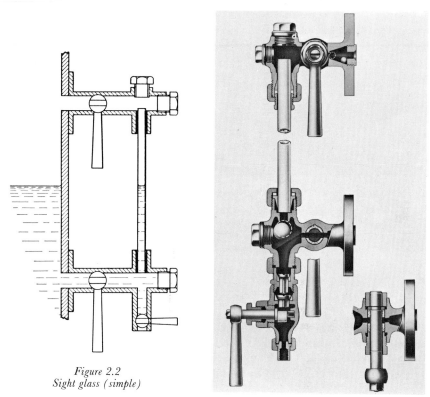

Figure 2.2
Sight glass (simple)

Figure 2.3
Sight glass with automatic cut-off.
(By courtesy, Hopkinsons Ltd.)

In many installations, the gauge glass is protected and the operator protected in case of breakage, by protector panels of toughened glass.

Similar gauges may be used to indicate the level of liquids in tanks. Gauges of this type are made to stand pressures of 24 bar of steam pressure at 250° C, or 70 bar of liquid pressure.

2.1.3 Float-actuated mechanisms and devices using other forms of interface detection

Introduction

Probably the commonest use of the equipment described in this section is for measuring the quality of crude oil or petroleum products which have been produced, processed, transferred or sold. Another common use is for ullage measurement in order to ensure that a container is not overfilled. When used on shore establishments, or sea-going tankers, for measuring the quantity of petroleum product entering or leaving a country, the Customs Authorities may be involved, and the equipment requires official approval. This usually means that the equipment must be capable of the highest accuracy, and provision has usually to be made for hand gauging, in order that the level equipment can be easily checked. Hand gauging, or dipping, is carried out by means of a graduated tape weighted at the end. The weighted end is lowered vertically into the tank until the weight, which is the zero end, just touches the bottom. The wetted portion of the tape will then be an accurate measure of the depth of the liquid. This method is often used as a final proof of the depth in sale and transfer in bulk, and for Customs purposes. However, the method is inconvenient and can be subject to errors owing to varying tape tension due to the method of lowering. In addition, if sediment accumulates on the bottom this may cause an error in reading. If the position of the surface of the liquid relative to a datum plate is measured, the problem of sediment is largely overcome. Where the product is toxic, hand gauging may be hazardous and should be avoided as far as possible.

If the product is contained in a vessel of known cross-sectional area and the depth of product is accurately measured, then the volume can be readily calculated. As liquids have a relatively large coefficient of expansion, the change of level with temperature is relatively large, being in the order of 1 part per 1000 of depth per °C rise of temperature in certain cases. In general, the coefficient of expansion of petroleum·products decreases with increasing density. Because of the temperature effect, 20° C is often accepted as the standard reference temperature for volume measurements, and volumes measured at other temperatures are corrected to this temperature when used as a basis for inventory or transfer records. The volume V_{20} at 20° C is related to the volume V_t at $t°$ by the equation:

$$V_{20} = V_t[1 + (20 - t)c]$$

where c is the mean coefficient of cubical expansion per °C of the liquid concerned.

The effect of changing temperature on the area of cross-section of the tank will, in general, be negligibly small unless there is considerable change in temperature. In cases where it is significant a correction may be made.

Providing the tank is adequately supported, changes in the dimensions of the tank owing to the weight of the contents are negligibly small in smaller tanks. In large tanks the effect may be significant, but can be allowed for in the calibration of the tank.

Where accurate measurement of volume is required, a calibration table is prepared showing the total volume of the storage tank for each mm in level. In these measurements the level is related to a fixed level or 'datum line', which is accurately known. The cheapest and least accurate method of making a calibration table is to prepare it from calculations based on detailed drawings of the tank. A common method which is more accurate but more expensive than the first method is called 'strapping'. In this method, the table is prepared from actual measurements of the dimensions of the finished tank. Where the greatest accuracy is required, a table is prepared by measuring water into the tank in appropriate quantities and noting the change of level. The table is checked as the tank is emptied by noting the changes of level as the water is measured out. This method is usually expensive for many readings must be taken under the most favourable conditions. The water in and out of the tank is measured with an accurate measuring vessel or by means of an accurate positive displacement meter (see section 3.1.2.2 on Flow Measurement). It must be realized however, that the calibration of the tank cannot be any more accurate than the method used to measure the water into or out of the tank during calibration. Complete details of the methods of calibration are given in documents such as *Petroleum Measurement Manual*, published by the Institute of Petroleum, and American Petroleum Institute Standard No. 2501 'Crude Oil Tank Calibration'.

Installation of equipment

In order that any level installation may yield the most accurate results possible, the system should be carefully designed and installed. With the best equipment described in this section, the position of the surface of the liquid in the tank may be measured with an accuracy of plus or minus 1·0 mm. In order that the contents of the tank may be known, this measurement of the position of the surface must be related to the datum line fixed on the tank.

In order that the volume of the contents may be corrected to the standard reference temperature it is necessary to measure the mean temperature of

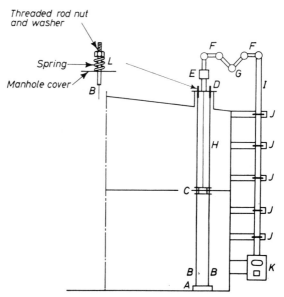

Figure 2.4   Installation of float type level indicator with guide wires on a fixed-roof tank
A, guide wire anchor. Wires may be anchored to a heavy weight or the bottom of the tank. *B*, float guide wires. *C*, float having sliding guides which move freely on the guide wires. *D*, manhole sufficiently large for float and anchor weight to pass through. *E*, flexible joint. *F*, pulley housings. *G*, vapour seal (if required). *H*, float tape. *I*, tape conduit. *J*, sliding guides. *K*, gauge head. *L*, guide wire tension adjustment

the contents. Methods of measuring the mean temperature are described in Temperature section. The nature of the installation used will depend upon the accuracy required. In general, the greater the accuracy required the more expensive the installation. However, in any installation, adopting the precautions described below will increase the accuracy and they should be adopted wherever possible.

The level-measuring equipment should be installed as far as possible from the inlet and outlet connections to the tank to minimize the effects of the turbulence caused when the tank is filled or emptied. This in itself is often

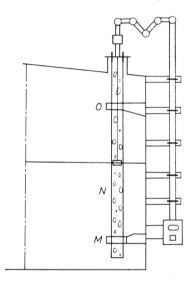

Figure 2.5   Installation of float type level indicator on fixed-roof tank with internal still pipe
M, sliding still pipe guide. N, still pipe sufficiently large for the float to slide freely up and down. Pipe to have two rows of holes or slots spaced at 0·3 m centres facing the tank shell. Internal surface must be free from burrs. O, still pipe support

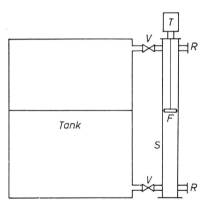

Figure 2.6   Installation of float type level indicator on external still pipe
V, isolating valves. S, external still pipe with removable flanges at each end. Pipe to be large enough to permit float to move freely. T, gauge head or level transmitter, if required head may be installed at ground level as shown. R, rodding-out points (if required)

insufficient to prevent the surges and swirling of the liquid from causing the float to wander, thus affecting the level indicator, so it is usually necessary to guide the float by means of guide wires as shown in *Figure 2.4*. If the surging or turbulence is particularly troublesome, it may be necessary to house the float in a still pipe having holes or slots extending to above the maximum level in the tank as shown in *Figure 2.5*. In cases where the contents of the tank are particularly corrosive or toxic, or there is a danger of the float or stillpipe accumulating scale or other solid matter, it greatly facilitates servicing if the stillpipe is fitted external to the tank as shown in *Figure 2.6* and isolating valves are provided. Where solid matter may be troublesome these valves should be of such a type that they can be rodded through if necessary. When it is required to service the float or still pipe, the top isolating valve may be closed, compressed air or nitrogen applied to the top of the still pipe forcing the liquid back into the tank, and the lower isolating valve closed. The top and bottom flanges may then be removed. The main disadvantage of the external still pipe is that the liquid in the still pipe may not be at the same temperature as that in the tank and it is not possible to make a simple correction for temperature changes.

The tank-side equipment should be installed on the shaded side of the tank to reduce the temperature changes to a minimum, and should be readily accessible for reading or servicing. Gauge heads must be rigidly fixed to the tank so that no relative movement can take place between them. The tape conduit must be adequate in size and correctly aligned so as to prevent the tape rubbing at any part. The conduit should be supported in sliding supports and must be adequately treated by galvanizing or other treatment in order to prevent corrosion. To facilitate hand-reference dipping a hand-gauging hatch should be provided near the level equipment; and a datum plate fixed to the tank shell in a convenient position. In this manner, the datum level for the level equipment can be fixed and calibration checks be carried out when necessary. Where it is advisable to keep the vapours from the tank out of the gauge head a liquid seal may be provided as shown in *Figure 2.4*. The depth of the seal is usually about 175 mm but by choice of suitable pipe legs any depth may be accommodated. Where freezing does not occur the sealing liquid is usually oil, but in colder climates some form of antifreeze is used.

The actual equipment will now be described under the following headings:
    Simple float-actuated mechanisms.
    More elaborate float-actuated systems.
    Methods using an interface detector.
    Other methods.

2.1.3.1 SIMPLE FLAT-ACTUATED MECHANISMS

(a) *Float and counterweight type*

This simple mechanism is illustrated in *Figure 2.7*. It consists simply of a large area float connected by a cable or chain to a counterweight which passes in front of a scale and acts as an index. The float should have the largest possible area in order to reduce the errors owing to friction and out-of-balance forces of the cable or chain. The force available to overcome the starting friction of the mechanism is determined by the product of the cross-sectional area of the float, the density of the liquid and the difference between the actual float position and the designed float position relative to the surface of the liquid. Thus, the error in the instrument owing to friction is inversely proportional to the cross-sectional area of the float. By the same reasoning, it can be seen that the error owing to out-of-balance of the cable is also inversely proportional to the area of cross-section of the float. The float may be suitably protected cork, or in the form of a shallow cylinder of metal with its axis at right angles to the surface. Where the tank contents are corrosive the float should be made of

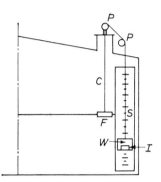

Figure 2.7   Simple float-operated level indicator
F, float having a large area and weight. *C*, cable, chain or tape. *P*, pulleys. *S*, level scale.
W, counterweights with index *I*

stainless steel or some other corrosion-resistant material or may be protected by a film of P.T.F.E. or a protective resin. As the depth to which the float sinks will depend upon the density of the liquid, it is necessary to take account of this in designing the system. In addition, the area of the float must be sufficiently large to reduce the errors produced by changes in depth of immersion owing to density changes to acceptable limits, i.e. well within the overall accuracy of the system.

(b) Magnetically coupled float and follower

When it is required to measure the level in a tank containing a volatile or toxic liquid the magnetically coupled level gauge shown in *Figure 2.8* has considerable advantage over the simple float-actuated gauge. The gauge consists of a tube closed at one end and open at the other. The tube is permanently fixed in the tank with the closed end immersed in the liquid. A float carrying a ring magnet floats in the liquid and is free to slide up and down the tube with the change of liquid level. The ring magnet is magnetically linked to the inner magnet so transmitting the rise and fall of liquid level to a calibrated rod, tape or gauge head, at which point the measurement is obtained.

The flanged tube and all wetted parts may be manufactured from F.M.B. stainless steel, polypropylene, aluminium alloy or brass whichever is suitable for the application.

Gauges suitable for working pressures up to 50 bar and temperatures up to 100° C are available. Where rod indication is used the depth measured is

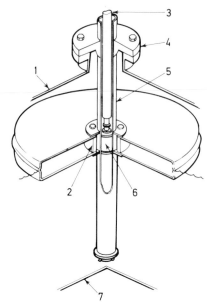

Figure 2.8   Magnetically coupled float and follower

1. Tank shell
2. Outer magnet
3. Calibrated rod or tape linked to gauge head
4. Mounting flange
5. Non-magnetic welded pressure tube assembly
6. Inner bar magnet transmits rise and fall
7. Tank shell

limited to 3·7 m but where calibrated tape is used this length may be increased to 14 m. Depths may be measured with an accuracy of ±0·5 mm.

(c) Simple float-operated contents gauges and level switches

An example of a simple float-operated contents or ullage gauge is illustrated in *Figure 2.9*. This type of gauge is frequently used on road, rail and marine vessels and on static applications particularly where the vessel may be sealed and subject to pressure or vacuum.

The method of operation is that the rise or fall of the liquid moves the float arm assembly through an arc about the pivot, P, which is located on the liquid side of the instrument. By means of a suitable gearing this movement rotates a driving magnet, B, which in turn rotates the follower magnet, A, so operating the pointer. The position of the gauge on the tank is determined by the circumstances. A position on the side or end of the tank, midway between the high and low level, with the dial vertical has the advantage that the float arm length is reduced to a minimum and the scale will be symmetrical.

The float is usually made of cork coated with an oil-resisting varnish, but metal floats are also available. Apart from the magnets, which are a high-permeability alloy, protected where necessary by silver-plating and rhodium coating, the instrument is manufactured from non-ferrous materials. The materials chosen depend upon the nature of the liquid measured. If required, mercury switches may be fitted to provide the initiation for a low level alarm.

Where it is required to initiate an alarm, start or stop a pump or open or shut a valve at a high or low level the magnetically operated switch or air pilot as shown in *Figure 2.10* may be used. The float assembly carries with it a permanent magnet which is opposed by a similar magnet which operates the switch, or air pilot valve. The adjacent poles of the two magnets are of the same polarity so that they repel each other, thus giving the mechanism a snap action.

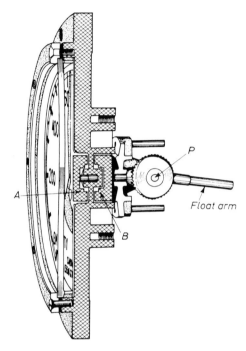

Figure 2.9   Simple float-operated contents gauge.
(By courtesy, Bayham Ltd.)

In the level switch mechanism the contacts change over with a snap action when the float passes the mid position. In the air pilot valve, a compressed air supply is led into the unit, and when the float is in its highest position the air valve permits the passage of air to the diaphragm or piston-operated valve causing the valve to close. A fall in liquid level causes the air valve to change over, shutting off the air supply and venting the air in the diaphragm valve to atmosphere, permitting the valve to open. If it is required to reverse the action of the pilot so that the diaphragm valve closes on fall in liquid level it is only necessary to change over the air supply and exhaust connections.

(d) The 'Ekstrom' level indicator

This type of level indicator is illustrated in *Figure 2.11*. It is frequently used for measuring the level of corrosive, toxic or inflammable liquids, and liquefied gases, because of its complete safety. It consists of a flanged external still pipe

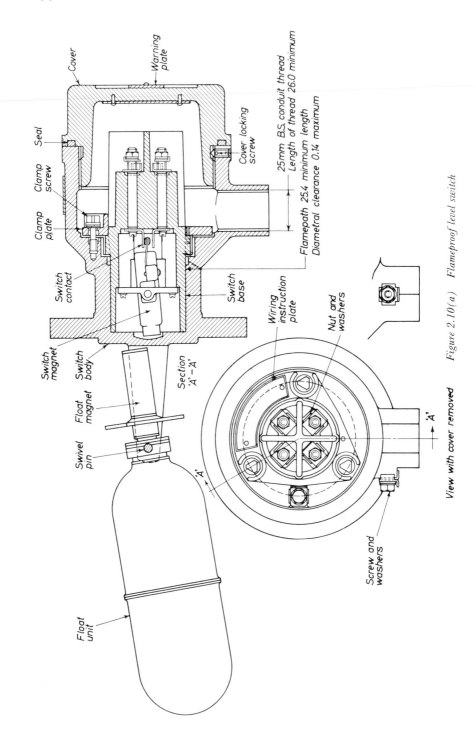

Cover

Warning plate

Seal

Clamp screw

Clamp plate

Cover locking screw

25mm B.S. conduit thread
Length of thread 26.0 minimum

Flamepath 25.4 minimum length
Diametral clearance 0.14 maximum

Switch contact

Switch base

Switch magnet

Switch body

Section 'A'-'A'

Float magnet

Swivel pin

Wiring instruction plate

Nut and washers

Float unit

Screw and washers

View with cover removed

Figure 2.10(a)   Flameproof level switch

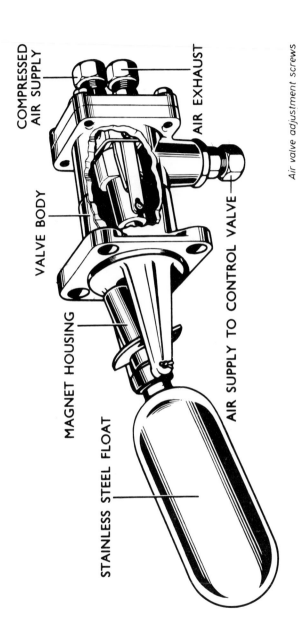

COMPRESSED AIR SUPPLY

AIR EXHAUST

VALVE BODY

AIR SUPPLY TO CONTROL VALVE

MAGNET HOUSING

STAINLESS STEEL FLOAT

Air valve adjustment screws

Magnet housing

Section through
air valve assembly

Figure 2.10(b)   Air pilot valve
'Mobrey' float-operated switch and air pilot valve.
(By courtesy, Ronald Trist & Co. Ltd.)

manufactured from a non-magnetic material, such as copper, aluminium, non-magnetic stainless steel or K monel having the same pressure rating as the vessel to which it is connected. The still pipe houses a float carrying a strong permanent magnet. The field of force of the magnet coincides with the level to which the float sinks. Parallel to the still pipe and close to it, is a graduated glass tube filled with butanol. In the glass tube there is a hollow iron sphere which has the same density as butanol so that its weight is exactly counterbalanced by the upthrust of the liquid so that it is weightless. The

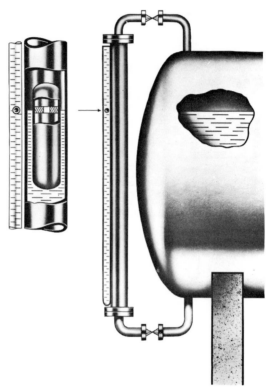

Figure 2.11   The 'Ekstrom' liquid level indicator. (By courtesy, G.E.C.-Elliott Process Instruments Ltd.)

sphere will therefore follow the magnet in the float, and hence indicate the position of the surface of the liquid. In addition to positioning the sphere, the magnet in the float may be arranged to tip a pivoted bar magnet and operate a mercury switch thus initiating an alarm when a preset level is reached. For use in flameproof areas, the bar magnet may be arranged to lift a flapper from a nozzle thus releasing an air pressure and hence initiating an alarm.

2.1.3.2   MORE ELABORATE FLOAT-ACTUATED SYSTEMS

(a) The Shand and Jurs liquid level gauges

In this type of gauge the stainless steel float may be in the form of a hollow disc for low pressures (i.e. up to 1 bar) or a raft formed of sealed tubes for tanks at pressures up to 2 bar. The float is attached to a flat precision graduated and perforated stainless steel tape which is guided over corrosion-resisting phenolic plastic pulleys having stainless steel shafts and P.T.F.E. sleeve bearings to reduce the friction to a minimum. The tape is attached to a take-up reel in the tank gauge-head shown in *Figure 2.12* which maintains a tension on the tape, and replaces the counterweight. The tension in the tape is maintained by means of a Neg'ator motor consisting of a spring with a torque characteristic which compensates for the tape pay-out and thus maintains a constant tension in the tape at the float. Thus the depth to which the float sinks is not influenced by the weight of the tape hanging free above it. The float is guided by means of stainless steel wires maintained at a constant tension by springs which are compressed sufficiently to produce a 23 kg guide-wire tension. The equipment is suitable for tanks up to 20 m high.

In one form of the instrument the depth of the liquid in the tank is read off from the tape through the window in the gauge head. On another type of display the perforations in the steel tape may engage on the pins of a

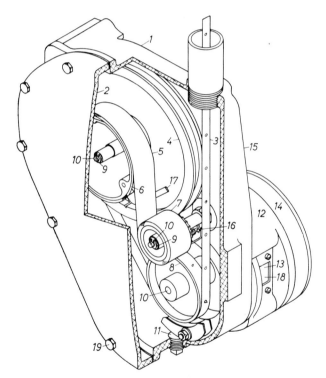

Figure 2.12   Tank gauge head with digital readout. (By courtesy, Whessoe Ltd.)
1, Precision cast main housing. *2*, Side cover. *3*, Perforated steel tape type 316 stainless. *4*. Moulded thermosetting phenolic tape drum. *5*, Broad Neg'ator Motor, stainless steel. *6*, Power drum. *7*, Storage drum. *8*, Precision made sprocket. *9*, P.T.F.E. bearings. *10*, Type 316 stainless steel shafts. *11*, Drain plug. *12*, Digital counter housing. *13*, Reading window. *14*, Stainless steel band covers adjustment slots. *15*, Operation checker handle (out of view). *16*, Operation checker internal assembly. *17*, Neg'ator motor guide, stainless steel. *18*, Counter assembly in chamber beyond tank pressure and vapours. *19*, Cap screws drilled for sealing

sprocket wheel and drive a reversible counter or indicator which indicates the level in feet, inches and tenths of an inch, or other convenient units.

The high pressure version of the instrument is suitable for pressures up to 20 bar. To avoid sealing problems the counter mechanism is magnetically coupled to the operating mechanism.

Where remote indication is required the level information may be transmitted by transmitters of the form described in Volume 3.

Tests indicate that gauging inaccuracies up to ± 5 mm may result from friction between the tape and seized pulleys or by contact between the tape and the walls of crooked or out-of-line piping or with small obstructions (scale, wax, etc.) in this piping. The most common method of checking is to move the system slightly by means of a manual operation checker. This method is not convenient for remote reading systems. A simple, effective and reliable method of reducing the effects of static friction is to use a vibro-checker which imparts a small vibrating motion to the tape.

The vibro-ckecker consists of a small electric motor, which may be flame-proof if required, driving an eccentric hub on which rides a loose-fitting nylon ring. When the motor is running the ring swings up and repeatedly strikes the tape. The resulting wave motion imparted to the tape is propagated throughout the system and is of sufficient intensity to overcome the friction which may exist at various bearing points. Consistent performance can thus be obtained over very long periods.

(b) The Kent/Enraf tank gauge

The principle of operation of this type of level indicator is illustrated in *Figure 2.13*. A solid displacer disc is suspended from a stainless steel wire, attached to a winding drum A. A two-phase servomotor winds or unwinds the wire until the tension in the weighing springs B, attached to the drum shaft, is in balance with the weight of the displacer minus the upwards pressure exerted on the displacer by the liquid. The balance contact C is then no longer touching either of the two servocontacts E or D.

The signal from the balance contact is first integrated by an electronic delay circuit to avoid unnecessary oscillation of the gauge on turbulent surfaces. During the rotation of the servomotor, the cam-driven transmitter K continuously changes the voltage-pattern in the receiver motor M causing the mechanical counter L to follow the liquid movement. The motor limit switches $I1$ and $H1$ or level alarm switches are actuated by cams I and H fixed to a rotating disc driven by the servomotor via a double worm drive. A switching arrangement to select the proper temperature element from a multi-element averaging thermometer can also be actuated by the same disc.

In addition to local indication of level the output will operate a remote indicator or feed into a telemetering system covering a complete tank farm.

In order that the operator may ensure that the operating circuits are live, a remote test facility is provided. This consists of a switch in an intrinsically safe circuit, which when operated causes the displacer to be raised from the liquid. When the switch is re-operated the level indicator returns to normal operating conditions and the displacer automatically returns to its former position thus providing an assurance that the operating circuits are operating satisfactorily.

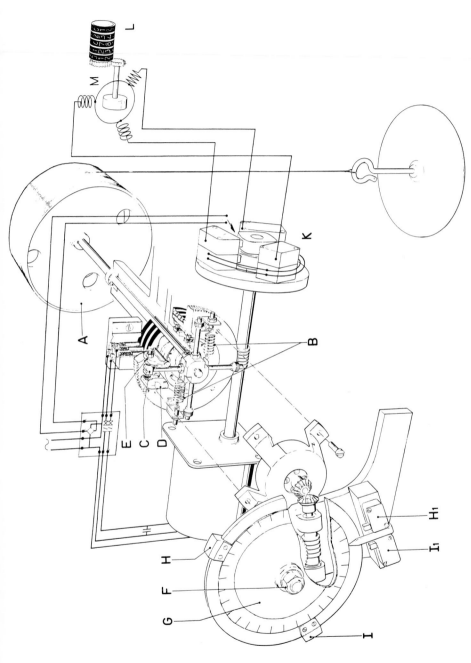

Figure 2.13   The Kent/Enraf tank gauge.   (By courtesy, George Kent Group)

The measuring range is up to 25 m, the accuracy $\pm(0\cdot5+0\cdot15L)$ mm, where L is the level in m, and the repeatability $\pm0\cdot2$ mm. The static error for a charge of 100 kg/m³ in product density is $0\cdot2$ mm.

The instrument is housed in an explosion-proof enclosure. The level readings are acceptable to Customs Authorities in the U.K. and many countries in Europe.

2.1.3.3   USE OF AN INTERFACE DETECTOR

(a) The Gilbarco–Elliott electronic tank gauge

This instrument differs from those already described in that a sensing element designed to detect its position in relationship to the liquid surface is raised or lowered by a servo-motor so that it always follows the liquid surface. To achieve this an 'up' signal of preset strength which would cause the servo-motor to raise the sensing element is compared with a 'down' signal from the sensing element which would cause the sensing element to be lowered. As the sensing element approaches the liquid surface the 'down' signal decreases, while increased distance between sensing element and the surface causes the 'down' signal to increase. Thus a balance between the 'up' and 'down' signals is always achieved with the same relative position of sensing element and liquid surface.

The sensing element consists of a hollow stainless steel cylinder $0\cdot38$ m long and 50 mm outside diameter. From the cylinder a stainless steel straight-wire antenna projects downwards 16 mm below the open end of the cylinder. This antenna is the only portion of the sensing element which is in contact with the liquid but the entire sensing element is designed to be unaffected by accidental immersion.

The sensing element hangs by a perforated stainless steel tape which links the sensing element to a sprocket on the servo-motor which raises or lowers the element. In addition the element is connected by a P.T.F.E.-protected radio-frequency cable to the electronic equipment which is housed with the servo-motor, the level indicator and any remote transmitters or switching mechanisms that may be required, in a tank-side explosion-proof box. The servo-motor is a slow-speed high-torque two-phase motor and raises or lowers the sensing element in accordance with the signals it receives from the servo-amplifier. In addition this motor drives the mechanical depth indicator calibrated in feet, inches and sixteenths of an inch, and remote transmission potentiometers if required. Because all the power required to operate the mechanism is provided by the motor, the indication will be independent of friction, density of the liquid, and other factors and a sustained accuracy of $\pm1\cdot5$ mm is possible in the level indication.

For readers who have an understanding of electronics the following more detailed explanation of the operation is given. The principle of the mode of operation is given in *Figure 2.14*. A radio-frequency oscillator modulated at 50 c/s is inductively coupled to the radio-frequency cable by the low impedance coupling loop C. The cable carries this radio-frequency signal to another coupling loop, A, located in the sensing element. This loop, A is inductively coupled to the antenna wire by the loop, B. Capacitors C_1 and

C_2 provide low impedance paths to earth for the radio-frequency signal but not for the 50 c/s signals. A silicon rectifier is connected to the antenna coupling loop, B, and demodulates the radio-frequency signal. This 50 c/s demodulated signal appears across the resistance R_2. Resistance R_1 offers a low-impedance path to this 50 c/s signal but capacitor C_2 offers a high impedance. Thus the 50 c/s signal is carried via the radio-frequency cable and appears across the capacitor C_1 which offers a high-impedance path. The

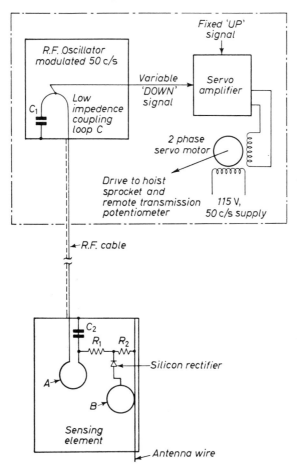

Figure 2.14 Principle of Gilbarco level indicator. (By courtesy, G.E.C.-Elliott Process Instruments Ltd.)

signal is then passed to the servo-amplifier where it is amplified sufficiently to energize the servo-motor winding. The phase of this signal is such that it tends to cause the servo-motor to turn in the direction which causes the sensing element to be lowered.

The servo-motor is also fed with a fixed signal phased such that the servo-motor tends to raise the sensing element. When the sensing element is above the level of the liquid relative amplitude of the two signals is such that the sensing element is lowered.

As the sensing element approaches the surface of the liquid 'losses' in the sensing element increase until, when the tip of the antenna enters the liquid to a depth of about 3 mm, the losses are sufficient to reduce the amplitude of the 50 c/s from the sensing element to such a level that the signal after amplification is smaller than the fixed 'up' signal. Thus, the sensing element will be raised. At one particular depth of antenna immersion both 'up' and 'down' signals will be equal resulting in zero signal to the servo-motor control winding. As the liquid level rises or falls so will the amplitude of the signal from the sensing element decrease or increase so that the sensing element will be raised or lowered to follow the liquid surface. The equipment is suitable for use on liquids at temperatures up to 220° C and pressures up to 17·5 bar g, and as the only materials exposed to the liquid or vapours in the tank are stainless steel or P.T.F.E., the standard gauge may be used on most corrosive and non-corrosive liquids. Special gauges are manufactured for use in extremely corrosive liquids. The equipment may be equipped with remote transmission and the servo-motor may be arranged to switch on the appropriate resistance bulbs for the measurement of the average temperature. In order that the operator may check that the instrument is functioning a switch is provided. Closing the switch causes the sensing element to be raised. When the switch is opened the indicator should return to its former reading.

(b) Whessoe servo-operated tank gauge

Another level gauge employing a servo motor is the Whessoe tank gauge in which the liquid surface detector takes the form shown in *Figure 2.15(a)*. This unit is suspended in the tank at the end of a perforated tape which runs over pulleys and down the outside of the tank into the gauge head where the servo motor, the control unit and the level indicator are located, see *Figure 2.15(b)*. If a level transmitter is provided, this is attached to the gauge head and driven by the servo system. The perforations on the tape ensure positive location on the sprocket wheel which drives the digital level indicator and the tape is recovered on a drum driven by the servo motor. The cable connecting terminals A, B and C with the control unit is clamped at its outer end to the level sensor and passes over pulleys and is connected to the control unit. Any slack in the cable as the sensor rises is taken up by the counterweight housed in the connecting tube as shown in *Figure 2.15(b)*. The sensor is guided by guide wires which pass through the space between the cotter pins and the sensor body. Under stable conditions the float (2) is on the surface of the liquid and supports the magnets in the mid position when both reed switches will be open circuit.

Should the level rise by a small amount the lower magnet will operate the lower reed switch closing the circuit between terminals A and C. This causes the control unit to operate the servo motor in a manner which raises the sensor until the position of the magnet relative to the reed switch is restored, thus winding in a length of tape equivalent to the change in level and changing the level indication by an equal amount.

A fall in the liquid level will cause the float to move downwards causing the upper magnet to close the circuit in the upper reed switch connecting terminals B and C. This causes the servo motor to be energised in a manner which results in a lowering of the sensor until the relationship of magnet and reed switch is

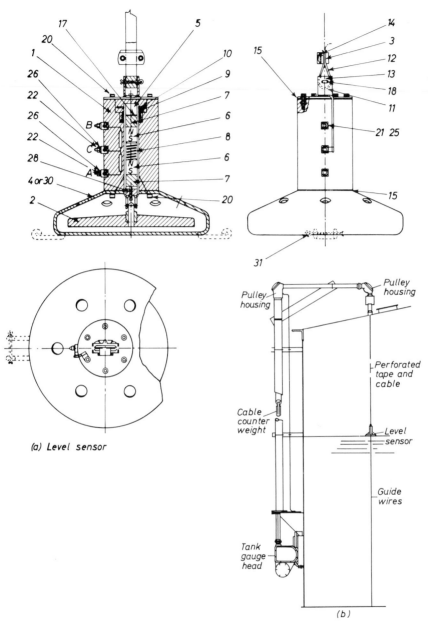

Figure 2.15   *Whessoe servo-operated tank gauge*

1. Body assembly	9. Float support spring	18. Cotter pin
2. Float assembly	10. Spring retaining washer	20. Cap screw
3. Tape clamp assembly	11. Hanging bracket	21, 22. Hexagon nut
4. Hood (less guide eyes)	12. Tape thimble	25, 26. Lock washer
5. Axial guide	13. Tape spool	28. Screw
6. Magnet	14. Cable clamp	29. Earthing cable assembly
7. Magnet retainer	15. Gasket (2)	30. Hood with guide eyes
8. Magnet adjusting spring	17. Circlip	31. Cotter pin

restored. Thus the indication of the gauge will again change by an amount equal to the change in level.

High and low level limit switches are provided to isolate the motor and prevent the sensor from being driven beyond the normal working range.

The sensor circuit is intrinsically safe to B.S. 1259: 1958 and the control unit and servo motor are housed in a flameproof enclosure to B.S. 229: 1957 for group II and III gases.

In order that the operation of the level gauge may be verified, a switch is provided at the remote level indicator which overrides the control system and cause the sensor to be driven upwards. It can then be allowed to find the liquid surface, and the level reading obtained should be the same as that before the switch was operated to within the working tolerance of the instrument.

The digital readout has a maximum range of 24·5 m, the gauge has a sensitivity of $\pm 1·0$ mm and the maximum sensor speed is 11 mm/s.

(c) Detection of level of solids

For detection of the level of solids in a silo or bin a device such as the Bin-Dex may be fitted to the top of the bin. The instrument consists of a metal weight or bob suspended by a stainless steel cable which runs over a weight-sensitive pulley system.

The pulley senses three load conditions:
1. The weight of the bob plus cable, when the bob is clear of the material in the bin.
2. The reduction in load when the weight or bob is partially supported by the material.
3. The increase in load if material is causing drag on the bob.

The cable reel is driven by a geared reversing motor and, at selected intervals determined by the setting of a timer, the motor is energised to raise the bob.

If the weight of the bob plus cable (condition 1) is sensed, the bob is lowered to the material surface, and the motor de-energised when the reduced load (condition 2) is sensed. If, on the other hand, the increased load due to drag (condition 3) is sensed, the bob is raised until it is in contact with the material surface but no longer covered, and the motor is again de-energised when the reduced load (condition 2) is sensed.

A d.c. electrical signal from a multi-turn potentiometer, gear driven from the motor shaft, operates a remote meter which indicates the level of the material in the bin.

2.1.3.4 OTHER METHODS

(a) Capacity bridge types—the Fielden Telstor

This equipment consists of an electrode, an electronic unit and an indicator. The electrode in the form of a long metal rod which reaches from the top to bottom of the vessel. The electrode is bare when the tank contents are electrically non-conducting, but is sheathed in polythene or P.T.F.E. when the tank contents are conducting. The electronic unit is merely a power supply and a highly stablised capacitance measuring bridge. One arm of the bridge is formed by the capacitance between the level sensing element and earth. A change in this capacitance owing to the rise and fall of the material around the electrode produces an out-of-balance current from the bridge which is measured by the indicating meter. On stable applications the meter can be calibrated in any desired units.

Both intrinsically safe and flameproof versions of the instrument are available. A simplified version may be used for level alarms or controls. In this form of the equipment the sensitivity of the device can be arranged so that the alarm is initiated by the proximity of material so that the probe will not be affected by the material in the vessel even if it is corrosive.

(b) Conductivity methods

When the liquid in a tank or sump is a conductor of electricity, electrodes may be inserted into the liquid which may be arranged to complete electrical circuits and so operate pump controls. Two patterns of relay units used in conjunction with such electrodes are shown in *Figure 2.16*. Each pattern can be supplied either for emptying a sump or filling a tank. *Figure 2.16* shows the connections for the type for emptying a sump. The type for tank filling is similar but the mercury switches are arranged at a different angle.

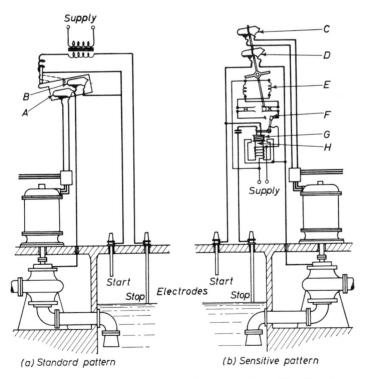

(a) Standard pattern   (b) Sensitive pattern

Figure 2.16   Pump control units. (By courtesy, George Kent Group)

The principle of operation is as follows:

Standard pattern relay (Figure 2.16a). When the liquid completes the circuit through the two electrodes, the solenoid is energised and closes the switches *A* and *B*. The former actuates the pilot coil of the motor starter while the latter completes a hold-on circuit through the tank, so that the pump continues to run until the liquid falls below the lower electrode.

Sensitive pattern relay (Figure 2.16b). Where the resistance of the electrode circuit including the liquid exceeds 125 ohms for 30 volt operation, or 500 ohms for 60 volt operation, as it may do when the liquid is water from mountain lakes, the sensitive pattern relay must be used. This is suitable for circuits where the electrode resistance does not exceed 3000 ohms, and may, if required, be arranged so that the electrode system is intrinsically safe for the methane and pentane class. The relay unit will, however, have to be located in a non-hazardous area or housed in a case which is certified flameproof. In this pattern, the electromagnet *H*, which is connected to the mains, acts also as the primary of a transformer of which the coil *G* is the secondary. When the liquid completes the circuit through the electrodes, the coil *G* is drawn into the gap of the magnet *H* and the contact *F* energises the appropriate coil of the secondary relay *E* thus closing the switches *C* and *D*. The former actuates the pilot coil of the motor starter while the latter completes the hold-on circuit.

(c) Method employing a radioactive source and detector

In this type of equipment a source of gamma radiation such as cobalt housed in a suitable lead shield and a detector element which may be a halogen-quenched Geiger–Müller tube of metal construction are used. The source is mounted on one side of a tank and the detector exactly opposite on the other side of the tank. The detector gives counts at a certain rate when the radiation from the source falls upon it. When the radiation is reduced by the tank contents passing between the source and detector, the count rate will be greatly reduced. A relay which can be set to trip at any count rate is fed from the detector and will change its state every time the contents surface passes the line joining source and detector. This type of equipment can therefore be used to initiate level alarms or controls. When it is required to measure level it can be arranged for a servo-motor to raise or lower the source and detector together in accordance with the state of the relay. Thus the equipment can be arranged to follow the level in a still pipe. In addition, the servo-motor is arranged to drive a counter so that the depth is indicated.

(d) Use of a thermistor

For detecting point levels to prevent tanks being overfilled, a thermistor may be mounted in the tank at a critical level. The thermistor is maintained at a temperature about 25° C above the tank ambient temperature by a constant current source. When the liquid touches the thermistor it is cooled and its resistance increases. This change of resistance is detected electronically and initiates the appropriate alarm or control action. The circuit is designed to be intrinsically safe so that the equipment may be used on petroleum products.

2.2 PRESSURE-OPERATED TYPES

2.2.1 Simple pressure-actuated mechanisms indicating one or more levels

The simple pressure-operated electrical indicator is an extremely useful type of instrument for indicating the level of the contents of a bin or bunker in which powdered or granular substances, such as pulverised fuel, are stored.

The transmitter, which consists of a cylindrical metal casting, is bolted on to the outside of the bin. It is fitted with a tough rubber diaphragm. The diaphragm has a definite movement, and operates a plunger, which in turn actuates up to three independent 'microgap' single pole change-over switches. The diaphragm is spring-loaded, and requires a pressure of about 0·15 bar to depress it. When the diaphragm is depressed, the plunger completes or breaks the electrical circuit of the alarm signal, indicating that the contents of the bin have reached the level of the transmitter. The internal connections of a transmitter are shown in *Figure 2.17.*

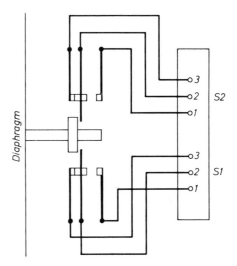

Figure 2.17   Level transmitter (By courtesy, G.E.C.-Elliott Process Instruments Ltd.)

By mounting a transmitter at the top, and another at the bottom of the bin, a red light may be switched on or off, to indicate the bin is full, while a green light may be switched off to indicate the bin is empty.

It is better for the alarm signal to be given by the light in the lamp being extinguished instead of the lamp lighting up. This is because, in the event of the lamp failing, the operator will assume that the signal has operated and will endeavour to attend to the necessary controlling of the supply. It will, however, soon become apparent that it is the lamp which has failed, and this can then be remedied before any alarm is caused.

When a lamp fails in the system where the lamp lighting up is the alarm signal, the operator will naturally assume that, with no light being visible, everything is in order, whereas the bin will continue filling up or discharging, thereby causing damage to equipment or the cessation of supply.

In many installations an audible warning is given as well as the visual one, in order to attract the operator's attention.

This type of installation is not suitable for liquids, for wet or damp materials liable to stick or 'pack', or for solids larger than about 25 mm cube.

2.2.2 Static pressure-actuated mechanisms including gas or liquid purge systems

2.2.2.1 LEVEL IN AN OPEN TANK

Since the pressure acting on any area at the bottom of a tank depends only upon the depth and the density of the liquid within the tank, the pressure measured at the bottom of the tank containing a liquid of known density is a measure of the liquid level in the tank.

This principle is used in many forms of liquid level measurement. The pressure may be measured by any of the pressure measuring instruments described in Section 1 provided it is suitable for the particular application, and may be a direct measurement or one using a gas or liquid purge system.

The order of choice of type of instrument from a practical point of view does not coincide with the order of treatment in this section. From the mathematical point of view the manometer is simple, but from the manufacturer's point of view it is a relatively complicated instrument, and from the purchaser's point of view it is comparatively expensive.

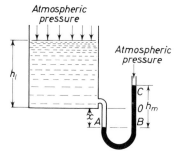

Figure 2.18
Static-pressure-actuated system (simple)

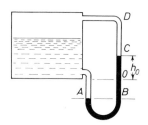

Figure 2.19
Static-pressure-actuated system (closed system)

From the manufacturer's and purchaser's point of view the order of simplicity would probably be:

(*a*) Diaphragm box and similar instruments as they are self-contained and require no auxiliary power. They cannot, however, be used in closed tanks.

(*b*) Air purge systems are almost as simple as the diaphragm box but they need in addition an auxiliary supply in the form of compressed air. This is sometimes provided by a hand-pump or foot-pump for remote locations.

(*c*) Manometers measuring differential pressure.

The simplest method, from the theoretical point of view, of measuring the pressure due to the liquid in an open tank is to use a mercury manometer.

One limb of the manometer is connected directly to the tank, while the other limb is open to the atmosphere, as shown in *Figure 2.18*.

Let A be the level of the interface between the mercury and the liquid in the tank, and let the mercury in the U tube stand at the level of the bottom of the tank, when the tank is empty. The tank contains a height h_e m of liquid of density d_e kg/m³, and the height of the mercury column of density d_m kg/m³ above B, on the same level as A, is h_m m. Let A be x m below the bottom of the tank. Then, since A and B are at the same horizontal level in the liquid:

$$\begin{aligned}
\text{Pressure at } A &= \text{pressure at } B \\
xd_eg + h_1d_eg + \text{atmospheric pressure} &= h_m d_m g + \text{atmospheric pressure} \\
\therefore \quad xd_e + h_1d_e &= h_m d_m \\
\text{or } h_1d_e &= h_m d_m - xd_1 \\
\text{or } h_e &= h_m \frac{d_m}{d_e} - x\frac{d_1}{d_1} \\
&= h_m \frac{d_m}{d_1} - x \qquad (2.1)
\end{aligned}$$

Now, since the mercury was originally at the level of the bottom of the tank, and the level in the open limb must have risen as much as the level in the other limb was depressed, $h_m = 2x$, or, $x = h_m/2$.

Substituting this value in equation 2.1

$$h_e = h_m \frac{d_m}{d_1} - \frac{h_m}{2} \qquad (2.2)$$

giving the depth of liquid in the tank

$$= h_m\left(\frac{d_m}{d_1} - \frac{1}{2}\right) = K\,h_m \qquad (2.3)$$

Thus, the limb BC may be calibrated in terms of level of liquid in the tank.

If O, at the level of the bottom of the tank, is made zero, then the height OC of the mercury above this zero is h_0 m, where $h_0 = \frac{1}{2}h_m$ or $h_m = 2h_0$

and

$$h_1 = 2h_0\left(\frac{d_m}{d_1} - \frac{1}{2}\right) = h_0\left(\frac{2d_m}{d_1} - 1\right) \qquad (2.4)$$

If the liquid in the tank is water and the mercury in the U tube has a specific gravity of 13·56 then:

$$h_1 = h_0\,(27·12 - 1) = 26·12h_0$$

Level measurement in a closed tank

The pressure of the air or vapour above the liquid in a closed tank may be different from that of the atmosphere. If the simple U tube described for open tanks is used for a closed tank, then: Pressure at A = pressure due to liquid + pressure of vapour or gas above the liquid.

Pressure at B = pressure due to mercury + atmospheric pressure.

Again since:

Pressure at A = pressure at B.

Pressure due to liquid + pressure of gas above the liquid

= pressure due to mercury + atmospheric pressure.

Thus the level of the mercury would not indicate the true level of the liquid, but the indication for a constant level would vary with the difference between the atmospheric pressure and the pressure in the tank above the liquid.

To eliminate this error, the pressure on both columns of liquid must be arranged to be the same. This may be achieved by connecting what was the open end of the U tube to the top of the tank as shown in *Figure 2.19*.

This practice has, however, one very serious defect. Vapours from the tank tend to condense in the limb $C\ D$, so the liquid formed will press down on the mercury, and tend to balance the pressure due to the liquid in the tank.

This will introduce an error, which will vary with the quantity of condensed liquid in $C\ D$. To avoid this difficulty, the arrangement shown in *Figure 2.20* using a condensing chamber E is used. The condensing chamber and the limb $D\ B$ are originally filled by allowing the level of the liquid in the tank to rise above the upper tapping, or by manual filling. The condensing chamber has a large area of cross-section in comparison with the U tube, so

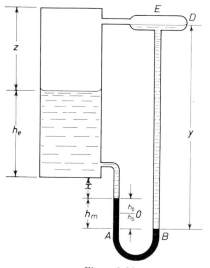

Figure 2.20
Static-pressure-actuated system
(with condensation chamber)

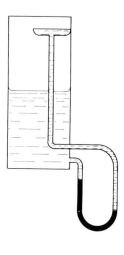

Figure 2.21
Static-pressure-actuated system
(with enclosed pipe)

that the level of liquid in it does not change much when the mercury moves in the manometer. In measuring the depth of cold water or heavy oil where the vapour pressure is low, there will be very little condensation in the chamber E. In such cases arrangements must be made to introduce a small bleed of vessel liquid into the chamber E if this type of installation is used.

The simplest way of looking upon this arrangement is that the two columns of liquid above A and B balance.

Suppose the level in the tank is z m below the level in the condensing chamber. The pressure due to the tank liquid at A is therefore less than that at B by the pressure due to a column of liquid of height z m, but the columns remain balanced because a length h_m m of liquid on the tank side is replaced by a similar column of mercury.

∴ Pressure due to column of height z m of liquid = Pressure due to column length h_m m of mercury − Pressure due to a column of length h_m m of liquid, or $z\, d_e\, g = h_m (d_m - d_e)\, g$ if the density of the vapour may be neglected,

or
$$z = h_m\left(\frac{d_m}{d_e} - 1\right) \tag{2.5}$$

If the density of the vapour cannot be neglected, and has a density d_v kg/m³,

then:
$$z(d_e - d_v)\, g = h_m(d_m - d_e)\, g$$

or
$$z = h_m\frac{(d_m - d_e)}{d_e - d_v} \tag{2.6}$$

$$= 2h_0\frac{(d_m - d_e)}{(d_e - d_v)}$$

Thus, the elevation above the datum of the level of the mercury in the left-hand limb will be an indication of how much the level of the liquid in the tank has fallen below the level in the condensing chamber, a level which may be regarded as being constant.

This system will introduce an error if the temperature of the liquid in the tank is different from that of the surrounding air. The liquid in the limb $B\,D$ will have a different temperature and consequently a different density from that in the tank. This difficulty may be overcome, to a large extent, by putting the major portion of the limb $B\,D$ in the tank as shown in *Figure 2.21*. This system is also useful when the liquid in the tank has to be kept hot to prevent it becoming too viscous to flow freely.

2.2.2.2   LIQUID SEALS

When the level of corrosive liquid, highly volatile liquid, tar, oil containing sulphur, or heavy viscous fluid is being measured, liquid seals should be used to prevent such liquids entering the mercury chamber. The sealing liquid must not mix with the vessel liquid, be attacked by it, or absorb corrosive elements from it. The arrangement in this case is shown in *Figure 2.22*. When there is vapour in the vessel, which would condense in the seal chamber and displace the sealing liquid, an intermediate condensing chamber may be placed between the seal chamber and the upper tank tap.

2.2.2.3   MEASURING ELEMENT AND COMPLETE INSTALLATION

The measuring element used in this type of instrument may have the same form as any of those already described for differential pressure measurement. This type of level measurement is, in effect, merely a measurement of dif-

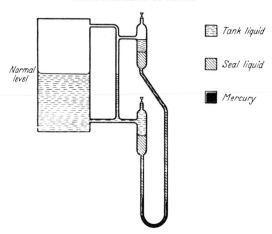

Figure 2.22   Static-pressure-actuated system (with liquid seals)

ferential pressure, the scale of the instrument being calibrated in units of depth instead of units of pressure. *Figure 2.23*, shows a diagram of a complete remote liquid level indicator used for indicating the level of liquid in boilers, de-aerating tanks, etc. Stainless steel is used in the construction wherever the water or mercury comes into contact with the instrument. A stainless steel float is carried by mercury in the well of the manometer, and its position is transmitted to the pointer by a magnetic armature. The position of this armature within the stainless steel tube is picked up on the outside of the tube by a yoke having strongly magnetized arms. Other details of its construction can be gathered from the diagram.

2.2.2.4   GAS PURGE SYSTEMS

In many installations it is not convenient to locate a manometer at the bottom of a tank, while in others, the contents of the tank may be very hot, highly corrosive, or contain solids in suspension which would block the pipes leading to the U tube. The simple manometer is not suitable in such cases, so that a gas purge or liquid system must be used.

Open tanks

Where the liquid level is required in an open tank, the pressure due to the tank contents may be found by using a purging gas. The purging gas, which may be air, carbon dioxide, or any suitable gas, is blown through the stand pipe which goes down almost to the bottom of the tank as shown in *Figure 2.24*.

As in the simple measurement of pressure, the pressure in the stand pipe will build up until it is equal to that due to the liquid above the level of the bottom of the pipe. If the flow of gas is small, say, 60 bubbles per minute, a pressure of gas equal to that in the stand pipe will be applied to the liquid level indicator and the recorder. They will, therefore, give an indication depending upon the pressure due to the depth of the liquid in the tank and so indicate the level. A correction can be made for the height of the end of the stand pipe above the bottom of the tank.

The liquid level indicator and recorder must be connected directly to the stand pipe as shown in *Figure 2.24* and not to any other point in the purge line. There must be a small fall of pressure along the gas line in order to produce a flow. If the rate of flow is small, this pressure gradient along the line will be very small, and, if the indicator is in the correct position, the error introduced will be so small that it may be neglected.

If an agitator is used in the tank, the stand pipe is placed in a larger pipe which protects it from turbulence. The larger pipe is open to the vessel liquid.

Where the depth of high consistency pulp on liquids containing sludge is being measured a combination of air and water purge is used. The stand

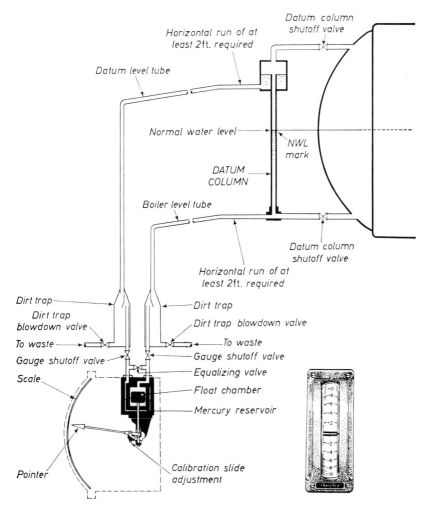

Figure 2.23   Boiler level indicator. (By courtesy, Bailey Meters & Controls Ltd.)

A. Air filter.
B. Reducing valve.
C. Needle valve for controlling flow of gas.
D. Bubbler, sight feed unit or rota-meter type of flow indicator, to indicate the rate of flow of purging gas.
E. Liquid level indicator.
F. 12 mm stand pipe, arranged so that its lower end is at least 75 mm above the sediment line in the tank. This pipe may be inside or outside the tank as shown.
G. Clean out plugs.
The recorder and the unit supplying the gas or air to the stand pipe may be as much as 300 m away from the tank.

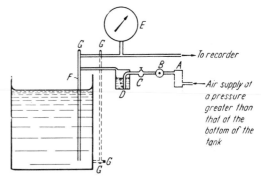

Figure 2.24   Gas purge systems

pipe is placed in a stilling well which is connected to the tank as shown in *Figure 2.25*. The flow of water is adjusted so that the level in the well is maintained slightly higher than the level in the tank, so that there is a continuous flow of water from the well to the tank, thereby keeping the connection free from solids. The depth of water in the well is then measured by the normal air purge system which gives a signal proportional to the level in the tank.

Closed tanks

The system used for closed tanks is shown in *Figure 2.26*.

Great care must be taken to avoid errors due to pressure gradients in the pipe line owing to flow along the line. These errors will be small provided the

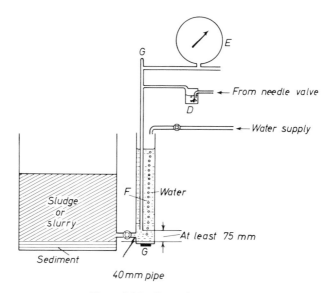

Figure 2.25   Gas and water system

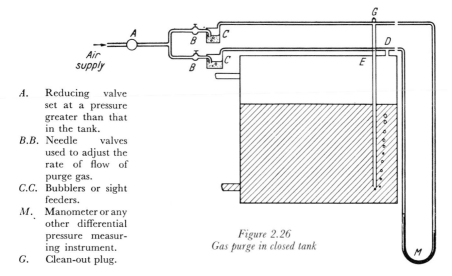

A. Reducing valve set at a pressure greater than that in the tank.

B.B. Needle valves used to adjust the rate of flow of purge gas.

C.C. Bubblers or sight feeders.

M. Manometer or any other differential pressure measuring instrument.

G. Clean-out plug.

Figure 2.26
Gas purge in closed tank

rate of flow in both purge lines is equal and small. The manometer or other differential pressure measuring instrument must be connected directly to the purge lines at the beginning of the stand pipe E, and at the tank connection D, and not to any other points in the lines.

Gas purge systems are very useful for measuring the depth of highly corrosive liquid in a tank. The stand pipe must be carefully chosen to resist the corrosive action of the tank liquid. Air is the most common purge gas used but on installations where the introduction of air into the system would cause a fire danger, carbon dioxide is used. In petroleum, oil, and similar plants, a fuel gas, ethane, methane, or butane purge is often used. The use of nitrogen as a purge gas is also very common in the petroleum industry, where large

F. Filter.

R. Regulating valves.

I. Flow indicators.

M. Measuring instrument

V. Isolating valves.

E. Equalizing valve.

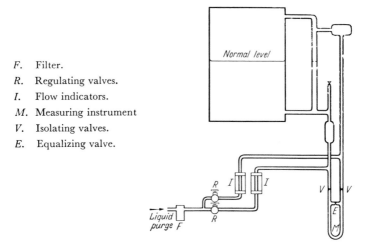

Figure 2.27 Liquid purge system

quantities of nitrogen are obtained by burning fuel gas to form flue gas which can be washed and stored.

2.2.2.5 LIQUID PURGE

When the depth of very hot liquids, asphalts, liquids containting solids in suspension, volatile or corrosive liquids, is required, a liquid purge is used as shown in *Figure 2.27*. Instead of gas, a suitable liquid is introduced into the manometer lead lines so as to maintain a very small flow into the tank. This flow prevents the tank contents from entering the lines and causing corrosion or blockage. The nature of the purging liquid must be such that the introduction of small quantities into the plant will not injure the product or process, it should be free flowing and not vaporise at the temperature of the lead lines. The purging liquid may be either soluble or insoluble in the vessel liquid. Often purified distillate from the vessel is suitable. Water or a light mineral oil is also used. The rate of flow of the purging liquid is adjusted to about 4·5 litres per hour through each connection, and the differential pressure measuring instrument is usually located below the level of the tank.

The measuring instrument used in purge systems is the same as that used for other similar forms of differential pressure measurement. The actual instrument used will depend upon the depth of liquid to be measured and the maximum differential pressure to be expected. Well type mercury manometers are used extensively for differential pressure measurement in level measurement, but other forms of measuring elements are also used. Whatever form of

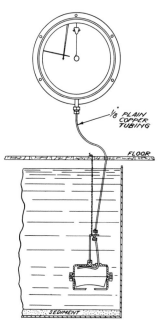

Figure 2.28
Diaphragm box installation.
(By courtesy, Foxboro-Yoxall Ltd.)

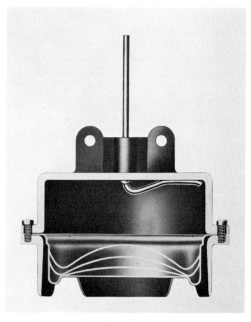

Figure 2.29
Diaphragm box installation.
(By courtesy, Foxboro-Yoxall Ltd.)

pressure measuring instrument is used, great care must be taken to see that liquid-filled pressure lines are free from gas, and that gas-filled lines contain no liquid. The maintenance of differential pressure measuring installations is described in the flow measurement section.

2.2.2.6 DIAPHRAGM TYPES

Where the provision or use of a continuous liquid or gas supply is difficult or undesirable, a diaphragm type of instrument may be used to measure the pressure at the bottom of an open tank. This diaphragm may be either flexible or stiff, and in certain cases may be replaced by a flexible synthetic bellows.

Flexible diaphragm

Figure 2.28 shows the installation of this type of depth measuring instrument. The diaphragm box, whose construction is shown in *Figure 2.29*, consists of a cylindrical box with a rubber diaphragm across its open end. The movement of this diaphragm is completely unrestrained. The diaphragm is merely a means of preventing the air in the box from dissolving but has no function other than that of a barrier. This diaphragm is protected by a further cylindrical box whose lower end is open to admit the tank liquid. The diaphragm box is suspended in the tank well above the sediment level. When the level of liquid in the tank rises, the pressure on the diaphragm increases, and the diaphragm

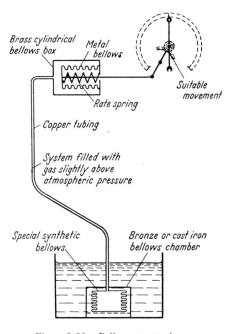

Figure 2.30   Bellows type tank gauge

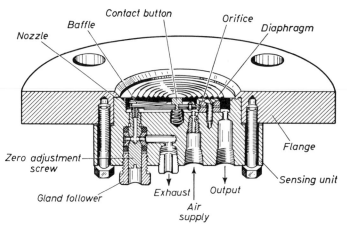

Figure 2.31   Force balance system

moves. This compresses the air within the closed system. The increased air pressure is transmitted by the capillary tube to the pressure measuring portion of the instrument which may be an indicator or a recorder.

In the depth gauge shown in *Figure 2.30*, the diaphragm box is replaced by a box containing a bellows of synthetic material. Changes of pressure within the bellows due to changing levels are communicated to the measuring portion of the instrument by copper tubing having a small bore.

Initially the diaphragm box and bellows are filled with air at a pressure slightly above atmospheric pressure. The diaphragm may be made from a variety of materials, from latex-dipped fabric, to oil- and heat-resisting rubber-like substances. The materials used for diaphragm and box depend upon the nature of the liquid whose depth is being measured, and they are chosen to resist the corrosive action of the tank contents.

Copper capillary tube is the standard connecting tube and is suitable for hot and cold water installations. Where the conditions are less favourable, the copper tube may be sheathed in lead, or armoured. Sheathing is necessary in sewers, and armouring where there is danger of the tube being damaged or fractured. Although the instrument depends upon pressure measurement, it is not influenced to a great extent by changes in the atmospheric pressure, for the measuring portion of the instrument measures the difference between the pressure within the closed system and the atmospheric pressure. Increase in atmospheric pressure will influence both these pressures so that the difference between them, which is a measure of the depth, will be the same.

Change in temperature will not influence the reading of this type of instrument any more than the reading of any other instrument which depends upon pressure measurement to indicate depth. In common with other instruments, its reading of depth will be in error due to the change in the density of the tank contents with change in temperature. If the diaphragm is truly flexible, then increase of temperature of the gas within the closed system will not influence the reading, for as soon as the pressure within the system increases due to the increased temperature, the diaphragm will allow it to expand until its pressure again balances that due to the liquid.

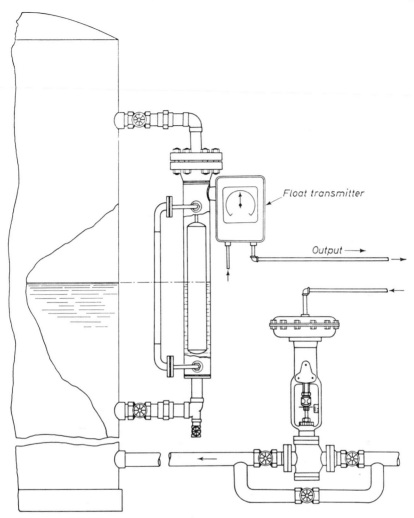

Figure 2.32   Level buoy (installation)

Stiff diaphragms

Another method of measuring the pressure due to the contents of a tank, is to use a diaphragm of corrugated metal. The movement of the centre of the diaphragm due to changing pressure is measured as in the Schaffer pressure gauge already described in the pressure measurement section. The d/p cell is also extremely useful for measuring level and in particular differences of level.

2.2.2.7   FORCE BALANCE SYSTEM

The pressure on a flexible diaphragm due to a liquid may be measured by balancing it with an air pressure on the other side. This forms a force balance system of the form used in the liquid level transmitter shown in *Figure 2.31*. It consists of a stainless steel diaphragm and a sensing unit. If the pressure on the diaphragm increases, the diaphragm will move towards the sensing unit and cause the baffle to move towards the bleed nozzle. This restricts the escape of air to the atmosphere, so that the air pressure behind the diaphragm builds up until it again balances the pressure due to the liquid. When the pressure falls, the diaphragm moves away from the sensing unit and an increased amount of air is allowed to escape to the atmosphere. The pressure behind the diaphragm therefore falls, until it again balances the pressure due to the liquid. The air pressure behind the diaphragm is transmitted via the receiver connection to an indicator or recorder which shows the level of the tank contents.

2.2.3 Buoyancy type

The fact that the upthrust on a body is equal to the weight of liquid it displaces is used in this type of level indicator. A long cylindrical body, known as a displacer, loaded to remain submerged in the densest liquid it encounters, is

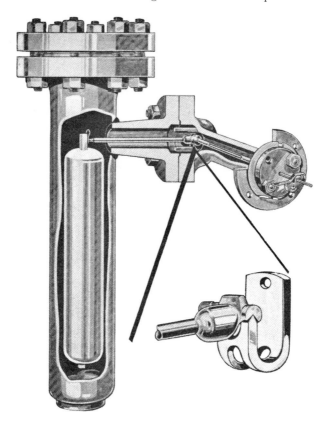

Figure 2.33   Assembly view of Fisher torque tube unit. (By courtesy, G.E.C. Elliot Control Valves Ltd.)

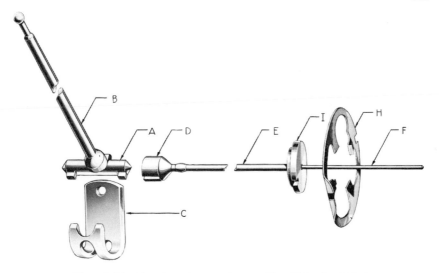

Figure 2.34   Top view of torque tube assembly—Fisher Level-Trol.
(By courtesy, G.E.C.-Elliot Control Valves Ltd.)

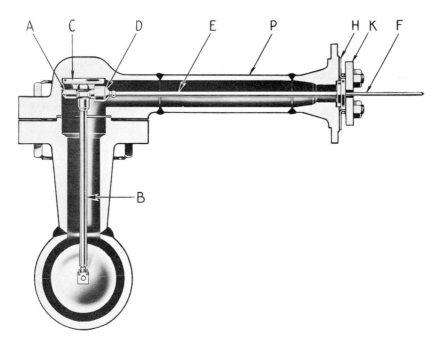

Figure 2.35   Fisher Level-Trol cutaway to show 'heart' of the device—the torque tube assembly.
(By courtesy, G.E.C.-Elliot Control Valves Ltd.)

suspended in the liquid in the tank or in a chamber outside as shown in *Figure 2.32*. The upthrust upon it will be proportional to the height on the float to which the liquid rises. The apparent weight of the float will decrease therefore as the level of liquid rises, and increase as the liquid level falls. When there is no liquid around the float W (*see Figure 2.33*), the entire weight of the float is carried by the free end of the float rod B. This causes a turning movement, equal to the product of the weight of the float and the length of the float rod, to be applied to the end of the torque tube. This turning movement or torque, is balanced by the torsional stress set up in the torque tube. Thus, the angle through which the shaft F is turned will be proportional to the apparent weight of the float (*Figure 2.34*). As the level of the liquid rises around the float, the apparent weight of the float will get less, and so will the turning movement acting on the torque tube. The torsional stress in the torque tube will therefore rotate the shaft F to a new angle proportional to the apparent weight of the float. The pointer or pen arm driven by the shaft F will therefore indicate or record the level of liquid in the tank, or may be used to operate a controller. The method of assembling the torque tube unit is shown in *Figure 2.34*, while *Figure 2.35* shows the complete unit.

The float is made of a corrosion-resisting alloy such as stainless steel, and is connected to the float rod by ball-and-socket joints. The length of the float determines the range of the instrument. The upthrust on the float will depend upon the specific gravity of the liquid, so that the instrument will indicate correctly only when the float is placed in the liquid for which it was designed. For accurate results, the change in liquid density for change in temperature must be taken into account, but this will to a large degree be compensated by the increase in dimensions of the float at the higher temperature. The great advantage of this type of installation is that the torque tube assembly is a very satisfactory method of transmitting the changes in the weight of the float to the outside of the float chamber when the pressure within the tank is high. It may be used on tanks containing liquids at pressures as high as 350 bar, and temperatures up to 400° C. Another advantage of the torque tube type of transmission is that the vertical motion of the float is small in comparison with the change in level.

3

MEASUREMENT OF FLOW

Instruments used in the measurement of flow can be divided into two main classes:

1. Quantity meters.
2. Rate-of-flow meters.

In the first class the total quantity which flows in a given time is measured; to obtain the average rate of flow this quantity must be divided by the time taken to flow. In meters of the second class, the actual rate of flow is measured; if the total quantity which flows in a given time is required, the quantities flowing in each small interval of time must be summed or integrated. If the rate of flow is reasonably steady, then the total quantity will equal the average rate of flow × time.

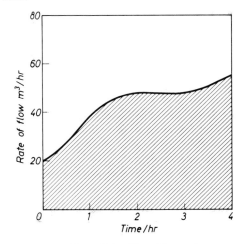

Figure 3.1   Flow-time graph

If the flow is not steady, however, the total quantity will be represented by the area under the graph obtained by plotting the rate of flow against the time as in *Figure 3.1*. This type of graph is obtained from recording flowmeters having square root compensation, and the total quantity will be represented in m³ by the shaded area as:

$$\text{Quantity} = \text{Rate of flow} \times \text{time} = \frac{\text{m}^3}{\text{hours}} \times \text{hours} = \text{m}^3$$

Finding the area under the curve is called 'integrating' the rate of flow, and is done mechanically in many instruments. The portion of the instrument which does this is called an 'integrator', and will be described later.

The instruments in these two sections are then further subdivided according to the nature of their primary elements. Each instrument may be considered as two parts, the primary element and the secondary element. The primary element may be defined as the portion of the instrument which converts the quantity which is being measured into a variable which operates the secondary element. The secondary element then measures this new variable which is arranged to depend upon the quantity being measured.

For example, in a weighing type of meter the weighing tank is the primary element. It converts quantity of water into a number of oscillations or rotations, whose number depends upon the quantity of liquid which has flowed. The secondary element is then a counter which adds up these oscillations or rotations and indicates them on a dial or chart. By arranging the correct relationship between the movement of the counter or pointer and the size of the divisions on scale or chart, an instrument is obtained which indicates or records the weight of liquid which has flowed.

In a differential type of flow meter, the primary element, which may be an orifice or a venturi tube, converts rate of flow into a differential pressure the value of which depends upon the rate of flow. The secondary element is then an instrument which measures this differential pressure and converts it into an indication of the rate of flow, on a scale or chart.

3.1   QUANTITY METERS

This type of meter gives an indication which is proportional to the total quantity which has flowed in a given time. The fluid passes through the primary element in successive and more or less completely isolated quantities by alternately filling and emptying containers of known fixed capacity. The number of times the container is filled and emptied is then indicated by the secondary element which consists of a counter with a suitably calibrated dial. The operation of the instrument is really the same as that of measuring the quantity of liquid transferred from one tank to another by counting the number of 'measuresful' transferred.

Quantity meters may be divided into two classes according to whether they indicate the weight or the volume which has flowed. They can then be further subdivided according to whether they are used for liquid or gas, and according to the nature of the primary element.

The division is therefore as follows:

Weighing meters.
Volumetric meters for liquids.
  Simple tank type.
  Positive displacement meters.
Volumetric meters for gases.
  Bellows type.
  Liquid sealed drum.
  Rotating impeller type.

3.1.1 Weighing meters

This type of meter is operated by the weight of the liquid.

In one type, the container is so arranged that when the liquid it contains reaches a predetermined height the container overturns and empties. Freed from the weight of the liquid it held, the container returns to its original position. As the container overturns when the centre of gravity reaches a predetermined height, the calibration will depend upon the density of the metered fluid. It will therefore be affected by changes in temperature, since this will affect the density of both liquid and container. The effect of the changed densities will in general be small. The number of times the container overturns is recorded on a counter which indicates the total weight of liquid which has flowed. Two containers may be used in another form of the instrument, the second container being brought to the top as the first is overturned by the weight of fluid it contains.

In another form, the container is suspended from a counter-balanced scale beam which tips up when the weight of the liquid in the container reaches a predetermined value.

3.1.2 Volumetric meters for liquids

3.1.2.1 SIMPLE TANK TYPE

In its simplest form, the instrument consists of a single tank which is alternately filled and emptied. The tank is allowed to fill, and when the liquid has reached the level of the top of a syphon it flows out. The number of times the liquid has syphoned-out is indicated by a float-operated mechanism.

When two tanks are used, and the level in one tank has reached a predetermined value, a float-operated mechanism cuts off the flow into that tank and directs it into a second tank. At the same time, a valve opens at the bottom of the first tank, allowing it to empty while the second tank is filling. When the second tank has filled, its outlet valve is opened, while that of the first tank is closed. The flow is then diverted from the second tank into the first. The number of times the tanks have been filled is indicated by a float-operated mechanism.

3.1.2.2 POSITIVE DISPLACEMENT METERS

Positive displacement meters are widely used on applications where the highest degree of accuracy and good repeatability is required. Their accuracy is not affected by pulsations in the flow rate, and accurate measurement is possible at higher liquid viscosities than with many other types of flow meters. Positive displacement meters are frequently used in oil and water undertakings for accounting purposes.

The principle of the measurement is that as the liquid flows through the meter it moves a measuring element which seals off the measuring chamber into a series of measuring compartments each holding a definite volume. As the measuring element moves, these compartments are successively filled and emptied. Thus for each complete cycle of the measuring element, a fixed quantity of liquid is permitted to pass from the inlet to the outlet of the meter. The seal between the measuring element and the measuring chamber is provided by a film of the measured liquid. The number of cycles of the measuring element is indicated by means of a pointer moving over a dial, a digital totaliser or some other form of register, driven from the measuring element through an adjustable gearing. This gearing ratio is adjusted during calibration so that the difference between the indicated and the actual quantity flowing is a minimum over the whole of the meters' rated capacity.

The measuring element may be regarded as an engine deriving its power from the flowing liquid stream and driving a load made up of the internal friction of the metering element and the operating torque of the register. Because there is a resistance to the movement of the measuring element a pressure difference exists between the upstream and downstream side of the measuring element and some liquid leaks through without being measured. The extent of the error, defined as the difference between the indicated quantity and the true quantity expressed as a percentage of the true quantity, due to this cause will depend upon many factors among them being:

(a) the size of the clearances between the measuring element and the measuring chamber. The closer the manufacturing tolerances, the smaller will be the errors owing to this cause. In a given meter, however, the clearances will be affected by wear. Changes of temperature in addition to changing the size of the measuring components can also affect the clearances for the measuring element and chamber may have different coefficients of thermal expansion. The total temperature effect is usually very small for suitably designed meters. When the meter is used at high static pressures it is necessary to ensure that the measuring chamber is sturdy enough to withstand the pressure without distortion. Another method of overcoming the problem of measuring chamber distortion is to allow the liquid from the upstream side to enter the space between the measuring chamber and the meter housing equalising the pressure on both sides of the chamber.

(b) the magnitude of the torque required to drive the meter register. The greater the value of the torque, the greater will be the pressure drop across the measuring element and hence the greater the quantity of liquid which will leak past unregistered. For this reason it is necessary to keep this torque to an absolute minimum. In recent accurate meters electronic or electromagnetic methods of detecting movement of the measuring element are used as these impose little or no load on the measuring element. Devices such as mechanical methods of temperature compensation, automatic batching systems etc., operated from the measuring systems, increase the load and should not be fitted when a high degree of accuracy is required. When they are fitted, the meter should be calibrated after the device has been fitted as the changed load will change the calibration. An increase in mechanical friction owing to corrosion or fouling of the bearings and in the counter will also cause an increased pressure drop and thus increase the leak rate and cause the meter to read low, particularly at low flow rates. Measuring the pressure drop across the meter will therefore give a very useful indication of when cleaning and recalibration has become necessary.

(c) the viscosity of the measured liquid. Increase in the viscosity of the measured liquid will increase the pressure drop across the measuring element but this is more than compensated for by the reduction in flow through the clearances for a given pressure drop. Thus, the error decreases with increase of viscosity at both low and high flow rates. The rangeability, or range of flows for which the meter error is less than a specified value, will therefore increase with increase of viscosity. Typical performance data for a film sealed positive displacement meter is shown in *Figure 3.2*. The meter factor defined as the ratio of the actual quantity passed to the volume indicated by the meter is plotted against the percentage of the rated flow.

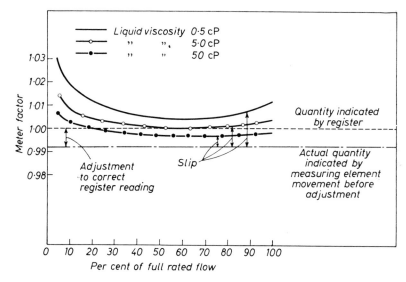

Figure 3.2   Typical performance data for film sealed positive displacement meter

It is important to note that the principal errors in positive displacement meters are due to the differences in the temperature, density and viscosity of the liquid at the working conditions from the values applying at the calibrating conditions. For this reason it is common practice where meters are used for measuring expensive liquids for accounting purposes to calibrate the meter *in situ* under working conditions by means of a meter prover. The prover may be a simple proving tank which has been previously calibrated by means of a standard test measure. The accuracy of a standard test measure is usually of the order of ±0·04 per cent, and the volume of the proving tank can be established with an error less than ±0·1 per cent.

Another form of meter prover consists of a calibrated length of pipe of uniform cross-section fitted with detectors a known distance apart. A piston in the form of a sphere fits closely in the pipe and is driven by the flowing liquid from one detector to the other displacing a known volume of liquid. The distance between the detectors is arranged to be sufficiently large to give the required accuracy, and an accuracy of ±0·02 per cent is claimed by one manufacturer.

The accuracy of a meter cannot be greater than the accuracy of calibration, but when accurate proving equipment is provided, the meter can be calibrated with the actual liquid and at the flow rate at which the measurement is to be made and the necessary adjustments made as often as required to give the necessary accuracy of measurement. Experience in the oil industry has shown that measurement by means of a well designed, calibrated and maintained meter can be more accurate than measuring quantity by level measurement in calibrated field tanks.

The accuracy of measurement attained with a positive displacement meter varies very considerably from one design of meter to another, with the nature and condition of the liquid measured, and with the rate of flow. Great care should be taken to choose the correct meter for the application. The most common forms of positive displacement meters are described under the following headings:

    (*i*)   Reciprocating piston type
    (*ii*) Rotating or oscillating piston type
  (*iii*) Nutating disc type
  (*iv*) Fluted spiral rotor type
    (*v*) Sliding vane type
  (*vi*) Rotating vane type
 (*vii*) Oval gear type

(*i*)   *Reciprocating piston type*

For use with cold water, this type of meter consists of a cast-iron cylinder lined with brass and fitted with a piston. In order to make a watertight contact with the cylinder wall, the piston consists of a sleeve rather than a disc and is machined to take piston rings which in cold water meters may be rubber, while when used in hot water the rings will be made of a metal suited to the use. For measuring corrosive liquids the materials from which the meter is made are chosen so that they resist the corrosive action of the liquid. The principle of the meter is shown in *Figure 3.3*.

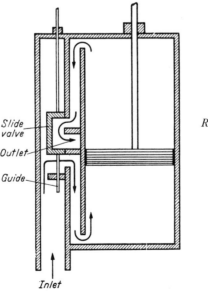

Slide valve

Outlet

Guide

Inlet

Figure 3.3
Reciprocating piston
type of meter

Suppose the piston is at the bottom of its stroke. The valve is so arranged that inlet water is admitted below the piston, causing it to travel upwards and the water above the piston to be discharged to the outlet pipe. When the piston has reached the limit of its travel, the top of the cylinder is cut off from the outlet side, and opened to the inlet water supply. At the same time the bottom of the cylinder is opened to the outlet side, but cut off from the inlet water. The pressure of the incoming water will therefore drive the piston downwards, discharging the water from below the piston to the outlet pipe. When the piston reaches the bottom of its travel the valve again operates and the process is repeated. The meter acts, therefore, rather like the conventional steam engine, the water pressure providing the motive power. At each stroke, a measured quantity of water is delivered at the outlet side.

As the piston reciprocates, a ratchet attached to the piston rod rotates a pinion which turns the counter. Adjustment of the quantity delivered per stroke can be effected by changing the length of the stroke or the dimensions of the cylinder.

In other forms of the meter more than one cylinder is used. In one form, four cylinders are used, arranged in a cross so that the axis of any one cylinder is at right angles to that of its neighbour. The cylinders are connected in turn by a rotating valve to inlet and outlet ports. The reciprocating motion of the pistons is, as described before, used to operate a counter.

These types of meters are used mainly for metering water, but they can be used for other liquids by using suitable corrosion-resisting materials in their construction. In general, the loss of head produced by this type of meter is relatively small, and by suitable choice of cylinder size meters can be made for a large range of flow rates.

(ii)   Rotating or Oscillating piston type

This type of meter, shown exploded in *Figure 3.4*, is widely used for metering household water supplies up to 50° C, but with the wide range of corrosion-resistant materials at present available it is also finding increasing use in chemical and oil industries. For use on cold water it consists of a meter body made of brass, and provided with union tail pieces of brass in the smaller sizes, or of cast-iron in the larger sizes. Fitted into the meter body is a two-piece graphited thermo-plastic working chamber [(10) and (7)] containing the ebonite piston (8) which is shown in *Figure 3.5*. This piston acts as a moving chamber, transferring a definite volume of liquid, from the inlet to the outlet port, for each cycle of its motion.

The piston has a peg through its centre and a pear-shaped groove cut out of one side. In the working chamber there is a shutter, and the piston rolls on the inside of the working chamber in such a way that the portion of the cut-out at the circumference of the piston is always in contact with the shutter, while the central peg on the underside describes a circle in a circular groove in the bottom of the working chamber. As the piston circulates, the pegs on the upper side engages in a cross-piece which is attached to a dog drive coupling which drives the combined reduction gear and counter. This consists of a leak-free transparent casing in which the number of rollers are immersed in a non-toxic fluid which acts as a lubricant and corrosion inhibitor. A sac

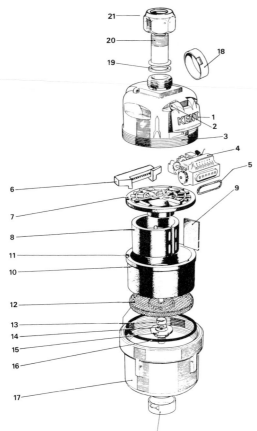

1. Lid.
2. Hinge pin.
3. Counter housing complete with lid and hinge pin.
4. Counter with worm reduction gear and washer.
5. Counter washer.
6. Ramp assembly.
7. Top plate assembly comprising top plate only; driving spindle; driving dog; dog retaining clip.
8. Piston.
9. Shutter.
10. Working chamber only.
11. Locating pin.
12. Strainer—plastic.
    Strainer—copper.
13. Strainer cap.
14. Circlip.
15. Non-return valve.
16. O ring.
17. Chamber housing.
18. Protective caps for end threads.

Figure 3.4   Domestic water meter, rotating piston type. (By courtesy, Kent Meters Ltd.)

attached to the counter casing acts as a balancing membrane and ensures that the pressure in the casing is equal to that of the water outside, ensuring there is no pressure difference across the O-ring seal at the end of the casing thus eliminating leakage.

Figure 3.6 shows the piston in four positions during its cycle of continuous operation.

Position 1. The piston is over the inlet port; the inlet water has entered the inside wall of the piston and is causing it to start its semi-rotary movement, sliding down the division piece, displacing the neutral water which becomes the exhaust water as it is expelled through the outlet port.

Position 2. The piston has rotated so that the central peg has moved through a quarter of a complete rotation. Incoming water is now both inside and outside the piston.

Position 3. The neutral water within the piston is cut off from both ports. Water from the inlet port is on the outside of the piston causing it to rotate further driving the outlet water through the outlet port. The peg has completed half a complete rotation.

Position 4. The water from the outside of the piston on the outlet side has

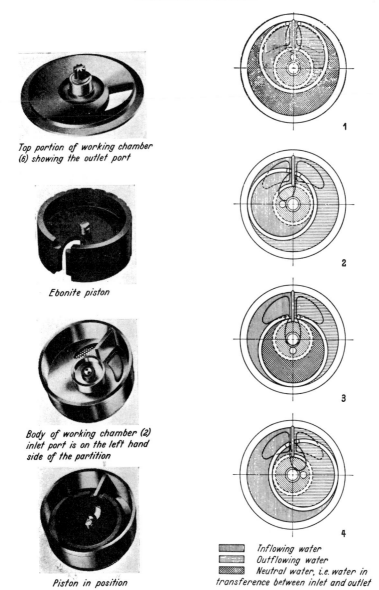

Top portion of working chamber
(6) showing the outlet port

Ebonite piston

Body of working chamber (2)
inlet port is on the left hand
side of the partition

Piston in position

Inflowing water
Outflowing water
Neutral water, i.e. water in
transference between inlet and outlet

Figure 3.5
Water meter; components.
(By courtesy, Kent Meters Ltd.)

Figure 3.6
Water meter; piston positions.
(By courtesy, Kent Meters Ltd.)

been almost all driven out; water is being driven out from within the piston, while the inside of the piston is now being connected to the inlet water. The peg has completed three-quarters of a complete rotation, and continues to rotate until the piston again reaches position 1.

For use on other applications a wide range of materials are available. The meter body may be made from gun-metal, acid-resisting bronze, monel metal, aluminium alloy or stainless steel, with measuring chambers of similar materials or materials such as moulded graphited polystyrene.

The piston may also be made from a range of materials such as aluminium, graphite and ebonite, while gears and bearings may be made of graphited nylon or graphited polystyrene.

Meters are available in a wide range of sizes which cover a flow range of 3·4 litre per hour in the 12 mm size to 13·6 litre per hour in the 30 mm size, but larger meters are available.

In general the simple type of meter has an accuracy of ±2 per cent over the whole recommended range or ±0·5 per cent when used at a constant flow at which it has been calibrated, i.e. at a fixed flow it has a repeatability of ±0·5 per cent.

When a greater accuracy is required this may be achieved by detecting the movement of the piston electromagnetically instead of requiring it to drive a gear train. This is achieved by having a piece of soft iron moulded into the piston and two coils with soft iron cores mounted adjacent to each other on the outside of the piston wall. One of the coils is energised with an a.c. supply and when the piston takes up a given position, the magnetic field from this coil is coupled to the secondary coil by the piece of soft iron. Thus for each rotation of the piston the secondary coil will receive an electrical impulse which after amplification is fed to a counter. Meters are available having a capacity per revolution which may be any value between 0·041 and 4·51, and rates of flow between 1 litre and 60 m³/h can be measured by meters of this type with an accuracy better than ±1 per cent and a repeatability at a constant flow of ±0·25 per cent.

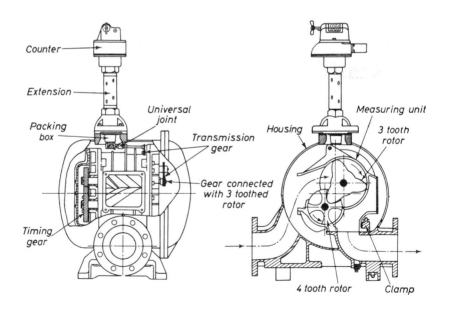

Figure 3.7   Fluted spiral rotor type meter. (By courtesy, Gilbarco Ltd.)

In another form of meter the piece of soft iron is replaced by a magnet which causes the contact on a reed-type switch in a counting circuit to close each time it passes.

(iii)   Nutating disc type

This type of meter is similar in principle to the rotating piston type. In this case, however, the gear train is driven not by a rotating piston but by a flat disc which is given a rocking or nutating motion by the water flowing through the meter. The disc is usually made of moulded hard rubber having a slot cut radially, into which the division plate of the measuring chamber fits. The disc is thus restrained so that it moves in such a way that one side slides up and down the division plate while the upper end of the disc spindle moves in a circle driving the counter mechanism. One edge of the disc is in contact with the upper edge of the measuring chamber while the opposite edge is in contact with the lower edge of the chamber. As the liquid flows the points of contact of disc and chamber move around the disc.

Meters are available having a rated capacity of 27 to 135 m³/h having a metering accuracy of the order of ± 2 per cent over the whole of the rated capacity of the meter.

(iv)   Fluted spiral rotor type

The principle of this type of meter is illustrated in *Figure 3.7*. The meter consists of two fluted rotors supported in sleeve type bearings, and mounted so as to rotate, rather like gears, in a liquid-tight case. The inner surface of the case and the impeller tips are carefully machined, and the impellers accurately centred, to reduce the clearance to a minimum. The rotors are designed so that they are always in complete static and dynamic balance, and their relative position is controlled by two helical timing gears. In this way, the rotation of the impellers is so synchronised, that there is little or no metal-to-metal contact, thus reducing friction and avoiding wear of the rotors. The shape of the rotors is such that a uniform uninterrupted rotation is produced by the liquid. The impellers in turn rotate the index or counter which shows the total quantity which has flowed.

An adjustment is provided which alters the relationship between the measuring element movement and the quantity registered in steps of 0·05 per cent. This type of meter is used largely for measuring crude and refined petroleum products and many chemicals. All meters having a capacity of 13·5 m³/h and larger are built with double case construction so that only the outer is subject to line stress and unbalanced liquid pressure. This type of construction also permits the measuring elements to be replaced without removing the meter body from the line. Meters are available for line sizes from 25 mm to 0·6 m to cover a range of flows from 1·6 to 3200 m³/h and pressure up to 80 bar. They are designed to operate over a general range from maximum capacity down to 20 per cent at an accuracy of the order of $\pm 0\cdot1$ per cent. In meters for use on liquid petroleum gas and other non-lubricating liquids, an automatic pressure lubricating system is provided for all bearings and gears.

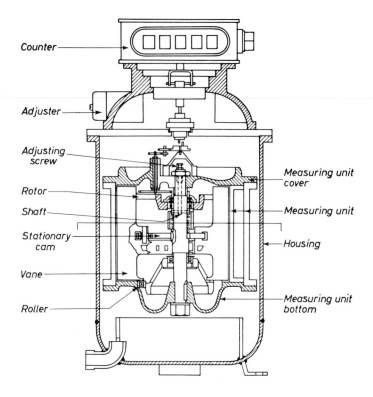

Counter

Adjuster

Adjusting screw

Rotor

Shaft

Stationary cam

Vane

Roller

Measuring unit cover

Measuring unit

Housing

Measuring unit bottom

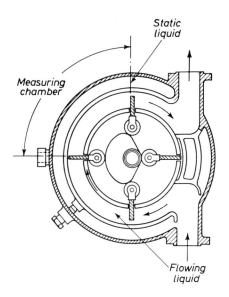

Static liquid

Measuring chamber

Flowing liquid

Figure 3.8   Sliding vane type meter. (By courtesy, Wayne Tank & Pump Co.)

Figure 3.9
Measuring chamber of oval gear meter
(By courtesy, Bopp & Reuther GmbH)

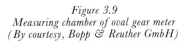

(v) Sliding vane type

The principle of this type of meter is illustrated in *Figure 3.8*.

The accurately machined body contains a rotor, revolving on ball-bearings, which has four evenly spaced slots forming guides for four vanes. Liquid, flowing through the meter, revolves rotor and vanes around a fixed cam. The four cam rollers follow the contour of the cam causing the vanes to move radially. The vane nearest the inlet port begins to move outwards towards the body wall, being fully extended by the time it reaches the beginning of the measuring chamber. The vane ahead is already fully extended; thus a measuring chamber of known volume is formed between the two vanes, the body wall and the top and bottom covers. A continuous series of these chambers is formed, four per revolution. The action is smooth and continuous. Neither vanes nor rotor touch any stationary parts, leakage past the vanes being prevented by the liquid film.

In general, this type of meter produces a comparatively low pressure loss, and is suitable for use on pressures up to 8·5 bar g. Models are available having a rated capacity of up to 80 m³/h, and are designed to give an accuracy of 0·1 per cent down to 20 per cent of the rated capacity.

In another form of sliding vane meter no cam is used but the opposite pairs of vanes are mounted on rigid tubular rods. The meter body is so designed that the vanes are guided to form the measuring chamber as the rotor turns. Models are available for pressures up to 10 bar g and rated capacities up to 180 m³/h.

(vi) Rotating vane type

This meter is similar in principle to the sliding vane type meter, but the measuring chambers are formed by four half-moon shaped vanes spaced equidistant on the rotor circumference. As the rotor is revolved the vanes turn to form sealed chambers between the rotor and the meter body. Meters are available in a wide range of materials, for working pressures up to 85 bar g, and models are available having rated capacities from 3·8 to 270 m³/h. As with other positive displacement meters an accuracy of ±0·1 per cent is possible down to 20 per cent of the rated capacity of the meter.

Figure 3.10
Pulse generator for oval gear meter
(By courtesy, Bopp & Reuther GmbH)

(vii)   Oval gear type

The measuring portion of this meter, illustrated in *Figure 3.9*, consists of two intermeshing oval gear wheels, which are rotated by the fluid passing through the meter. The rotation of the oval gears is transmitted to the counter mechanism by means of a spindle passing through a packed gland. When the meter is required for use on corrosive or toxic liquids the rotation of the gears is transmitted through the hollow spindle of one oval gear to a follower, by means of magnets embedded in the gear wheel and follower. Thus, the only portion of the meter in contact with the metered liquid are the oval gears, the measuring chamber and the meter body, and these can be manufactured in stainless steel or other corrosive-resistant material.

The follower, in turn, rotates the counter or pointer by means of an adjustable gear train. In addition, the follower may rotate an electronic pulse generator, which may be used to operate a remote counter, a blending controller and a rate of flow indicator.

When used to operate a remote counter, two pulse generators may be used as shown in *Figures 3.10* and *3.11* to ensure that the failure of the transmission system is immediately indicated. A pulse generator consists of a transistorised oscillator having a control head. An aluminium toothed wheel driven by the metering mechanism passes between the two halves of the control head, causing the oscillator to oscillate, or not, depending on whether a tooth is present or not. This causes the current taken from the amplifier unit by the oscillator to change from 2 mA to 0·2 mA. The amplifiers and the impulse-former increase the energy content of the pulses to a sufficiently high level to operate a synchronous motor. If the motors driven by the two pulse generators get out of phase, a relay initiates an alarm, allows a shutter to cover the meter register, and can shut down the system completely.

Photoelectric pulse generators, consisting of a disc driven from the meter mechanism and having a large number of radial lines, reproduced photo-graphically, rotating between a light source and a photocell, and capable of producing a frequency of 2000 Hz at the meter maximum are also available. Where necessary the mechanism can be enclosed in an explosion-proof housing to V.D.E. standards.

Oval gear meters are available in a wide range of materials, such as bronze, aluminium, stainless steel etc., in sizes from 10 to 400 mm, suitable for pres-sures up to 60 bar g and flows having rated maxima from 0·06 m³/h to 1200 m³/h at temperatures up to 250° C. Special meters have been designed for temperatures down to $-190°$ C, and for pressures up to 350 bar g.

In the small meters between 10 mm and 65 mm nominal size the guaran-teed accuracy down to 20 per cent of the rated flow is 0·25 per cent. In the larger sizes, 80 mm nominal size and above, the guaranteed accuracy is 0·1 per cent.

Installation and maintenance of positive displacement meters

When in store before installation the meter must be handled with care and the openings covered to ensure that no foreign material can enter the meter.

Before the meter is installed the lines upstream of the meter must be very thoroughly flushed to ensure that they are free from solids such as pipe scale,

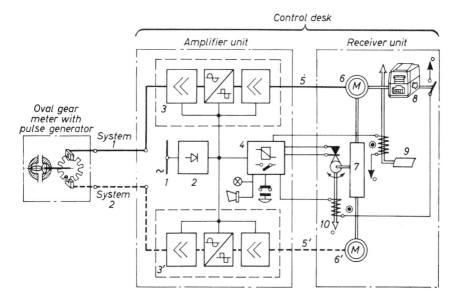

Figure 3.11   Oval gear meter with remote control. (By courtesy, Bopp & Reuther GmbH)

1, main supply; 2, d.c. supply; 3, amplifier, pulse modifier and main amplifier for system 1; 4, relay circuit with sound and light signal; 5, receiver motor cable; 6, motor for system 1; 7, differential gears with switch contacts for the control of the relay, symmetrical setting for synchronous operation; 8, receiver (meter register with ticket printer); 9, register screen with crank; 3′, amplifier, pulse modifier and main amplifier for system 2; 5′, cable between 3′ and 6′ of the control system; 6′, motor for system 2 (control system)

etc. A filter of mesh size sufficiently small to ensure that all particles large enough to stop the meter are prevented from entering, should be installed on the inlet side of the meter as close to the meter as possible. The filter should be cleaned on a routine basis, the frequency of cleaning being determined by the amount of solid matter accumulated in a given time.

The meter will measure not only liquid but any entrained gas. If, therefore, there is any possibility of gas or vapour entering the meter, an air eliminator should be installed immediately upstream of the meter.

An air eliminator consists of a vertical tank of the required pressure rating fitted with baffles. The liquid enters at the bottom of the tank, and in flowing upwards its velocity is considerably reduced, and the gas or vapour bubbles rise to the surface. As gas leaves the liquid, it collects in the space above, forcing down the level of the liquid surface. A float which operates a discharge valve follows the liquid surface, and when the level falls sufficiently, the discharge valve opens, allowing the gas or vapour to escape. When the liquid again rises to a predetermined level, the valve closes again preventing escape of liquid. From this brief description of the air eliminator, it will be seen that it must never be installed in a line containing liquid at a pressure below atmospheric if it is to function properly.

When installed the meter should be adequately supported and not subjected to stresses by the pipework. The counter should be conveniently located for easy reading. When used on volatile liquids the meter should be fitted on the

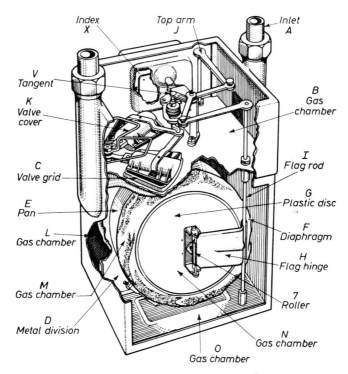

Figure 3.12 Dry gas meter; portion of outer case removed. (By courtesy, George Wilson (Gas Meters) Ltd.)

discharge side of the pump in order to eliminate the possibility of liquid vaporising as it passes through the meter. The pipework should be adequately sized and the number of tees, elbows, etc. should be reduced to a minimum especially on gravity fed meters. The liquid must flow through the meter in the direction indicated on the meter body and any other maker's instructions about correct positioning must be observed. The liquid pressure must never exceed the maximum indicated on the meter, and the rate of flow must be limited to the rated capacity of the meter if excessive wear is to be avoided. Soft closing valves should be used to avoid subjecting the meter to excessive shock pressures, and the meter should not be subject to excessive temperature changes.

When meters are provided with electronic methods of pulse generation for signal transmission, adequate shielding of conductors connecting the pulse generator to the counter must be provided. The conductors must be kept well away from all sources of magnetic induction, kept as short as possible, and the shield earthed at the counter end only. In addition the conductors should be run in a separate conduit, never in a conduit with other conductors carrying large currents.

Meters are calibrated and adjusted for accuracy by the manufacturer, but if possible they should be calibrated on the metered liquid after installation, and adjusted if necessary. To obtain the maximum accuracy the meters should be calibrated at regular intervals, say every six months, and readjusted as required.

3.1.3 Volumetric meters for gas

The volume of a gas is usually measured in terms of the standard cubic meter which is the volume measured at 1013·25 m bar, 15° C and dry. Correction for departure from these conditions are explained in section 3.3.

3.1.3.1 BELLOWS TYPE (DRY GAS METER)

This type of meter has remained fundamentally the same for over 100 years, and is probably the most common kind of meter in existence. It is found in every home where gas is used, and may merely record the quantity of gas consumed in a given period, or be of the pre-payment type, which will deliver gas only after it has been paid for by the insertion of a coin.

The performance of these meters is governed by the Regulations of the Department of Trade and Industry. Extracts from the General Directions, Gas Meter Examiners', 1968, converted to S.I. units are given below:

1. No meter examiner shall stamp any meter unless he is satisfied that it conforms to the standards prescribed by this regulation.
2. No air or gas will escape from the meter.
3. No meter, when measuring either air or gas passing through it at between 5 per cent and 100 per cent of its badged flow rate, will register more than 2 per cent over or 3 per cent under the actual quantity of such air or gas.
4. The mean difference between the pressures at the inlet and outlet of any

meter through which air or gas is passing at any rate of flow for which it is designed will not exceed:

(a)   in the case of any such meter having a capacity not exceeding 141·5 m³/h, 12·7 mm water column;

(b)   in the case of any such meter having a capacity exceeding 141·5 m³/h but not exceeding 566 m³/h, 20·3 mm water column;

(c)   in the case of any such meter having a capacity exceeding 566 m³/h, 25·4 mm water column.

5. No meter will cause such an oscillation of pressure, when measuring either air or gas passing through it at any rate of flow for which it is designed, that the difference between the highest and the lowest pressure at the outlet thereof is greater:

(a)   in the case of a meter to which paragraph (a) of the last foregoing paragraph applies, than 7·6 mm water column;

(b)   in the case of any other meter, than 10·2 mm water column.

The quantity of gas delivered is measured by the number of times the measuring chambers are filled and emptied. This is done by counting the number of horizontal movements of two diaphragms. Although the gas is delivered in separate measured quantities, by having four measuring chambers each filling and exhausting at staggered intervals, the flow through the meter is continuous and free from fluctuations. *Figure 3.12* shows a bellows type of meter, a portion of the outer case having been removed to show the inside.

The metal case comprises an upper and lower section, the upper being divided into two compartments one of which is free from gas and contains the integrating mechanism X, whilst the other has a constant supply of gas from the pipe A to the gas chamber B. A simple valve arrangement permits this supply to pass at the predetermined intervals into the four measuring chambers below. The lower part of the case is divided vertically by a metal division D, to each side of which is fixed pan E in such a manner that when the diaphragm F is circumferentially attached thereto the lower case becomes four distinct chambers $LMNO$.

The diaphragms are made of a leather prepared from sheep-skin imported from the East Indies, which has been found to have the necessary flexibility, close-grained texture, and wearing qualities required for meter leathers. This sheepskin is tanned with bark, then with standard chrome liquid, the excess acid neutralised and the skin treated with a suitable preservative in order that it may resist the scouring action of the gas. The preservative also seals the pores in the skin and prevents it from stretching or shrinking while in use. Great care is exercised in fixing these diaphragms as 'puckers' and creases must be removed, otherwise wear will occur through the constant rubbing of the creases. The actual size is also very critical, as upon this the accuracy of the meter will depend. Where meters are used on natural gas synthetic diaphragms are used.

The diaphragm is composed of flexible leather F attached centrally to a plastic disc or discs G. The leather permits a horizontal movement to the disc and the reciprocation of the latter serves to partly rotate the flag rods I by means of the hinge H. The flag rods pass through stuffing boxes to the upper gas chamber and have fixed to their top extremity links J which serve to revolve a vertically positioned crank that actuates the covers K in their movement over the valve grids, C. The crank carries a means of adjusting the

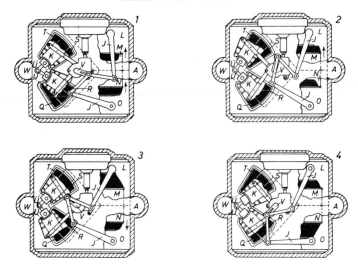

Figure 3.13   Action of dry gas meter. (By courtesy, George Wilson (Gas Meters) Ltd.)

distance swept by the discs (Tangent V) and is also conveniently geared to the index. *Figure 3.13* illustrates the action of the meter.

1. Gas is entering the chamber O via port Q, diaphragm moves as arrow. Gas in chamber N exhausts through centre port via gasway U to outlet pipe W. In the other grid S opens permitting gas to enter chamber M which is now exhausted. The movement of the disc follows the gas from L through to port T into the gasway U and the outlet.
2. Chamber M is half filled. Chamber L half exhausted. Discs travelling in direction of arrow. Chamber N is completely exhausted whilst port R is about to open gas supply and the movement of the cover will allow the gas in chamber O to flow back through port Q in the centre port and out.
3. Chamber M is full. L has just finished exhausting. Chamber N is half full exhausting through port Q.
4. Chamber M exhausting through port S. Chamber L is filling in through port T. Chamber N is full. Chamber O is empty.

Materials

Case and partitions are made of tinplate which must be free from pinholes and heavily coated on the outside with cellulose or paint to protect it from corrosion. Where the gas contains light oils, it tends to preserve the tinplate; but the diaphragms are likely to suffer. The most detrimental constituent of the gas is sulphur, which tends to oxidise and dissolve in the moisture present, producing sulphurous acid which will readily attack metals, and these must therefore be protected. The meter must also be protected externally from the ravages of the atmosphere, particularly as they are aggravated by the position in which the meter is placed; such as in a passage, under a sink, in a damp cellar or a closed cupboard. Great care must be taken, therefore, to see that the protective coating is complete and effective.

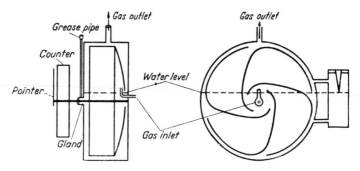

Figure 3.14 Liquid sealed drum type gas meter

Valves

The valve grids are usually made from a specially prepared alloy of tin and antimony, whilst the valve covers are now manufactured from graphite impregnated plastic. Both components are ground and faced with particular care and the walls and base of the slide valves are reduced to a minimum in order to reduce friction and surface deposits.

Installation

Effects of temperature. If the meter is to read accurately it must be installed in a position where it is not subject to large variations of temperature. As gases expand by 1/273 of their volume at 0° C for every Celsius degree rise in temperature, the mass of gas, and hence the calorific value of the gas, delivered for every cycle of operation of the meter, will be reduced if the temperature rises. Hence the meter tends to over-record in terms of mass and calorific value of gas, and to favour the suppliers when the temperature rises. When the temperature falls it tends to under-read and favour the consumer.

The meter should not be located in a damp position, for this tends to increase the danger of corrosion.

Calibration

The meter is calibrated by drawing through it a known volume of air, measured by collecting it in a container. The change in meter reading is noted, and the tangent adjusted until the meter reads correctly. This is repeated until the meter is within the limits of accuracy required by the Ministry of Power regulations. The pressure drop across the meter and the oscillation is also measured.

3.1.3.2 LIQUID SEALED DRUM

This type of meter differs from the bellows type of meter in that the sealing medium for the measuring chambers is not solid but is water or some other suitable liquid.

The instrument is shown in section in *Figure 3.14*. It consists of an outer chamber of tinned brass, tinplate, or Staybrite steel sheeting, containing a rotary portion. This rotating part consists of shaped partitions forming four measuring chambers, made of light-gauge tinplate, or Staybrite steel, balanced about a centre spindle so that it can rotate freely. Gas enters by the gas inlet near the centre, and leaves by the outlet pipe at the top of the outer casing. The measuring chambers are sealed off by water or other suitable liquid, which fills the outer chamber to just above the centre line. The level of the water is so arranged that when one chamber becomes unsealed to the outlet side, the partition between it and the next chamber seals it off from the inlet side. Thus, each measuring chamber will, during the course of a rotation, deliver a definite volume of gas from the inlet side to the outlet side of the instrument. The actual volume delivered will depend upon the size of the chamber and the level of the water in the instrument. The level of the water is therefore critical, and is maintained at the correct value by means of a hook type of level indicator in a side chamber which is connected to the main chamber of the instrument. If the level becomes very low, the measuring chambers will become unsealed, and gas can pass freely through the instrument without being measured; while if the level is too high, the volume delivered at each rotation will be too small, and water may pass back down the inlet pipe. The correct calibration is obtained by adjusting the water level.

When a partition reaches a position where a small sealed chamber is formed, connected to the inlet side, there is a greater pressure on the inlet side than on the outlet side. There will therefore be a force which moves the partition in an anti-clockwise direction, and so increases the volume of the chamber. This movement continues until the chamber is sealed off from the inlet pipe but opened up to the outlet side; while at the same time the next chamber has become open to the inlet gas but sealed off from the outlet side. This produces continuous rotation; the rotation operates a counter which indicates complete rotations and fractions of a rotation, and can be calibrated in actual volume units. The spindle between the rotor and the counter is usually made of brass, and passes through a grease-packed gland. The friction of this gland, together with the friction in the counter gearing, will determine the pressure drop across the meter, which is found to be almost independent of the speed of rotation. This friction must be kept as low as possible, for if there is a large pressure difference between inlet and outlet sides of the meter, the level of the water in the measuring chambers will be forced down, causing errors in the volume delivered; and at low rates of flow the meter will rotate in a jerky manner.

It is very difficult to produce partitions of such a shape that the meter delivers accurate amounts for fractions of a rotation; consequently the meter is only approximately correct when fractions of a rotation are involved.

The mass of gas delivered will depend upon the temperature and pressure of the gas passing through the meter. The volume of gas is measured at the inlet pressure of the meter, so if the temperature and the density of the gas at s.t.p. are known, it is not difficult to calculate the mass of gas measured. The gas will, of course, be saturated with water vapour, and this must be taken into account in finding the partial pressure of the gas.

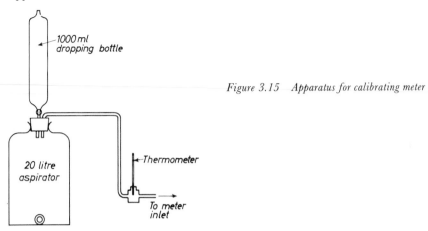

Figure 3.15 Apparatus for calibrating meter

Calibration

The meter may be calibrated by using a 20-litre aspirator and a 1000 ml dropping bottle as shown in *Figure 3.15*. The actual volume of gas delivered is known, and, therefore, the water level in the meter can easily be adjusted so that the calibration is correct. If the volume of gas measured is required at s.t.p., allowance must be made for the actual temperature and pressure of the incoming gas.

Application and installation

This type of meter is useful for measuring small flows of gases for analytical tests, and in the measurement of the calorific value of a fuel gas.

For special purposes, meter bodies may be made of other materials such as ebonite or Perspex. The material of the meter body and the sealing liquid used in the meter must be such that they are not affected in any way by the gas which is being metered, and the sealing liquid should not attack either the material of the meter body or the sealing grease of the packed gland.

The meters are designed for flows between 6 litres and 90 litres per hour, and should not be used for gases at high temperatures, or installed in a hot position, for this will cause a great deal of sealing liquid to be lost by evaporation. A catchpot should be used upstream of the instrument if the gas is very wet, otherwise the meter water level will be increased by condensation. In all uses there should be routine calibration and topping up.

Care and maintenance

The meters must be washed out from time to time to remove any accumulation of dirt and grease. The gland is then repacked with grease, and the meter refilled and calibrated. It should then be tested for leaks. If a small pressure is applied to the side while the outlet side is closed, a leak will make itself obvious by the fact that the pointer will tend to drift. After cleaning, the instrument must be mounted in a level position, and the water level adjusted to give a correct calibration.

Recent developments

In a form of the meter developed recently the measuring chambers are formed by partitions of helical form in order that the measuring spaces may have a constant cross-sectional area throughout their lengths: thus fractions of a revolution pass volumes of gas directly proportional to the amount of turning. The partitions are also designed to be always perpendicular to the surface of the water when they enter or leave it, so that the meter may be run at speeds up to 1000 revolutions per hour without undue loss of pressure or accuracy.

The sectional area of the case at the waterline is designed to be large in relation to the area of the outer chamber of the drum to ensure that differential pressure produces only a small change of water level in the measuring chambers, and has only a small influence on the accuracy of the meter over a wide range of speeds. These meters are available in half litre size, but they have a large hourly capacity since they may be run at speeds up to 1000 revolutions per hour.

3.1.3.3 ROTATING IMPELLER TYPE

This type of meter is similar in principle to the rotating impeller type meter for liquids, and could be described as a two-toothed gear pump. A typical example is shown in *Figure 3.16*. Although the meter is usually manufactured almost entirely from cast iron, other materials may be used if desired. The meter basically consists of two impellers housed in a casing and supported on rolling element bearings. A clearance of a few thousandths of an inch between the impellers and the casing prevents wear with the result that the calibration of

Figure 3.16   Rotating impeller meter for gases. (By courtesy, W. C. Holmes Ltd.)

the meter remains constant throughout its life. The leakage rate through the meter under operating conditions is very small because both gas and impeller tips are moving in the same direction and at similar velocities. The leakage rate is only a small fraction of 1 per cent and this is compensated for in the gearing counter ratio. Each lobe of the impellers has a scraper tip machined onto its periphery which prevents deposits forming in the measuring chamber. The impellers are timed relative to each other by gears fitted to one or both ends of the impeller shafts.

The impellers are caused to rotate by the decrease in pressure which is created at the meter outlet following the use of gas by the consumer. Each time an impeller passes through the vertical position, a pocket of gas is momentarily trapped between the impeller and the casing. Four pockets of gas are therefore trapped and expelled during each complete revolution of the index shaft. The rotation of the impellers is transmitted to the meter counter by suitable gearing so that the counter reads directly in either cubic feet or cubic metres. As the meter records the quantity of gas passing through it at the conditions which prevail at the inlet, it is necessary to correct the volume indicated by the meter index for various factors. These are normally pressure, temperature and compressibility. Correction can be carried out manually if the conditions within the meter are constant. Alternatively, the correction can be made continuously and automatically by small mechanical or electronic computers if conditions within the meter vary continuously and by relatively large amounts. Meters can also drive through external gearing, various types of pressure and/or temperature recording devices as required.

Meters of this type are usually available in pressures up to 60 bar, and will measure flow rates from approximately 12 m³/h up to 10 000 m³/h. Within these flow rates, the meters will have a guaranteed accuracy of $\pm 1 \cdot 0$ per cent over a range of from 5–100 per cent of maximum capacity. The pressure drop across the meter at maximum capacity is always less than 50 mm wg. These capacities and pressure loss information are for meters operating at low pressure: the values would be subject to the effects of gas density at high pressure.

3.2   RATE OF FLOW METERS

3.2.1   Direct measurement of velocity

If a pipe has a cross-section of A m², and the velocity of a liquid through it is uniform across the section and has a value V m/s, then $A \times V$ m³ of liquid will flow through the pipe per second. In practice, the velocity is not uniform across a section of the pipe, being greater at the centre and less where the liquid is in contact with the pipe walls. If, however, a velocity is measured which bears a constant relationship to the mean velocity across the section, then the volume flowing will be $K \times A \times V$ m³/s, where K is a constant for a particular pipe, and represents the relationship between the true average velocity and the measured velocity. The actual value of this constant will vary with the form of the section through which the liquid flows, but can be found experimentally for any particular type of section. The rate of flow type of meter is based on this principle.

In this section, therefore, meters will be described in which the quantity of liquid flowing through the meter is measured by measuring the velocity of the liquid through the meter. They are subdivided according to whether they measure liquid or gas, and according to the nature of the primary element. The subdivision is therefore as follows:

Rate of flow meters for liquids
    Deflecting vane type
    Rotating vane type
    Helical vane type
    Turbine type
    Combination meters
    Magnetic flow meters
    Ultrasonic flow meters
Rate of flow meters for gases
    Deflecting vane type
    Rotating vane type
    Thermal type

3.2.2 Rate of flow meters for liquids

3.2.2.1 DEFLECTING VANE TYPE

If a rectangular vane is freely pivoted about its upper edge, and allowed to hang in a stream of liquid, it will tend to obstruct the liquid stream and to deflect the liquid downwards, as shown in *Figure 3.17(a)* so that the liquid will flow beneath the vane. Since to every action there is an equal and opposite reaction, the liquid will tend to deflect the vane backwards and upwards.

The forces acting on the vane will be those shown in *Figure 3.17(b)*, and the vane will be in equilibrium under the action of these three forces. Since the weight of the vane is constant, the angle through which it is deflected will depend upon the size of the force due to the impinging liquid. The size of this force will depend upon the rate of change of momentum of the impinging liquid, and will therefore depend upon the velocity of the stream. For a liquid of given density the angle through which the vane is deflected is, therefore, a measure of the velocity of the stream in which it is placed. The relationship between the angle of deflection and the velocity of the stream will not be a linear one, but the correlation between the two may be found experimentally.

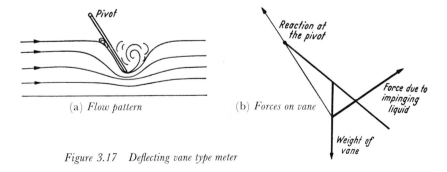

(a) *Flow pattern*   (b) *Forces on vane*

Figure 3.17   Deflecting vane type meter

This is the principle of the deflecting-vane type of meter, whether it be for liquids or gases. Owing to the obstruction the vane offers to the stream, there will be an appreciable loss of head within the meter.

Meters based on this principle are used by water boards to measure the flow at water, particularly at night when the flow is an indication of the losses due to leakage. The meter is particularly suitable for this purpose as the movement of the vane per unit change of flow is greatest at the lowest flow.

3.2.2.2 ROTATING VANE TYPE

If, instead of a single vane pivoted at the upper edge, a number of wings are arranged around the circumference of a disc, or merely attached radially at intervals around a pivoted spindle, in such a way that at least one wing is in the stream at any time, the force which produces deflection in the deflecting vane will now produce continuous rotation. The rate of this rotation will be a measure of the velocity of the liquid through the meter.

This is the principle of the meter shown in *Figure 3.18*. Liquid passing through the meter is directed on to the wings of the vane, and rotates it at a rate which depends upon the velocity of the liquid, so that the total rotation

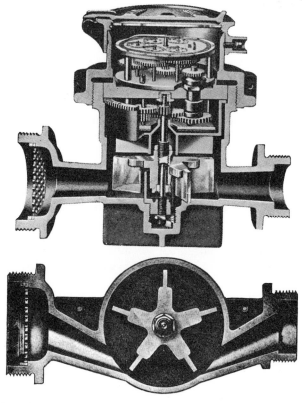

Figure 3.18   Rotating vane type meter. (By courtesy, Kent Meters Ltd.)

of the vane will be an indication of the total quantity of liquid which has flowed through the meter. Liquid out of this directioned stream is not stationary and influences the rate of rotation of the moving assembly. Regulation of the meter is, therefore, achieved by moving an adjustable baffle which affects the drag on the rotating vane. Both forward and return flows will be measured, and the resultant flow registered on the dial. The secondary element, or index, of the meter may be of two types. It may be a simple revolution counter giving a straight reading; or it may incorporate a number of dials. In other cases, especially in current meters for streams, it may operate an electric or acoustic transmitter arranged to drive some form of indicator each time the rotor makes a certain number of revolutions. On the other hand, it may incorporate a tachometer which is calibrated to give the rate of rotation of the vane, or the speed of the liquid.

The meter is constructed of corrosion-resisting material. For use on water all working parts are made of nickel, while the rotary vane is made of a special plastic. Two body designs are available, one for use in horizontal pipes and the other for use in vertical pipes, the latter being suitable for either upward or downward flow. Details of the accuracy of this type of instrument may be obtained from the manufacturer's literature.

The meter may be adjusted if it does not register accurately. It causes only a small loss of head.

3.2.2.3 HELICAL VANE TYPE

For larger rates of flow in closed channels, the fan type of rotating vane is replaced by a helical vane mounted centrally in the body of the meter, with its axis along the direction of flow as shown in *Figure 3.19*. The vane consists of a hollow cylinder with accurately formed wings. Owing to the effect of the buoyancy of the liquid on the cylinder, friction between its spindle and the sleeve bearings is small. The water is directed evenly onto the vanes by means of guides.

Transmission of the rotation from the undergear to the meter register is by means of a ceramic magnetic coupling.

The body of the meter is cast iron and the mechanism and body cover is of a thermoplastic injection moulding. The meter causes only a small loss of head and has a large capacity. It is, therefore, suitable for measuring the flow through district mains.

The meter is available in sizes from 40 mm having a maximum continuous rate of flow of 17 m³/h to the 150 mm size having a flow rate of 209 m³/h.

3.2.2.4 THE TURBINE TYPE METER

This type of meter consists of a practically friction-free rotor pivoted along the axis of the meter tube and designed in such a way that rate of rotation of the rotor is proportional to the rate of flow of fluid through the meter. This rotational speed is sensed by means of an electrical pick-off coil-mounted in the meter housing as shown in *Figure 3.20*. As the rotor spins an alternating current is generated in the coil by a magnet housed in the rotor. The number

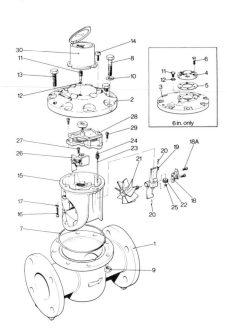

1 Body
2 Top cover with regulator plug and
   regulator sealing ring
3 Top cover plate
4 Joint plate
5 Joint plate gasket
6 Joint plate screws
7 Top cover sealing ring
8 Body bolt
9 Body bolt unit
10 Body bolt washer
11 Regulator plug
12 Regulator plug sealing ring
13 Joint breaking screw
14 Counter box screw
15 Measuring element
16 Element securing screw
17 Element securing screw washer
18 Back bearing cap assembly
19 Back vane support
20 Tubular dowel pin
21 Vane
22 Worm wheel
23 Vertical worm shaft
24 First pinion
25 Drive clip
26 Regulator assembly
27 Regulator assembly screw
28 Undergear
29 Undergear securing screw
30 Register

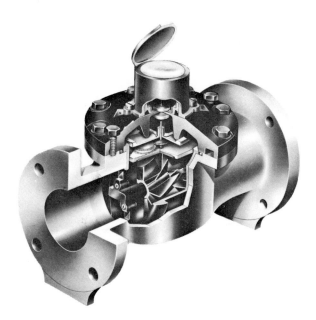

Figure 3.19   Helical vane type 2000 meter. (By courtesy, Kent Meters Ltd.)
(above) exploded view
(below) cut-away view

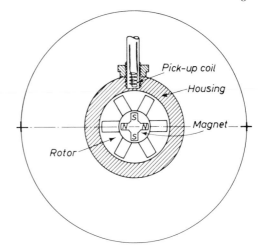

Figure 3.20
Principle of turbine type flow meter

of cycles per revolution may be varied by design to suit the application.

Another design uses the rotor blades to vary the reluctance of a magnetic circuit; in this case the magnet is housed in the pick-up coil. Alternatively the coil may be a part of a radio frequency circuit and the presence of a blade modulates the circuit producing a flow signal.

In some designs of turbine meter the rotor is pivoted on ball bearings or a fluid bearing is produced by drilling a hole through the nose cone and allowing process liquid to flow through and lubricate sleeve type bearings. In others the rotor contains a tungsten carbide bush running on tungsten carbide or fluorosint platform bearings, or the rotor and its bearings are supported by a stainless steel shaft passing through both upstream and downstream rotor support and guide vanes.

In the Pottermeter type of turbine meter shown in section in *Figure 3.21* the rotor is designed so that the pressure distribution of the process liquid is used to suspend the rotor in an axial 'floating' position.

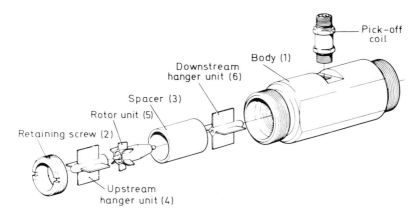

Figure 3.21   Exploded view of small Kent Pottermeter. (By courtesy, Kent Instruments Ltd.)

Owing to the conical shape of the downstream end of the rotor there is a decrease in fluid velocity with a corresponding pressure recovery tending to force the rotor upstream, thus under certain flow conditions, balancing the downstream viscous drag. In the larger sizes, the design of the rotor may vary and the venturi be built into a spacer which fits between the upstream and downstream guide vanes.

Meters are usually calibrated on water but may be calibrated on other liquids such as kerosene. They have a linearity of better than ± 0.5 per cent over the normal working range and better than ± 0.25 per cent over a restricted range. They have a repeatability of ± 0.1 per cent of the meter reading and a pressure drop of from 0.3 to 0.9 bar at the maximum linear flow rate on water depending on the size of the meter.

The effects of changes in calibration owing to difference in metered fluid viscosity or density may be eliminated by calibrating the meter under conditions exactly similar to those under which the meter will be used.

A turbine meter with viscosity compensation over a limited range is available. In this type of meter a filtered sample of the measured fluid flows between the stationary case and a drum which is rotated by the rotor. Increase in liquid viscosity increases the drag on the drum and hence compensates for the influence of the changed calibration on the meter reading.

The Pottermeter is manufactured in non-magnetic stainless steel and may be made suitable for use at high static pressures. Standard meters are available in sizes from the 5 mm having a linear flow range of 0.109 to 0.341 m³/h to the 500 mm size having a linear flow range of 654 to 6540 m³/h.

The a.c. output from the meter may be integrated when a rate of flow indication is required but the main use of this type of meter is for flow totalising or for feeding flow information into in-line blending controllers. The small a.c. current is amplified and the waveform modified and the square waveform output used to operate an electromagnetic counter which indicates the total quantity which has flowed through the meter.

Owing to the low energy level of the output of the sensing element it is intrinsically safe to B.S. 1259 provided it is connected by Zener barriers to the receiving equipment.

A reverse flow through the sensing element produces the same form of signal as a forward flow, so that a reverse flow results in a measuring error.

Installation

The meter should always be handled with care. It may be mounted in any attitude but for maximum meter rangeability it should be mounted with the rotor axis horizontal. The flow should be in the direction shown on the body. In order to reduce the risk of vaporisation in the meter, the meter should be installed downstream of the pump so that the liquid passes through at a pressure well above its vapour pressure. Although the bearing supports act as flow straighteners it is necessary to provide a straight pipe section of 10 pipe diameters upstream of the meter and 5 pipe diameters downstream.

Although solid particles may pass through the meter provided they are of such a size and shape that they move freely past the rotor blades, it is advisable to install a filter upstream of the meter. Any ferromagnetic material in the line must also be removed as it will adhere to the magnet and interfere with the accuracy of the meter.

The meter should be kept away from interfering signals from motors, solenoid valves, transformers, etc. The wiring between the sensing element and the amplifier should be a twisted twin cable with a screen earthed at the amplifier end only.

Meter provers

A meter prover consists of a calibrated length of pipe of highly uniform cross-section which may be a straight length as shown in *Figure 3.22* or where a larger volume is required consist of a series of loops. While the liquid is flowing through the line and meter a sphere, which completely fills the pipe, is launched into the line. When it passes the first sphere detector, usually a pressure-operated switch, it starts the counter counting the pulses from the positive displacement or turbine meter. When the second sphere detector is

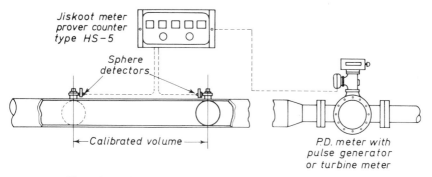

Figure 3.22   Meter provers. (By courtesy, Jiskoot Autocontrols Ltd.)

actuated it stops the counter. The internal finish of the pipe is usually of a very high standard and may be coated with a glossy epoxy resin and the system calibrated and certified to better than 0·02 per cent accuracy. The simple straight prover may be bi-directional and the loop assembly form usually has facilities to enable the sphere to be returned to the starting point.

It is possible to obtain the number of meter pulses for a given volume very accurately thus facilitating adjustment of the meter calibration should this prove necessary.

The meter prover may be located in a meter testing bay, or more commonly installed permanently on a bypass to meters used for custody transfer. The prover is frequently mounted on a trailer or skid unit so that it can be used to calibrate meters 'in-line' at any part of a refinery.

3.2.2.5   COMBINATION METERS

Where widely fluctuating flows are encountered, a combination meter should be used. The combination consists of a large meter (which may be helical, fan or other type) in the main, with a small rotary meter, or other suitable type, in the by-pass. Flow is directed into main or by-pass, according to quantity of flow, by an automatic valve. Small flows are registered by the by-pass meter, the valve in the main being closed. When the flow increases above the accurate minimum for the main meter, the automatic valve in the main opens fully and the by-pass is cut off. When the flow again falls below the accurate minimum for the main meter, the main valve closes and the by-pass is opened. Thus, flows down to the minimum for the by-pass meter may be accurately measured, while the delivery capacity of the meter is the maximum for the main meter. Combination meters are also made with both meters in operation at rates of flow near the maximum capacity. The total flow is given by the sum of the registrations on both dials. Meters are also available in which the quantity which has flowed through both meters is indicated on one dial.

Recorders

In addition to indicating the total quantity of water which has flowed in a given period, these meters may be used to record the total flow on a chart carried by a drum which is rotated at a constant speed by a clockwork or electrical mechanism.

The pen is positioned on the chart by means of a spindle driven from the pointer of the meter. This spindle has a right-handed helical groove machined in it, so that as it rotates the height of the pen above the zero is increased and records the total flow. The slope of the curve so drawn will be an indication of the rate of flow at the time considered.

When the pen reaches the top of the chart, the pen carrier is transferred to a left-handed helical groove, so that, in further recording, the total flow is indicated by the distance of the line from the top of the chart. On reaching the bottom of the chart the pen is returned to the right-handed helix; and so on.

3.2.2.6 MAGNETIC FLOW METERS

Principle of operation

In this type of meter the velocity of the stream through a tube or pipe is measured by measuring the electromagnetic effect it produces. The meter consists of two portions, the transmitter or primary element, which is mounted in the pipe line and generates the electrical variable, and the receiver or secondary element which measures the magnitude of this electrical variable.

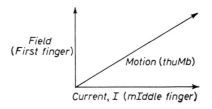

Field (First finger)

Motion (thuMb)

Current, I (mIddle finger)

Figure 3.23   Fleming's Right Hand rule

The operation of the transmitter is based on Faraday's well-known law of electromagnetic induction:

Whenever a conductor cuts lines of magnetic force an e.m.f. is generated, and the magnitude of this e.m.f. is proportional to the rate at which lines of force are cut. The e.m.f. is generated in a plane which is mutually at right angles to the velocity (V) of the conductor and the direction of the magnetic field (H), and the direction will be given by Fleming's Right Hand rule (*Figure 3.23*).

If the right hand is held with thumb, first finger and middle finger mutually at right angles, and the First finger represents the Field direction, thuMb the direction of Motion, then the mIddle finger gives the direction of the induced current I.

The magnetic flowmeter uses the above principles to measure the average flow velocity in the manner shown in *Figure 3.24*. A uniform magnetic field

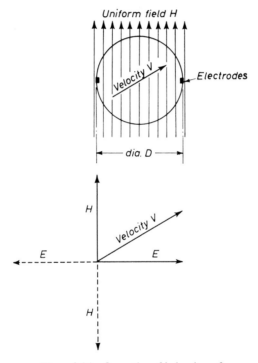

Figure 3.24   Generation of induced e.m.f.

is induced through a portion of the flow tube consisting of a non-magnetic material, lined if necessary with insulating material. The flowing liquid itself is the conductor moving through the magnetic field. As the disc of liquid passes between the electrodes inserted in, and insulated from, the tube wall, an e.m.f. is generated which is then measured. The conducting disc of liquid is equivalent to a conductor of length D, the distance between the electrodes which is the same as the tube diameter. The steady succession of discs moving through the magnetic field past the electrodes yields a continuous voltage.

When the magnetic field is reversed, the polarity of the generated e.m.f. is also reversed as shown by the dotted lines. Thus if H is an alternating magnetic field as produced by an electromagnet through which an alternating current is flowing, the e.m.f. E will be alternating at the same frequency as the magnetic field.

The magnitude of the generated e.m.f. E will be given by

$$E = CHDV \text{ where } C \text{ is a dimensionless constant}$$

or
$$V = \frac{E}{CHD}$$

Thus the volume rate of flow

$$Q = VA = \frac{EA}{CHD}$$

A, D and C are constant for a given tube size.

If the maximum value of the alternating magnetic field can be maintained constant, the maximum value of the alternating e.m.f. will be proportional to the volume flow rate of the liquid. This may be measured directly by an instrument of the type described in section 4.7.5 such as the Dynalog, which may be calibrated in flow units. Alternatively, the signal may be applied to a converter which converts the a.c. voltage input into a proportional d.c. output of the required range, e.g. 10–50 m.a. or 4–20 m.a. which may then be indicated or recorded on an instrument scaled in flow units, or applied as the measured value to a controller. If required, the converter can be arranged to give a pulse output in addition to the analogue output, each pulse representing a discreet volume. The pulse output is fed into a totalizing counter scaled in engineering units thus giving the total quantity that has flowed. If necessary the converter may be enclosed in a watertight and flameproof housing.

Application of the meter

In order that the meter may function the liquid flowing through the transmitter should have a conductivity of not less than 0·02 microsiemens per metre. This compares with a conductivity of about 1 mS/m for most tap waters. Thus most process liquids may be measured except non-conducting liquids such as hydrocarbons.

The transmitter will function in any position provided the line is always completely full. As the meter actually adds the incremental velocities across the tube from one electrode to the other, neither turbulence nor variations in the flow profile seriously affect the accuracy. Straightening vanes and straight runs of pipe are not therefore necessary. As the passage through the transmitter is line size, the pressure drop through the transmitter is the same as that for the equivalent length of straight pipe having an internal diameter equal to that of the transmitter.

As there are no pockets or crevices to trap foreign material the transmitter is suitable for use on heavy slurries and food products. If a concentric residue of 'slime' which has the same conductivity as the flowing fluid builds up on the inside of the transmitter tube it behaves in the same way as a ring of stagnant liquid and the meter reading will be unaffected. Removal of the residue in these circumstances is necessary only when the pressure drop

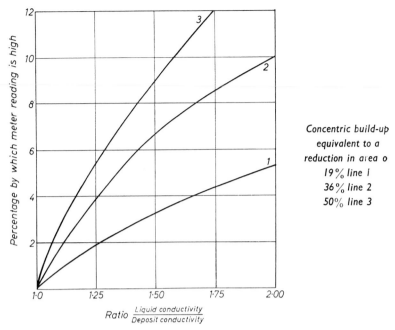

Figure 3.25 *Effect of build-up on transmitter accuracy. (By courtesy, Foxboro-Yoxall Ltd.)*

through the meter becomes prohibitive. Where the conductivity of the deposit differs from the conductivity of the flowing liquid the affect on the accuracy is shown in *Figure 3.25*. Where the build-up has an extremely low conductivity a thin film will cause the meter to read zero because the electrodes will no longer be in contact with the liquid. As an optional extra an electrode cleaning assembly may be installed. When actuated, this assembly interrupts the signal leads to the transmitter and applies the supply voltage to the electrodes thus boiling away process deposits from the electrodes.

The tube lining and electrodes are the only portion of the meter to come into contact with the process stream and these can be made from materials which are highly resistant to corrosion. The meter can therefore be used to measure the flow of liquids which are too corrosive to be measured by other means.

For pulsating flows of low frequency, say less than one cycle per second, the pen of the indicating instrument will tend to follow the pulsations and read the instantaneous value. If a slow pen speed is used to obtain a straight line record, since the indication is directly proportional to the flow, the indication will be the average value with negligible error when the amplitude of pulsations is no greater than plus or minus 10 per cent of the instrument span and they are sinusoidal in nature.

If air bubbles or solid matter are entrained in the liquid the meter will read the volume flow rate of the total mixture. It is only necessary to have the tube completely filled with a homogeneous mixture at all times. For example, if every hour 50 m³ of water and 20 m³ of suspended solid pass through the meter then the meter will read 70 m³/h.

through the meter then the meter will read 70 m³/*h*.

Effect of liquid conductivity and lead resistance on the instrument accuracy

In order to measure the e.m.f. generated at the transmitter it is necessary to have a connecting cable and a voltage measuring instrument. When a current flows through the transmitter and the remainder of the circuit, the sum of the voltage drops in each part of the circuit will be equal to the e.m.f. (E). Using Ohm's law.

$$iR_1 + iR_2 + iR_3 = E$$

where R_1 is the impedance of the transmitter

R_2 is the impedance of the connecting cable

R_3 internal impedance of the measuring instrument.

Thus the voltage drop measured by the measuring instrument is the voltage drop across itself *viz.* iR_3

$$\text{where } iR_3 = E - i\,(R_1 + R_2)$$

Thus to measure the e.m.f. within the standard accuracy

1. The measuring instrument must have a sufficiently high input impedance to keep the magnitude of the current i flowing below a certain very small value.

2. The sum of the internal impedance of the transmitter and the lead impedance must not exceed a certain maximum value. Thus, for a given measuring system, the maximum lead length is fixed by the transmitter size and the liquid conductivity. For example to measure with the standard accuracy with the 2 m lined metal metering tube meter, the maximum length of signal cable between transmitter and converter is 1·5 m when the liquid conductivity is 0·02 mS/m and 305 m when the liquid conductivity is 5 mS/m.

   In addition, since the e.m.f. is alternating, the cable acts as a capacitative shunt across the transmitter output and is in parallel with the measuring instrument input terminals. As the cable length increases the capacitative reactance drops and more current tends to flow through the transmitter internal resistance increasing the voltage drop within the transmitter. Increasing the cable length, would therefore, tend to make the indicating instrument read low.

   When used with liquids of low conductivity where this error would be significant, the low-level millivolt signal from the transmitter is protected from these capacitance losses by surrounding the conductors by shields, power driven from the converter, so that the conductor is surrounded by a screen at its own potential thus preventing electrostatic effects from interfering with the measurement. This power driven shield exists between the converter and the transmitter as well as inside the transmitter housing. The electrodes also have power driven cup shields. The shields are then vinyl insulated and surrounded by a screen earthed to the transmitter, and this surrounded by vinyl insulated screen earthed to the converter.

Form of the instrument

Figures 3.26a and *b* show the external views of the transmitter with housing removed. Two basic measuring tubes exist. The fibre glass reinforced epoxy

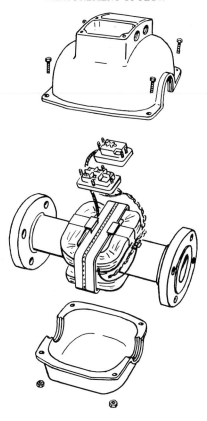

Figure 3.26(a)   The magnetic flowmeter—exploded view. (By courtesy,
Foxboro-Yoxall Ltd.)

tube for chemical applications, which being an insulator requires no insulating
liner, and the type 316 stainless steel tube lined with P.T.F.E. for corrosive
liquids, polyurethane for abrasive liquids, or polyvinylidene fluoride (Kynar)
where the tube is in contact with food or other similar sanitary applications.
The fibre glass reinforced measuring tube is suitable for process pressures up
to 17 bar and temperatures of 95° C.

The 316 stainless steel tubes can be designed for pressures up to 175 bar
and the P.T.F.E. liner is suitable for process temperatures up to 175° C, the
polyurethane up to 70° C and polyvinylidene fluoride up to 60° C.

The uniform magnetic field is produced by two saddle-shaped coils of
insulated copper wire. The coils are wrapped in fibreglass tape for added
protection, vacuum impregnated with plastic, shaped and baked to form
rigid permanent bundles. Electrostatic shielding prevents capacitatively
induced voltage between the coils and the electrodes. The magnetic field is
rendered uniform and everywhere at right angles to the flowing fluid by the
laminated soft-iron core. The core is clamped rigidly in place by mounting
brackets welded to the tube.

The two button type electrodes are mounted in and insulated by P.T.F.E.
sleeves from the metering tube 180° apart and in a plane at right angles to the

Figure 3.26(b)   The magnetic flowmeter with case removed.
(By courtesy, Foxboro-Yoxall Ltd.)

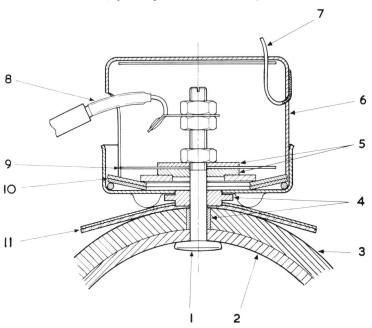

Figure 3.27   Form of electrode

1. Electrode.
2. Tube lining.
3. Metal tube.
4. Electrode insulators.
5. Insulators for electrode tensioning system.
6. Screening can.
7. Earth connection.
8. Signal lead.
9. Metal washer.
10. Belleville washers for electrode tensioning.
11. Metal washer and insulator.

direction of flow and the magnetic field. Type 316 stainless steel electrodes are standard but where process conditions demand other metals such as platinum 20 per cent, iridium, Hastelloy C, Monel, Titanium, Carpenter 20 and Tantalum are available. The form of the electrode assembly is shown in *Figure 3.27.*

The components are assembled in the sequence suggested by the diagram but with the Belleville washers in their unstressed condition, when they are deflected in the opposite direction of that shown in the diagram. The first nut on the electrode stem is then tightened so that the Belleville washers are deflected through the 'flat' position until they almost touch the base part of the screening can. The electrical connection to the electrode is then completed via the solder tag and finally the outer portion of the screening can is placed in position and its earth connection made.

The transmitter components are housed in two rugged removable cover sections constructed from low copper-content cast aluminium, selected as the most suitable for resistance to corrosion. The housing is protected by several layers of primer and baked-on textured vinyl finished coatings. Heat dissipation is also accomplished through the housing which ensures cooler operation of the field coils. The housing joints are gasketed with silicone rubber sealants rendering the housing watertight. The transmitter meets the American Intrinsic Safety Standard for use in Class 1 Groups C and D, Division 2, areas.

The system accuracy of transmitter and converter is $\pm 1\cdot0$ per cent of full scale flow rate and the repeatability is $0\cdot25$ per cent full scale.

3.2.2.7 ULTRASONIC FLOW METERS

This flowmeter is based upon the fact that sound waves travel through a material medium and if the medium moves the sound waves are carried with it. Thus if a transmitting transducer is located on one side of a pipe, and two receiving transducers are fitted on the other side of the pipe, one upstream and the other downstream of the transmitter, then the sound waves will be received earlier by the downstream receiver than by the upstream receiver. This difference in transit time may be measured in one of two ways. One way is to measure the phase difference of the sound waves arriving at the receivers. The other way is to measure the time difference between the time of arrival of a burst of waves. In either case, the measured difference of transit time bears a linear relationship to the flow rate and the equipment may be calibrated to read directly in units of flow rate, or by the use of an integrating motor, total flow may be recorded. Equipment capable of giving an indication with an accuracy of $\pm0\cdot5$ per cent is available for pipe sizes of from 40–400 mm at flow rates of from 3 mm/s upwards. The main advantage of the method is that equipment may be designed with a response time of a few milliseconds so that pulsating flow may be measured. In general, this type of equipment is used for measuring flow of liquids, but models of measuring flow of gas are also available.

In another form of the instrument, bursts of sound are propagated alternately in opposite directions between a pair of transducers situated diagonally along the pipeline. Because the signal travelling upstream is delayed, and the downstream moving signal is speeded up by the moving fluid, the alternate bursts yield waves of different frequency and this difference is a measure of fluid velocity.

3.2.2.8 — VORTEX FLOWMETERS

It has been discovered that when a cylindrical rod is placed in a flowing stream the fluid flow across the cylinder will result in the formation of two unsymmetrical rows of vortices rotating in opposite directions as shown in *Figure 3.28 (c)*. The production of a vortex produces a reaction force on the cylinder which if the cylinder has the form shown in *Figure 3.28 (b)* causes fluid to flow through several sensing ports and the single cross flow hole. The fluid flows alternately in one direction and then in the other with a frequency

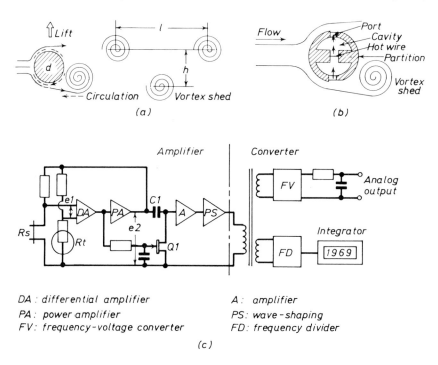

DA : differential amplifier   A : amplifier
PA : power amplifier   PS : wave-shaping
FV : frequency-voltage converter   FD : frequency divider

(c)

Figure 3.28   The Vortex Flowmeter. (By courtesy, Foxboro-Yoxall Ltd.)

which is dependent only upon the ratio of the velocity of the fluid to the diameter of the probe provided the Reynolds number lies between 2×10^3 and $1 \cdot 0 \times 10^5$ for gases and 4×10^3 and $1 \cdot 4 \times 10^5$ for liquids.

A platinum detection wire located in the cross flow hole is maintained at a constant temperature, such that frequency of the fluctuating heating current is proportional to the flow velocity.

The diameter of the detector is not changed for pipe diameters ranging from 400 mm to 1200 mm and the associated fittings are so designed that the detector can be installed or removed from the line without leakage.

The method of obtaining the output is shown in *Figure 3.28 (c)*. The detection wire (Rs) and a compensating resistor (Rt) form a pair of bridge arms in the circuit. The circuit is designed in such a way that the detection wire is maintained at approximately 20° C above process fluid ambient temperature. A change in cooling flow from vortex shedding changes the value of (Rs) momentarily, thereby inducing an unbalance voltage input to the differential amplifier (DA). The output of (DA) provides a signal to the power amplifier (PA) which adjusts the bridge voltage or feedback current to maintain (Rs) at a constant value for bridge balance (by holding the wire temperature fixed). The variation in feedback current is amplified and shaped through the amplifier (A) and pulse shaper (PS) for transmission to the converter.

The output of the meter is independent of the density, temperature and pressure of the flowing fluid and represents the flow rate to better than ±1 per cent of full scale.

3.2.2.9 THE SWIRLMETER

Another meter which depends on the oscillatory nature of vortices is the Swirlmeter shown in *Figure 3.29*. A swirl is imparted to the body of flowing liquid by the curved inlet blades which give a tangential component to the fluid flow. Initially, the axis of the fluid rotation is the centre line of the meter, but a change in the direction of the rotational axis (precession) takes place when the rotating liquid enters the enlargement causing the region of highest velocity to rotate about the meter axis. This produces an oscillation or precession, the frequency of which is proportional to the volumetric flowrate. The sensor, which is a bead thermistor heated by a constant current source converts the instantaneous velocity changes into a proportional voltage which

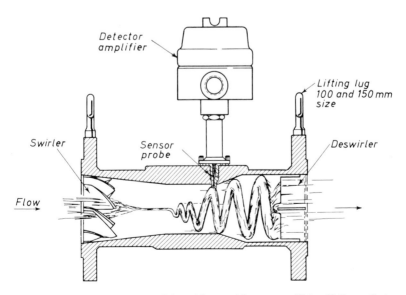

Figure 3.29   Cutaway view of the swirlmeter. (By courtesy, Fisher & Porter Co.)

is amplified, filtered and shaped to form a 15 V peak-to-peak frequency signal proportional to the volumetric flow rate. Other forms of sensor are available for special applications.

The operating range of a Swirlmeter depends upon the specific application, and the rangeability may vary between 10 to 1 and 100 to 1.

Typical gas flow ranges are 8 to 50 m³/h in the 25 mm size to 100 to 2500 m³/h in the 150 mm size. Corresponding water flow ranges are 0·36 to 3·6 m³/h and 10·8 to 180 m3/h.

The frequency range is 20–1300 Hz for gases and 2–150 Hz for liquids. The linearity is ⩽ ±1 per cent of rate and the repeatability ⩽ ±0·25 per cent of rate. Normal operating pressure is up to 40 bar and operating temperatures from −50° C to +10° C or 0° C to 70° C.

3.2.3   Rate of flow meters for gases

3.2.3.1   DEFLECTING VANE TYPE; VELOMETERS

The principle of this type of instrument is similar to that of the same instrument for liquids. The construction, however, has to be different, for the density of a gas is usually considerably less than that of a liquid. As the force per unit area acting on the vane depends upon the rate of change of momentum, and momentum is mass multiplied by velocity, the force will depend upon the density and upon the velocity of the impinging gas. The velocity of gas-flow in a main is usually very much greater (6 to 10 times) than that of liquid-flow but this is not sufficient to compensate for the greatly reduced density. (Density of dry air at 0° C and 760 mm = 0·001 3 g/ml while density of water = 1 g/ml.)

The vane must therefore be considerably larger when used for gases, or be considerably reduced in weight. The restoring force must also be made small if an appreciable deflection is to be obtained.

The simple velometer consists of a light vane which travels in a shaped channel. Gas flowing through the channel deflects the vane according to the velocity and density of the gas, the shape of the channel, and the restoring torque of the hairspring attached to the pivot of the vane.

The velometer is usually attached to a 'duct jet', which consists of two tubes placed so that the open end of one faces upstream, while the open end of the other points downstream. The velometer then measures the rate of flow through the pair of tubes, and, as this depends upon the lengths and size of connecting pipes and the resistance and location of the pressure holes, each assembly needs individual calibration.

The chief disadvantages of this simple velometer are the effects of hot or corrosive gases on the vane and channel. This disadvantage may be overcome by measuring the flow of air through the velometer produced by a differential air pressure equal to that produced by the 'duct jet'. In this way the hot gases do not pass through the instrument so that it is not damaged.

3.2.3.2   ROTATING VANE TYPE

Anemometers

As in the case of the deflecting vane type, the force available from gases to produce the rotation of a vane is considerably less than that available in the measurement of liquids. The vanes must therefore be made light, or have a large surface area. The rotor as a whole must be accurately balanced, and the bearings must be as friction-free as possible.

The rotor may be of two kinds. In the cup anemometer (*Figure 3.30*) the rotor consists of three light conical-shaped cups, attached to arms carried by

Figure 3.30 (above)
Cup counter anemometer.
(By courtesy, Casella
and Co. Ltd.)

Figure 3.31 (right)
Air meter.
(By courtesy, Casella
and Co. Ltd.)

a vertical spindle, which in the more sensitive type of instrument runs on a jewel bearing, and is geared into a light counting mechanism. The counter wheels being made of a self-lubricating styrene based plastic. The instrument has a counter range of 0–9999·99 Km and an accuracy better than ±1·8 Km/h from 9–150 km/h. The number of cup revolutions in a period of, say, 3 minutes is obtained from the counter reading by timing with a stop watch. The air velocity may then be obtained from an appropriate graph. The rotor revolves because the resistance offered to the wind by the open end of the cup is greater than that offered by the closed end. There is, consequently, a greater force exerted by the wind on a cup with its open end facing the wind than on a cup with its closed end facing the wind. The rotor therefore turns with the apices of the cups leading. This type of instrument is used extensively in meteorological work.

In the generator form of the instrument the spindle of the anemometer carries a 12-pole permanent magnet rotor whose motion within a wound stator induces in it an alternating voltage proportional to the wind speed. This operates indicators consisting of a metal rectifier, a temperature correcting element and a moving coil D.C. voltmeter calibrated in wind speed. A recorder may also be operated from the same signal.

In the air meter type *Figure 3.31,* the rotor consists of a fan or light vane wheel with flat or slightly curved vanes, made of light metal or mica, mounted on slender arms, with its axis in the direction of the flow of the gas. The rotor drives a light counting-train, which gives the distance moved by the air in feet. By timing over a period and determining the number of revolutions, the actual velocity of the gas may be found. If the actual volume of gas flowing is required, the cross-section of the stream should be traversed and the average velocity of the stream found as is described in the section on the pitot static tube dealt with in Section 3.4.1.1.

The instruments are made for various ranges of velocities within the limits of 16–3000 m/min having a sensitivity of from ±1 to ±2·5 per cent.

Rotary gas meter

The rotary meter is a development of the air meter type of anemometer, and is shown in *Figure 3.32.* It consists of three main assemblies, the body, the measuring element and the multipoint index driven through the inter-

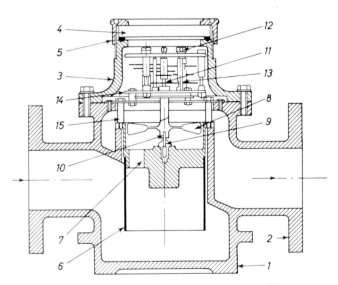

Figure 3.32   Diagrammatic section of a rotary gas meter. (By courtesy, Parkinson and Cowan Compteurs)

gearing. The lower casing (1) has integral in line flanges (2) and is completed by the bonnet (3) with index glass (4) and bezel (5).

The measuring element is made up of an internal tubular body (6), which directs the flow of gas through a series of circular ports (7) on to a vaned anemometer (8). The anemometer is carried by a pivot (9) which runs in the sapphire/agate bearing assembly (10), the upper end being steadied by a bronze bush (11).

The multipointer index (12) is driven by an intergear (13) supported between index plates (14). The index assembly is positioned by pillars (15) which are secured to the top flange of the internal tubular body.

Materials of construction

All meter casings are cast iron and, in the small size, the internal tubular body cast as an integral part of the lower casing. The larger sizes have a separate internal tubular body, made from cast iron with a brass or mild-steel skirt, which forms part of the overall measuring element.

The anemometer is made from aluminium and is mounted on an aluminium spindle. The lower end of the spindle houses a sapphire thrust bearing which runs on a steel pivot. Journal loading is accepted by an agate bearing adjacent to the sapphire. The upper end of the fan spindle is fitted with a stainless-steel pinion which transmits the rotary movement of the anemometer to the intergearing.

The intergearing is made up from a series of brass/stainless-steel spur gears and pinions. The spindles of these gears are located in steatite bearings which are housed in aluminium index plates.

The index dial is viewed through the index window which is manufactured from polished plate glass, and is secured by a brass bezel to the meter bonnet.

As the interiors are made to standard dimensions, a replacement may be fitted while the normal interior is being cleaned, repaired and recalibrated. The interior may be recalibrated without its parent casing, so that it may be returned to the maker for recalibration if this is found to be necessary.

Application

This type of meter is used for secondary measurement of flow in industrial and commercial installation at pressures up to 1·5 bar and flows up to 200 m³/h.

Calibration and maintenance

These meters are calibrated by passing gas through the meter from a cali-brated gas holder at definite known rates of flow. For instance, a meter which registers up to 200 m³/h is tested in steps of 20 m³/h. Any tendency to under-register or over-register is corrected by slightly bending the blades of the fan. Meters of this type are accurate to within ±2 per cent of actual flow rate over a flow range of 10:1.

This type of meter does not require a great deal of maintenance unless the gas which it is measuring is very dirty or contains a great deal of tar. In such circumstances the measuring element will require frequent and regular cleaning by immersion in a bath of solvent.

When it is required to measure gas at higher flow rates and higher pressure a turbine meter similar to that used for liquids may be used. The rotating turbine may drive an index directly through a magnetic coupling which isolates the gas from the index system.

Meter sizes up to 400 mm capable of standing pressures up to 80 bar and having flow rates up to 6500 m³/h are available with an accuracy of ± 1 per cent over a flow range of 8:1 in the small sizes to 20:1 in the larger sizes. Equipment may be added to compute the mass flow by measuring the temperature and density of the flowing gas.

3.2.3.3 THERMAL TYPE

This type of meter depends upon the fact that if a known quantity of heat is given to a gas whose specific heat at constant pressure is known, the rise in temperature will be inversely proportional to the mass of gas present. Thus, if there is a closed channel through which gas is flowing, and it is arranged to supply to the gas a known quantity of heat, say every second, then the rise in temperature will be inversely proportional to the mass of gas passing through the heated section every second. In the practical form, the average temperature of the gas before it is heated must be measured. This may conveniently be done by placing across the cross-section of the channel a mesh of wire whose resistance will depend upon its temperature, as in the resistance thermometer described later. The gas then takes up a known quantity of heat from a heating element which consumes a known quantity of electrical energy. This element is stretched across the whole cross-section of the channel so as to heat the gas as uniformly as possible. The rise in temperature is then found by placing a resistance element, similar to the first one, in the path of the gas. If the two resistance elements are made two arms of a Wheatstone Bridge, then the amount of unbalance produced, or the change of resistance required in one of the other arms of the bridge to produce balance, will be a measure of the rise in temperature.

As heat is supplied at a constant rate, this rise of temperature will be inversely proportional to the mass of gas flowing through the closed channel. The bridge may, therefore, be calibrated in terms of mass flow of gas.

This method has the disadvantage that it is difficult to integrate the rate of flow from an inverse ratio chart, and that the rate of change in rise of temperature with rate of change of flow will be small when the flow is large. A meter has therefore been designed which has a variable heat input. The heat input is arranged to give a constant rise in temperature. The amount of heat required to produce this constant rise in temperature will be proportional to the mass of gas flowing, so that a direct reading instrument is produced. This instrument also has the advantage that the operation of the instrument is more stable owing to the fact that the temperature relationship of all parts of the instrument is constant.

Because a great deal of heat is absorbed in the form of latent heat when a liquid is changed to a vapour, the results will not be accurate if the gas measured contains entrained liquid. This liquid must therefore be removed or evaporated by previous heating of the gas, if accurate results are to be obtained.

By using the Wheatstone Bridge method of measuring the rise in temperature, a rise of 1° or 2° C can be measured and maintained, and the instrument is more stable owing to the fact that the temperature relationship of all parts of the instrument is constant.

3.3 PRESSURE DIFFERENTIAL METHODS

Probably the most widely used type of flowmeter is that which depends upon differential pressure measurement. Whatever the construction of the meter, the principle involved is the same. The net cross-sectional area of the stream is reduced, causing an increase in the velocity, and hence an increase in the energy of motion or kinetic energy. Since energy cannot be created or destroyed, this increase in kinetic energy is obtained at the expense of pressure energy, so that the pressure of the fluid is reduced. A known fraction of this reduction of pressure or pressure differential is measured, and enables the velocity of the stream to be calculated. In order to understand this type of flow measurement clearly, it is necessary to be familiar with the underlying theory which is, therefore, dealt with here.

Streamlined flow

Consider the flow along a straight, smooth-walled channel of uniform cross-section. If the paths of all particles of fluid are exactly parallel to the channel, the flow is described as streamlined or viscous flow. The velocity of all the particles need not be equal but there must be no transverse motion; i.e. there must be no motion of the particles across the stream. This type of flow occurs at low velocity and in liquids having a high viscosity. In this type of flow, the flow pattern is governed to a large extent by the viscous forces in the liquid. Viscosity will be defined a little later in this section.

Turbulent flow

If the velocity of flow is increased, the flow pattern changes, and some particles have a velocity across the channel as well as their velocity along the channel. This transverse velocity may vary from slight wavering to violent swirls and eddies. Such flow is called turbulent flow.

The swirling in the turbulent flow may become so violent that the paths of the particles become helical; i.e. the particles move in spirals along the channel. This type of flow is described as swirling or helical flow.

The most common type of flow is the simple turbulent flow, and head meters are designed to measure this. If applied to helical flow, a head meter will not give accurate results. Helical flow may, however, be suppressed by allowing a sufficient length of straight pipe before the meter is reached, or by fitting straightening vanes.

For accurate measurement by head meters, the flow should not vary by more than plus or minus 20 per cent unless the fluctuations are slow enough to permit a readable record of the actual variations to be made. In pulsating flow such as that produced by a single-acting reciprocating pump both the pressure and the velocity vary rapidly. These variations travel along the channel in the form of waves. In open channels they rapidly die out, but in closed channels they may persist for a considerable time. Before measurement, the flow should be made uniform by including in the main, between the pump and the measuring element, a cushioning chamber in the form of a reservoir of large capacity.

If the differential pressure fluctuations due to the varying flow are greater than 40 per cent of the operating differential pressure the readings on the flowmeter will be far from accurate even if the record is made readable by the inclusion of a restriction in the lines or pressure measuring instrument. This is because the instrument, when damped, will measure the mean of the varying differential pressure. The mean flow is, however, measured by the mean of the square root of the differential pressure, so that the indication of the meter is not an accurate indication of the mean flow. This point is illustrated mathematically by a simple example in the appendix.

Energy of a fluid in motion

Consider a fluid flowing along a smooth walled channel of uniform cross-section. It will have several forms of energy. The most important of these will be potential energy, kinetic energy, pressure energy and heat energy.

Potential energy

This is the energy the fluid has by virtue of its height above some fixed level. For example, 1 m³ of liquid of density ρ_1 kg/m³ will have a mass of ρ_1 kg and would require a force $9\cdot81\rho_1$ N to support it at a point where the gravitational constant (g) is $9\cdot81$ m/s. Therefore if it is at a height of Z_1 metres above a reference plane it would have $9\cdot81\rho_1 Z_1$ joules of energy by virtue of its height.

Kinetic energy

This is the energy a fluid has by virtue of its motion. 1 m³ of fluid density ρ_1 kg/m moving with a velocity V_1 m/s would have a kinetic energy of $\frac{1}{2}\rho_1 V_1^2$ joules.

Pressure energy

This is the energy a fluid has by virtue of its pressure. For example a fluid having a volume v_1 m³ and a pressure of p_1 N/m² would have a pressure energy of $p_1 u_1$ J.

Other forms of energy

In addition to the above forms of energy the fluid will have energy by virtue of its temperature, that is, heat energy. If there is resistance to its flow in the form of friction, other forms of energy will be converted into heat energy.

Viscosity

The frictional resistance which exists in a flowing fluid is called viscosity. The particles of fluid which are actually in contact with the walls of the channel are at rest, while those at the centre of the channel will be moving with the maximum velocity. Thus the layers of fluid near the centre, which are moving at the maximum velocity, will be slowed down by the slower-moving layers around; while the slower-moving layers are being speeded up. It is well known that some liquids flow more easily than others: petrol flows more easily than lubricating oil. The measure of this resistance to flow in a fluid is called the coefficient of viscosity or simply viscosity. More precisely, viscosity of a fluid is defined as the tangential force per unit area at either of two horizontal planes at unit distance apart, one of which is fixed while the other moves with unit velocity, the space between the planes being filled with the fluid. It is a factor of which account must be taken in measurement of flow with head meters. It may be measured by measuring the rate of flow of the fluid through a tube whose length is very great in comparison with its diameter, when a known difference of pressure is applied. It may also be measured by measuring the terminal velocity of a small sphere when falling through a long column of the fluid.

In SI units the dynamic viscosity of a fluid will be expressed in units of Ns/m^2, i.e. a fluid has a dynamic viscosity of Ns/m^2 if a force of 1 N is required to move a plane 1 m^2 in area at a speed of 1 m/s parallel to a fixed plane the moving plane being 1 m away from the fixed plane and the space between the planes being completely filled with the liquid. In practice the dynamic viscosity is often expressed in centipoise (cP) where 1 cP $= 10^{-3} Ns/m^2$.

It is sometimes useful to know the ratio of the dynamic viscosity of a fluid to its density at the same temperature. This is called the kinematic viscosity of the fluid.

$$\text{Kinematic viscosity at } t° \text{ C} = \frac{\text{dynamic viscosity at } t° \text{ C}}{\text{density at } t° \text{ C}}$$

In SI units the unit is the m^2/s but in practice the kinematic viscosity may be expressed in centistokes (cSt) where 1 cSt $= 10^{-6} m^2/s$.

Viscosity is not affected very much by pressure unless the pressure is very high but it is affected by temperature. For liquids, the viscosity decreases with increase of temperature, but for gases it increases. Viscosity does not affect the accuracy of quantity meters unless they are used for very viscous oils but it does influence rate-of-flow meters. The correction to be applied is, however, well defined and the conditions for which the correction is required are limited to liquids of higher viscosities.

It is viscosity which is responsible for the damping out, or suppression, of disturbances due to bends and valves in a pipe. The energy which existed in the swirling liquid is changed into heat energy, but the rise of temperature is generally too small to be measured.

3.3.1 Energy equation for streamlined flow of liquids

3.3.1.1 BERNOULLI'S THEOREM

Energy cannot be created or destroyed, it can only be changed in form. It is also true that matter or mass cannot be created or destroyed. Consider 1 kg of fluid flowing through a channel with rigid walls which will allow neither fluid nor heat to pass through. Consider two sections, section 1 and section 2, and let the conditions at each section be represented by the following symbols.

	At section 1	At section 2	Units
Area	A_1	A_2	m²
Velocity	V_1	V_2	m/s
Pressure	p_1	p_2	N/m²
Density	ρ_1	ρ_2	kg/m³
Specific volume, volume of 1 kg	$v_1 = 1/\rho_1$	$v_2 = 1/\rho_2$	m³
Height of centre of gravity of section above reference plane	z_1	z_2	m
Internal energy per kg	I_1	I_2	J

If 1 kg of fluid enters at section 1 and there is no accumulation of fluid between section 1 and section 2 then 1 kg of fluid must leave at section 2. The energy of the fluid at section 1

= potential energy + kinetic energy + pressure energy + internal energy

$$= 1.z_1.g + \tfrac{1}{2}.1.V_1^2 + p_1 v_1 + I_1$$

Energy of fluid at section 2 $= 1.z_2 g + \tfrac{1}{2}.1.V_2^2 + p_2 v_2 + I_2$. Since no energy can leave the channel and since energy cannot be created or destroyed,

Total energy at section 1 = Total energy at section 2

$$z_1 g + V_1^2/2 + p_1 v_1 + I_1 = z_2 g + V_2^2/2 + p_2 v_2 + I_2 \qquad (3.1)$$

Now if the temperature of the fluid remains the same, the internal energy remains the same and

$$I_1 = I_2 \qquad (3.2)$$

and equation (3.1) reduces to

$$z_1 g + V_1^2/2 + p_1 v_1 = z_2 g + V_2^2/2 + p_2 v_2 \qquad (3.3)$$

This equation applies to liquids and ideal gases.

Now consider liquids only. These can be regarded as being incompressible and their density and specific volume will remain constant along the channel, and

$$v_1 = v_2 = 1/\rho_1 = 1/\rho_2 = 1/\rho \qquad (3.4)$$

Thus equation (3.3) may be written

$$Z_1 g + V_1^2/2 + p_1/\rho = Z_2 g + V_2^2/2 + p_2/\rho \qquad (3.5)$$

Dividing by g this equation becomes

$$Z_1 + \frac{V_1^2}{2g} + \frac{p_1}{\rho g} = Z_2 + \frac{V_2^2}{2g} + \frac{p_2}{\rho g} \qquad (3.6)$$

3.3.1.2 EQUATIONS IN TERMS OF HEAD

Let a side hole be drilled in the channel at the level of the centre of gravity of section 1, and connected to the lower end of a vertical tube from which the air has been removed. Liquid will rise in the tube to a height such that the pressure due to the column of liquid in the tube will balance the static pressure p_1; i.e. to a height $p_1/\rho g$ above the side hole, or $p_1/\rho g + Z_1$ above the horizontal level taken as the reference plane.

Since the liquid inside the channel is the same as the liquid in the gauge tube the location of the holes does not matter. The level of the liquid in the gauge tube will be $p_1/\rho g + Z_1$ above the reference plane wherever the hole is located. Similarly if a gauge tube is connected to section 2, the level in this tube will be $p_2/\rho g + Z_2$ above the reference plane.

Thus the difference of level in the gauge tubes, or differential head, will be given by h m where:

$$h = \left(\frac{p_1}{\rho g} + Z_1\right) - \left(\frac{p_2}{\rho g} + Z_2\right) \qquad (3.7)$$

But from equation (3.6) we have:

$$\left(\frac{p_1}{\rho g} + Z_1\right) + \frac{V_1^2}{2g} = \left(\frac{p_2}{\rho g} + Z_2\right) + \frac{V_2^2}{2g}$$

or

$$\left(\frac{p_1}{\rho g} + Z_1\right) - \left(\frac{p_2}{\rho g} + Z_2\right) = \frac{V_2^2}{2g} - \frac{V_1^2}{2g}$$

or

$$h = \frac{V_2^2}{2g} - \frac{V_1^2}{2g} \qquad (3.8)$$

or

$$2gh = V_2^2 - V_1^2$$

or

$$V_2^2 - V_1^2 = 2gh \qquad (3.9)$$

Now the volume of liquid flowing along the channel per second will be given by Q m³ where

$$Q = A_1 V_1 = A_2 V_2 \qquad (3.10)$$

or

$$V_1 = \frac{A_2 V_2}{A_1} \qquad (3.11)$$

Now substituting this value in equation (3.9)

$$V_2^2 - V_2^2 \frac{A_2^2}{A_1^2} = 2gh$$

or

$$V_2^2 \left(1 - \frac{A_2^2}{A_1^2}\right) = 2gh \tag{3.12}$$

or dividing by $\left(1 - \frac{A_2^2}{A_1^2}\right)$ equation (3.12) becomes

$$V_2^2 = \frac{2gh}{1 - \dfrac{A_2^2}{A_1^2}} \tag{3.13}$$

or taking the square root of both sides

$$V_2 = \frac{\sqrt{2gh}}{\sqrt{1 - \dfrac{A_2^2}{A_1^2}}} \tag{3.14}$$

Now

$$\frac{A_2}{A_1}$$

is the ratio:

$$\frac{\text{area of section 2}}{\text{area of section 1}}$$

and is often represented by the symbol m and

$$\left(1 - \frac{A_2^2}{A_1^2}\right) = 1 - m^2$$

and

$$\frac{1}{\sqrt{1 - \dfrac{A_2^2}{A_1^2}}}$$

may be written as

$$\frac{1}{\sqrt{1 - m^2}}$$

this is termed the velocity of approach factor, often represented by E. Therefore equation (3.14) may be written

$$V_2 = E\sqrt{2gh} \tag{3.15}$$

and

$$Q = A_2 V_2 = A_2 E\sqrt{2gh} \text{ m}^3/\text{s} \tag{3.16}$$

Mass of liquid flowing per second $= w = \rho Q = A_2 \rho E\sqrt{2gh}$ kg   (3.17)

Or, in terms of differential pressure

$$\Delta p = h\rho$$

$$Q = A_2 E \sqrt{\frac{2g\Delta p}{\rho}} \text{ m}^3/\text{s} \tag{3.18}$$

$$\omega = A_2 E \sqrt{2g\rho\Delta p} \text{ kg/s} \tag{3.19}$$

3.3.1.3 WORKING EQUATIONS

Discharge coefficient

The equations developed above apply to streamlined flow. In practice, the flow is rarely streamlined, but is turbulent. Thus, in addition to its velocity along the channel, a particle of liquid will have a velocity across the stream. The velocities of the particles across the stream will be entirely random, and will not affect the rate of flow very considerably. Thus, if the quantity of liquid entering at section 1 at any instant can be regarded as being equal to that leaving at section 2, the equations may be applied to the flow. Owing to viscosity there will be a tendency for these random motions to be smoothed out.

In developing the equations, effects of viscosity have been neglected. In an actual fluid the loss of head between sections will be greater than that which would take place in a fluid free from viscosity. Thus, the actual mass flow, whether it be turbulent or streamlined, will be less than that given in terms of head by equation (3.17), and the amount by which it is less will depend upon the viscosity of the liquid, and the nature of the channel and of the flow.

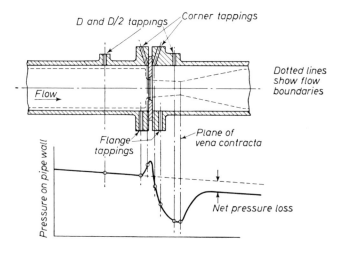

Figure 3.33   Section of square-edged orifice plate showing variation of pressure along the pipe wall. (By courtesy, BS Code 1042, Part 1:1964)

If the area of cross-section at section 2 is less than at section 1, it is usual for the pattern of the flow and the corresponding pressures to be of the form shown in *Figure 3.33*. Owing to the inertia of the fluid particles the area of cross-section of the stream continues to contract even after the fluid has passed through the orifice plate. Thus, the actual area of the stream at a pressure represented by p_2 may be less than the area of cross-section of the orifice A_2. The volume- and mass-flows through section 2 will, therefore, be less than those given by equations 3.16 and 3.17. In order to correct for this contraction and for other effects such as those due to viscosity and the positioning of the pressure tappings, another factor is introduced into the equations for flow.

This factor is the discharge coefficient, C, and is given by the equation

$$\text{Discharge coefficient, } C = \frac{\text{Actual mass rate of flow}}{\text{Theoretical mass rate of flow}}$$

or, if the conditions of temperature, density, etc., are the same at both sections, it may be written in terms of volume,

$$C = \frac{\text{Actual volume flowing}}{\text{Theoretical volume flowing}}$$

The area of cross-section of the jet of fluid passing through an orifice plate is about 0·6 of the area of the orifice, so that the discharge coefficient for an orifice plate has a value of about 0·6. In a venturi tube or nozzle the lines of flow closely follow the contour of the primary device and the area of the stream at the downstream tap is the area of the throat so that the discharge coefficient is close to unity.

Determination of discharge coefficients

In certain cases it is possible to determine the discharge coefficient experimentally, by measuring the actual volume of liquid discharged through the meter in a given time by means of a weighing tank or a volumetric tank. At the same time, the indication of the meter is noted. As all the factors in the equation, with the exception of C, are known, C can be calculated. This procedure can only be adopted in a limited number of cases, and under strictly limited conditions. Fortunately, individual calibration is not very often necessary, as recognised coefficient values are now known for certain definite and reproducible forms of restrictions. The discharge coefficient depends upon the Reynolds number for the installation, and upon other factors.

Dynamic similarity: Reynolds number

If one wishes to estimate the flow through one installation by comparing it with the flow through a geometrically similar installation, the flow patterns must be similar in both installations. In these circumstances, the systems are said to be dynamically similar, and the paths traced by corresponding particles in the systems are geometrically similar, i.e. are in the same scale ratio as is involved for the two installations.

Two installations will be dynamically similar if the Reynolds number is the same for both. Reynolds number R_d depends upon

1. The diameter at the 'throat' of the installation 'd'.
2. The mean velocity 'V' corresponding to d.
3. The density of the fluid 'ρ'.
4. The absolute viscosity 'η'.

and is defined by the equation

$$R_d = \frac{dV\rho}{\eta} \tag{3.20}$$

R_d is dimensionless, and the numerical value will be the same for any given case provided the factors are expressed in self-consistent units.

Provided the value of Reynolds number is above a certain value, known as the 'limiting value', it is found that the flow coefficients remain substantially constant for standardised nozzles and orifice plates, i.e. they are unaffected by the velocity, density and viscosity of the flowing liquid. This limiting value is about 40,000 for the nozzle, 20,000 for an orifice and even as low as 50 for specially designed floats in area-meters.

Thus, if the 'coefficient of discharge' is determined for one installation it will be the same for all other dynamically similar installations; and if the value of Reynolds number for the installation is above the critical value, it will be the same for all geometrically similar installations. To be geometrically similar, however, the systems must be similar in all details such as similarity of installation, of pressure-tappings and surface conditions in channels and in the throttling device.

This brings out a very important point in the installation and use of a differential pressure device in practice. If the instrument is to read correctly, then the conditions laid down when it was designed must be fulfilled, if this condition of similarity is to be satisfied. The relationship between the pressure tappings and the throttling device must be that for which the instrument was designed. If, during the course of use, the corners of an orifice plate wear, then the flow pattern will change, and so will the coefficient of discharge. If there is an accumulation of dirt or any foreign matter in the pipes, or on, or around, the device, then the flow pattern will change, and so will the effective pipe diameter and the coefficient of discharge. Damage to the orifice plate, venturi tube or whatever primary element is used, burring of the pressure taps, or allowing the pressure taps to protrude into the pipe, can upset the effective coefficient of discharge of the instrument to a very marked degree. This is because they all tend to change the flow pattern within the installation from that for which the instrument was designed or calibrated.

For lower Reynolds numbers and for smaller or rougher pipes the basic coefficient is multiplied by a correction factor Z whose value depends upon the area ratio, the Reynolds number and the size and roughness of the pipe. Values of the basic discharge coefficient and the correction factor Z for each type of device and position of pressure tappings are given in the appropriate tables in B.S. 1042:Part 1:1964.

Consider a practical differential pressure-producing device having the following dimensions:

Internal diameter of upstream pipe	D mm
Orifice or throat diameter	d mm
Pressure differential produced	h mm water gauge
Density of fluid at the upstream tapping	ρ kg/m³
Dynamic viscosity of fluid at the upstream tapping	μ poise
Absolute pressure at upstream tapping	P bar

Then introducing the discharge coefficient C, correction factor Z, and the numerical constant, the equation for quantity rate of flow $Q\ \mathrm{m^3/h}$ becomes

$$Q = 0.012\,52\ CZEd^2\sqrt{\frac{h}{\rho}} \qquad (3.21)$$

and the weight or mass rate of the flow $W\ \mathrm{kg/h}$ is given by

$$W = 0.012\,52\ CZEd^2\sqrt{h\rho} \qquad (3.22)$$

Modification of flow equations to apply to gases

Gases differ from liquids in that they are compressible. If the gas under consideration can be regarded as an ideal gas (and most gases are ideal at temperatures well away from their critical temperatures and pressures), then the gas obeys several very important gas laws. These laws are stated in the pages that follow.

Dry gases

1. *Boyle's law.* The volume of any given mass of gas will be inversely proportional to its absolute pressure provided the temperature remains constant, e.g. if a certain mass of gas occupies a volume v_0 at an absolute pressure p_0 and a volume v_1 at an absolute pressure p then

$$p_0 v_0 = p v_1 \quad \text{or} \quad pv = \text{constant}$$

or $\qquad\qquad\qquad v_1 = \dfrac{v_0 p_0}{p} \qquad\qquad\qquad (3.23)$

   i.e., if the absolute pressure on a gas is doubled its volume is halved, provided the temperature remains constant.
2. *Charles' law.* The volume of a given mass of gas at constant pressure is proportional to its absolute temperature, e.g., if a certain mass of gas occupies a volume v_1 at a temperature T_0 kelvin, then its volume v at T° Kelvin is given by

$$\frac{v_1}{T_0} = \frac{v}{T} \quad \text{or} \quad v = \frac{v_1 T}{T_0} \qquad (3.24)$$

   i.e., if the temperature Kelvin of a gas is doubled its volume is doubled provided the pressure remains constant.
   The temperature Kelvin may be obtained from the temperature in $^\circ$ Celsius by adding 273.15
   It is often convenient to use Charles' law in another form. If instead of the gas being allowed to change in volume, the volume is kept constant; the pressure will change and Charles' law may be written: The pressure of a given mass of gas at constant volume is directly proportional to its absolute temperature.
3. *The ideal gas law: general case where pv and T change.* Suppose a mass of gas at pressure p_0 and temperature T_0 Kelvin has a volume v_0, and the mass of gas at pressure p and temperature T has a volume v; then if the change from

the first set of conditions to the second set of conditions takes place in two stages:

(a) Change the pressure from p_0 to p at constant temperature T_0. Let the new volume be v_1.
From Boyle's law

$$p_0 v_0 = p v_1 \quad \text{or} \quad v_1 = \frac{v_0 p_0}{p}$$

(b) Change the temperature from T_0 to T at constant pressure p.
From Charles' law

$$\frac{v_1}{T_0} = \frac{v}{T} \quad \text{or} \quad v_1 = \frac{v T_0}{T}$$

Hence, equating the two values of v_1

$$\frac{v_0 p_0}{p} = \frac{v T_0}{T}$$

or

$$\frac{p_0 v_0}{T_0} = \frac{p v}{T} = \text{constant} \tag{3.25}$$

If the quantity of gas considered is one mole, i.e. the quantity of gas which contains as many molecules as there are atoms in 0·012 kg of Carbon 12 (nuclide ^{12}C), this constant is represented by R the gas constant and equation 3.25 becomes

$$p v = R_0 T \tag{3.26}$$

where $R_0 = 8·314 \text{ J/mol K}$ and p is in N/m^2 and v is in m^3.

Adiabatic expansion. When a gas is flowing through a primary element the change in pressure takes place too rapidly for the gas to absorb heat from its surroundings. When it expands owing to the reduction in pressure it does work, so that if it does not receive energy it must use its own heat energy, and its temperature will fall. Thus, the expansion which takes place owing to the fall in pressure does not obey Boyle's law, which applies only to an expansion at constant temperature. Instead, it obeys the law for adiabatic expansion of a gas *viz*:

$$p_1 v_1^{\gamma} = p_2 v_2^{\gamma} \quad \text{or} \quad p v^{\gamma} = \text{constant} \tag{3.27}$$

where γ is the ratio of the specific heats of the gas

i.e.

$$\gamma = \frac{\text{Specific heat of a gas at constant pressure } (C_p)}{\text{Specific heat of a gas at constant volume } (C_v)}$$

and has a value of 1·40 for dry air and other diatomic gases, 1·66 for monatomic gases such as argon and helium, and about 1·33 for triatomic gases such as sulphuretted hydrogen, carbon dioxide, sulphur dioxide, etc. The value of γ is most conveniently found from the velocity of sound in a gas. Sound vibrations in a gas take place so rapidly that the pressure changes which accompany them are adiabatic, and therefore the velocity of sound in a gas depends upon γ for the gas.

Thus, if a fluid which is being metered cannot be regarded as being incompressible another factor is introduced into the flow equations to correct for the change in volume due to the expansion of the fluid while passing through the restriction.

This factor is called the 'expansibility factor' ε and has a value of unity for incompressible fluids. For ideal compressible fluids expanding without any change of state the value can be calculated from the equation

$$\varepsilon = \sqrt{\left(\frac{\gamma r^{2/\gamma}}{\gamma-1} \quad \frac{1-m^2}{1-m^2 r^{2/\gamma}} \quad \frac{1-r^{(\gamma-1)/\gamma}}{1-r}\right)} \tag{3.28}$$

where r is the ratio of the absolute pressure at the downstream tapping to that at the upstream tapping.

i.e. $r = p_1/p_2$

or can be obtained from a nomagraph in B.S. 1042: Part 1:1964. Where the value of $r > 0.97$ the expansibility factor for practical purposes may be taken as 1. In order that the working equations may apply to both liquid and gases the factor ε is introduced and the equations become

$$Q = 0.01252\,CZ\varepsilon Ed^2\sqrt{h/\rho} \text{ m}^3/\text{h} \tag{3.29}$$

and

$$W = 0.01252\,CZ\varepsilon Ed^2\sqrt{h\rho} \text{ kg/h} \tag{3.30}$$

$$\varepsilon = 1 \text{ for liquids}$$

Critical flow of compressible fluids

It can be shown theoretically for flow through a convergent tube such as a nozzle, the value of r at the throat cannot be less than a critical value r_c. When the pressure at the throat is equal to this critical fraction of the upstream pressure, the rate of flow is a maximum and cannot be further increased except by raising the upstream pressure. Reduction of the downstream pressure does not reduce the throat pressure and cannot increase the rate of flow. The velocity of the gas is the velocity of sound in the gas for the temperature and pressure prevailing at the throat and variations in pressure cannot be transmitted upstream. The critical pressure ratio r is given by the equation

$$2r_c^{1-\gamma/\gamma} + (\gamma-1)m^2 r_c^{2/\gamma} = \gamma-1 \tag{3.31}$$

The value of r is about 0.5 but it increases slightly with increase of M and with decrease of specific heat ratio. Values of r are tabulated in B.S. 1042: Part 1:1964.

When the throat is followed by a divergent outlet as in a venturi nozzle or venturi tube, critical conditions may exist at the throat when the exit pressure is larger than the critical pressure at the throat owing to the pressure rise in the divergent outlet. In all circumstances, however, the mass rate of flow cannot exceed the maximum determined by the pressure conditions at the throat.

Where the fluid separates from the walls to form a vena contracta, as in the square orifice plate, critical flow does not occur and the rate of flow appears to increase indefinitely as the pressure ratio r decreases.

Critical flow may be used for measuring rate of flow provided a pressure difference of more than half the absolute upstream pressure is acceptable. The method may be used, for example, for measuring the output of a compressor by discharging to the atmosphere.

The basic equation for critical flow is obtained by substituting $(1-r_c)P$ for Δp in equation 3.19, substituting r_c for r in equation 3.28 and inserting the basic coefficient C applicable to flows at high Reynolds number. Thus, provided r is less than r_c, the diameter of the upstream pipe does not enter the calculation and the equation reduces to

$$W = 1 \cdot 252 \, U d^2 \sqrt{\rho P} \text{ kg/h} \tag{3.32}$$

where

$$U = C \sqrt{(\gamma/2) r_c \frac{\gamma - 1}{\gamma}} \tag{3.33}$$

For any given specific heat ratio U is found to be virtually constant if the area ratio is less than 0·4 and the value is given graphically in B.S. 1042: Part 1:1964.

The volume rate of flow (in m³/h) is obtained by dividing the weight-ratio of flow by the density (in kg/m³) of the fluid at the reference conditions.

Departure from gas laws

At room temperature and at absolute pressures less than 10 bar most common gases except carbon dioxide behave sufficiently like an ideal gas that the error in flow calculations brought about by departure from the ideal gas laws is less than 1 per cent. In order to correct for departure from the ideal gas laws a deviation coefficient K is given in B.S. 1042: Part 1:1964 which is used in the calculation of densities of gases where the departure is significant. For ideal gases $K = 1$.

Wet gases

This modification of the equation applies to dry gases. In practice, however, most gases are wet, so that they are really a mixture of pure gas and water vapour. Thus, if a gas contains water vapour, its pressure will be partly the pressure of the gas and partly the pressure of the water vapour present. The partial pressure due to the saturated water vapour does not obey Boyle's law. In fact, it is independent of the volume, and depends only upon the temperature. If the gas is in contact with water, and its temperature rises, water will vaporise until the new pressure of the water vapour present is the saturation vapour pressure for the new temperature. On the other hand, if the pressure on a certain volume of gas saturated with water vapour is increased, the pressure of the gas will increase, and so will the partial pressure of the water vapour, but only temporarily, since water vapour will condense to reduce the partial pressure of the vapour to the saturated vapour pressure for the existing temperature. The saturated vapour pressure of water is given in the graph in *Figure 3.34*.

The presence of water vapour in the gas affects the rate of flow corresponding to a given differential pressure in two ways: (1) The density of the moist gas may differ from that of the dry gas, and (2) the gas being measured is only a portion of the mixture passing through the primary element.

If the gas is saturated with water vapour, a correction can be made in its density for the water vapour present. Whereas, in Imperial units, standard conditions for measuring the volume of wet gas existed no such agreement has been reached for expressing gas volume in SI units. It is therefore essential to state the volume and the conditions of temperature, pressure and humidity under which the gas was measured.

If we wish to express the weight of dry gas at 15° C and 1013·25 m bar (corresponding to standard atmospheric pressure or 760 mm Hg) although the gas was measured in a saturated condition a density correction can be made.

Density if dry air at 1013·25 m bar and 15° C $\quad$ = 1·2256 kg/m
Specific gravity of dry gas relative to air $\quad$ = δ
Partial pressure of saturated water vapour at 15° C = 17·05 m bar
Partial pressure of gas in saturated gas $\quad$ = 1013·25 − 17·05 m bar
$\qquad\qquad\qquad\qquad\qquad\qquad\qquad\qquad$ = 996·2 m bar

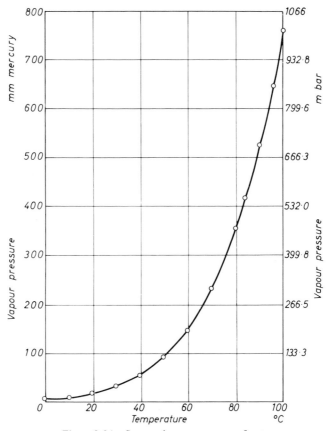

Figure 3.34   Saturated vapour pressure of water

Weight of dry gas in m³ of saturated gas = *Sp Gr* of gas × weight of 1 m³ air

$$\times \frac{\text{partial pressure of gas}}{\text{total pressure of gas}}$$

$$= \delta \times 1\cdot2256 \times \frac{996\cdot2}{1013\cdot25} = 1\cdot205 \, \delta \, \text{kg} \qquad (3.34)$$

The density of any dry gas under any particular condition of temperature and pressure may be obtained from the density under standard conditions by applying the gas equation 3.25.

If the gas is not saturated, then its actual state must be found. The actual state of humidity of a gas can be expressed in two ways. The relative humidity may be given. This is defined as the ratio of the mass of water vapour present in a given volume of gas, to the mass of water vapour necessary to saturate the same volume of gas at the same temperature. If it is assumed that unsaturated vapours obey Boyle's Law then the relative humidity may be written as the ratio

$$\frac{\text{pressure exerted by the vapour}}{\text{saturation vapour pressure at the same temperature}}$$

The relative humidity may be found by determining the dew point for the gas. This is the temperature at which water vapour in the form of dew is deposited on a cooled surface. The dew point determined, the relative humidity may be found, for:

$$\text{Relative humidity} = \frac{\text{pressure exerted by vapour}}{\text{saturation vapour pressure at working temperature}}$$

$$= \frac{\text{saturation vapour pressure at the dew point}}{\text{saturation vapour pressure at the working temperature}}$$

These saturation vapour pressures are obtained from the graph relating vapour pressure of water to its temperature *Figure 3.34*.

Thus, if the temperature, the absolute pressure at the upstream tapping, and the state of humidity of the gas are known, a correction factor can be worked out and applied to obtain the actual mass of gas flowing.

If the temperature is T Kelvin and the pressure P m bar at the upstream tapping and the partial pressure of the water vapour p_v m bar are known the gas density ρ is given by the equation

$$\rho = 6\cdot196 \left[\delta \frac{(P-p_v)}{KT} + \frac{0\cdot622}{T} p_v \right] \text{kg/m}^3$$

Where K is the gas law deviation for the gas at a temperature T, and 6·22 is the density of water vapour relative to that of dry air and applies in practice in the temperature range 0–50° C.

For dry gas p_v is zero and the equation becomes

$$\rho = 6\cdot196 \frac{\delta P}{KT} \text{kg/m}^3 \qquad (3.35)$$

Choice of suitable dimensions

In designing a primary element for a particular job, it is usual to find the maximum flow to be measured. A pitot static tube is often very useful for this purpose. The differential pressure which may be obtained without introducing either corrections for compressibility of the measured fluid, or too great a permanent loss of pressure head is then decided upon.

When a fluid has passed through a differential pressure device, the reduction of pressure is partially restored, but the pressure recovery is never complete. This loss of pressure across the primary element must be kept low, and, in the case of fluids flowing under a small head of pressure, it often determines the type of pressure differential device to be used. The pressure loss curves for nozzles, orifices and venturi tubes are given in *Figure 3.35*.

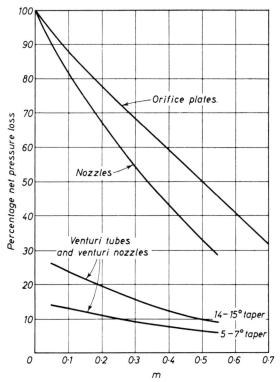

Figure 3.35   Net pressure loss as a percentage of pressure difference. (By courtesy, BS Code 1042: Part 1:1964)

When details such as the size of the main, the type of primary element to be used, the range of the secondary element, and the physical conditions of the measured fluid have been decided, the required throat size of the primary element may be calculated. The method is given in B.S. Code No. 1042: Part 1:1964. The method of successive approximations is used, and all the necessary graphs and tables are available in the Code. Numerical examples are also given. In measurement of gases and vapours an arbitrary rule for the selection of differential pressure ranges is to select a range in mm w.g. which does not exceed 271 times the numerical value of the absolute pressure of the gases or vapours in N/m^2.

3.4 CLASSIFICATION OF PRESSURE DIFFERENTIAL METHODS OF MEASURING FLOW

From the equations it can be seen that the quantity and mass flow in any pressure differential method of measuring flow depends upon several constants, together with the area through which the stream is flowing and the pressure head responsible for the flow. The flow may therefore be measured in several ways.

In the first place the area through which the fluid is flowing may be kept constant and the flow measured by measuring the variation in head. If a constant rate of flow is required, this may be obtained by maintaining both area and head constant. Thirdly, the head may be kept constant, and the area required to maintain the necessary flow measured; or finally, both head and area may be allowed to vary as in an open stream, and, by measuring both, a measure of flow may be obtained. Differential pressure methods will, therefore, be described under the following headings:

Constant area, variable head methods
   Primary elements;
   (a) Pitot static tubes
   (b) Venturi tubes and flow nozzles
   (c) Pitot venturi tubes
   (d) Dall flow tube
   (e) Orifice plates
   (f) Centrifugal type
   Secondary elements. Methods of measuring pressure differential
Constant area, constant head methods
Variable area, constant head methods
   Gate type of area meter
   Orifice and plug type
   Tapered tube and 'float' type
Variable area, variable head methods
   Weirs
   Flumes
Target flow meters
Integrators
   Mechanical integrators
   Electrical integrators
   Pneumatic integrators

3.4.1 Constant area: variable head methods

3.4.1.1 PRIMARY ELEMENTS

(a) Pitot static tubes

The pitot static tube is an extremely useful device for making temporary measurements of flow and is used a great deal in measuring the velocity of aircraft relative to the air. It actually measures the velocity of the stream at a point, but by traversing the stream and measuring the velocity at several points it is possible to obtain the average velocity of the stream, and hence, by multiplying this by the area of cross-section, the volume flowing.

Principle. If a tube is placed with its open end facing into a stream of fluid, then the fluid impinging on the open end will be brought to rest, and its kinetic energy converted into pressure energy. Thus, the pressure built up in the tube will be greater than that in the free stream by the 'impact pressure' or pressure produced by the loss of kinetic energy. This increase in pressure will depend upon the square of the velocity of the stream. The difference is measured between the pressure in the tube and the static pressure of the stream. The static pressure is measured by a tapping in the wall of the main, or by a tapping incorporated in the pitot static tube itself. The difference between the pressure in the tube and the static pressure will be a measure of the 'impact pressure', and therefore of the velocity of the stream.

Consider the small stream of liquid flowing on to the tip of the pitot tube. Originally it will be moving with a velocity V_1. and finally it is at rest.

In equation 3.8, h, the pressure differential or impact pressure developed, is given by

$$h = \frac{V_2^2}{2g} - \frac{V_1^2}{2g} \quad \text{where} \quad V_2 = 0$$

therefore

$$h = -\frac{V_1^2}{2g}$$

i.e. the pressure increases by

$$\frac{V_1^2}{2g}$$

The negative sign indicates that it is an increase in pressure and not a decrease.

Increase in head

$$h = \frac{V_1^2}{2g} \quad \text{or} \quad V_1^2 = 2gh$$

$$V_1 = \sqrt{2gh} \qquad (3.36)$$

As the whole of the stream flowing on to the end of the tube may not be brought to rest, since some may be deflected around the edge, this value of V_1 may vary from the true velocity.

This variation will depend upon the design of the tube. To compensate for this variation, a coefficient C, called 'the pitot tube coefficient', is intoduced. In a correctly designed tube this coefficient may be unity.

C departs from unity according to the degree of imperfection of the tube and the conditions of use, and the correction can be applied to the individual readings used to obtain an average velocity or as an overall correction to the average value. Where an overall accuracy of ± 1 per cent is required stringent conditions of measurement are required and the reader is referred to B.S. 1042: Part 2:1973. In industrial situations where these conditions cannot be met and a less accurate estimate of flow rate is acceptable some of these conditions may be relaxed but in these circumstances an estimate of the accuracy of measurement is not possible but it will probably be not better than ± 5 per cent and possibly much worse.

Equation 3.36 will therefore become

$$V_1 = C\sqrt{2gh}$$
                                                                      (3.37)

and the flow rate in a circular duct can be calculated from the equation

$$Q = \frac{\pi}{4} D^2 \overline{V}_1$$

where D is the mean diameter in the plane of measurement and $\overline{V}_1$ the mean velocity obtained as described later.

Likewise in a rectangular duct

$$Q = ab\overline{V}_1$$

where a and b are the lengths of the sides of the duct. Details of the location of the traversing points are given in the B.S. mentioned above.

Constructional details. Pitot static tubes may be of two kinds:

(i) Elementary type which consists of a single hole pitot tube with a separate static tapping as shown in *Figure 3.36*. The tube must be stiff enough to withstand the impact of the fluid stream and any vibration set up by the stream.

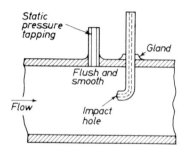

Figure 3.36   Single hole pitot tube

This may be achieved by giving the tube a streamlined section, so that the impact force and the vibration are reduced, and the resistance to bending of the tube in the direction of the stream increased.

(ii) The N.P.L. modified ellipsoidal nosed standard pitot-static tube which is shown in *Figure 3.37* which gives the proportions in terms of the outside diameter of the head. The pitot-static tube consists of two concentric tubes, the inner one transmitting the total pressure (i.e. static and dynamic) while the annular space between the tubes transmits the static pressure. The nose profile consists of two quarter ellipses separated by a distance equal to the diameter d, of the total pressure hole. The major and minor axes of the full ellipses are d and $(d-d_1)$ and the nose length $2d$. The diameter of the head must not exceed 15 mm and the static pressure holes are located $8d$ from the tip of the nose and not less than $8d$ from the axis of the stem.

The diameter d_1 of the total pressure hole is between $0.1d$ and $0.35d$ and coincides with the axis of the head. The static pressure holes are equally spaced around the circumference of the head. They are six in number unless the head is less than 3 mm when the number must not be less than four. The holes must be sharp and free from burrs or dents.

Method of use. If the pitot static tube is to be used as a permanent device for measuring the flow in a main, the relationship between the velocity at the point of its location to the mean or average velocity in the main must be known. In order to find this relationship, two tubes should be used. One tube is located in the main and kept fixed and used as a 'control', to make certain that the rate of flow through the main remains constant while the velocity distribution is being found.

The method of finding the average velocity across the pipe is based on the assumption that the velocity profile follows a logarithmic law in which case the mean velocity passing a cross-section is accurately equal to the numerical average of the local velocities measured at certain specified positions across a traverse. *Figure 3.38* specifies the positions for a ten-point traverse. Experience shows that the log-linear method gives good results even when the velocity profiles do not follow the logarithmic form provided a sufficient number of diameters are traversed. Thus the average velocity is obtained by finding the sum of the individual velocities and dividing the result by ten. Traverses should be made on at least two diameters at right angles. If the results of the two traverses do not agree to within ± 4 per cent of their mean two additional traverses must be made. The four must agree to within ± 10 per cent of the mean to be acceptable.

Precautions in use

(1) The axis of the pitot tube must be parallel to the axis of the pipe and it must be free from vibration.
(2) The fluid shall behave as a single-phase fluid, i.e. be a permanent gas, vapour or liquid and have a velocity of flow between 3 m/s and 30 m/s for gases or vapours, and between 0·1 m/s and 2·4 m/s for liquids.
(3) The length of unobstructed straight pipe upstream of the plane of measurement should be not less than 20 pipe diameters unless a flow straightener is fitted 5 D upstream of the measuring plane when the distance may be reduced to 10 D.
(4) The swirl angle shall not exceed $\pm 5°$ from the direction of the axis of the pipe and there must be no appreciable pressure fluctuations.
(5) The diameter of the head of the pitot static tube must not exceed one-twenty-fifth of the diameter of the duct which shall have a constant diameter to within $\pm 0·5$ per cent.
(6) No sudden changes of diameter or protruberance shall occur for a distance of $\frac{1}{2} D$ downstream or upstream of the measuring plane and the minimum distance of the pitot-static tube from the wall is three-quarters of the tube diameter.
(7) In order to keep the blockage effect to a minimum, a traverse on a diameter should be carried out from two diametrically opposite holes so that the pitot-static tube is never inserted more than half a pipe diameter.

Advantages of the pitot-static tube

(1) It produces no appreciable pressure loss in the main unless it is made large in comparison with the size of the main.
(2) It can be inserted through a comparatively small hole into the main without

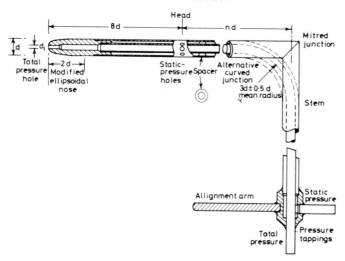

Figure 3.37   N.P.L. modified ellipsoidal-nosed standard pitot-static tubes.

the necessity for shutting down the main. It is, therefore, very useful for estimating the flow through a main in order that a more permanent type of flow measuring device such as an orifice type may be designed for the main.

(3) It can be used to find the distribution of velocities in a main or flue. This is useful, for example, when it is required to site a sampling tube in a flue for waste gas analysis.

(4) Its cost is low.

Disadvantages

(1) Unless elaborate precautions are taken a high degree of accuracy is not obtainable.

(2) The differential pressure produced particularly in gas measurement is small. A manometer sensitive to pressure changes of about 5×10^{-3} mm H_2O will be required to measure the velocity head corresponding to a speed of 3 m/s with an accuracy of ± 1 per cent of the minimum pressure difference to be observed. Suitable manometers are described in section 1.3.1.

Because of this limitation a modified type of pitot tube such as pitot-venturi tube is used and the pitot tube used mainly for exploratory purposes.

The Annubar

For the more permanent type of pitot-tube form of installation an Annubar may be used as shown in *Figure 3.39(a)* and (*b*). The pressure holes are located in such a way that they measure the representative dynamic pressure of equal annuli. The dynamic pressure obtained at the four holes facing into the stream

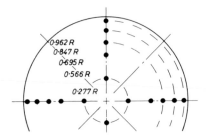

Figure 3.38   Location of measuring points according to 10-pt log-linear rule

is then averaged by means of the 'interpolating' inner tube (*Figure 3.39(b)*) which is connected to the high pressure side of the manometer.

The low pressure side of the manometer is connected to the downstream element which measures the static pressure less the suction pressure. In this way a differential pressure representing the mean velocity along the tube is obtained which enables the flow to be obtained within ±0·55 to ±1·5 per cent of actual flow on the 50 mm to 600 mm sizes for most liquids and gases over the makers specified range. Accuracy for sizes 25 to 40 mm and 650 mm to 4·6 m sizes is ±1·1 to ±2·3 per cent.

(a) *Venturi tubes and flow nozzles*

The Venturi tube was first proposed by Clemens Herschel in 1887. It can have two forms, the conical type and the nozzle type.

The conical type. This type consists of five portions (*Figure 3.38*):

1. The short cylindrical inlet, the same diameter as the upstream pipe line, machined inside, or cast smooth, so that its diameter can be accurately determined. It has a side hole, or several side holes joined to a ring, called a piezo ring, for measuring the static pressure.
2. The entrance cone which includes an angle of 21°, joined to the inlet by a smooth curve.
3. The short cylindrical throat, accurately machined and fitted with a single hole, or several holes and a piezo ring, to measure the pressure at the throat. This is joined to the entrance cone by a smooth curve tangential to the surface of the cone and cylinder. This smooth curve avoids the resistance brought about by a sharp corner, and prevents the liquid from breaking away and not filling the throat completely at high velocities. The throat diameter must be between 0·224 and 0·742 of the entrance pipe diameter, and not less than 19·3 mm. It may have an inspection hole so that the condition of the throat may be examined.
4. The end of the throat leads by an easy curve into the exit cone which includes an angle of from 5° to 15°.
5. The exit cone ends in an outlet flange for connecting the venturi tube to the pipe line. A pressure tapping is sometimes provided at the outlet flange, so that the overall loss of pressure may be measured.

Venturi tubes for pipes of diameter 50 mm or less are usually made of

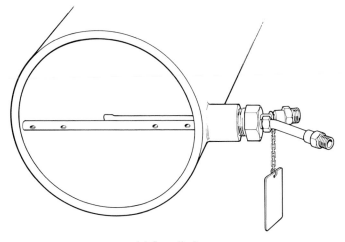

(*a*) Installation.

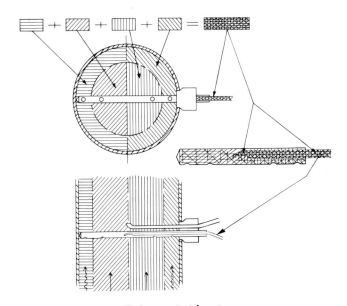

(*b*) Interpolating tube

Figure 3.39   The Annubar. (By courtesy, A. E. D. International Ltd.)

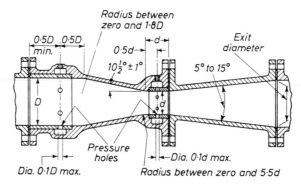

Figure 3.40   Venturi tube. (By courtesy, B.S. Code 1042: Part 1:1964)

brass or bronze and smoothly finished all over inside. Larger venturi tubes are made of cast iron. The straight entrance portion and the throat are sometimes lined with brass, bronze or any non-corrodible material, such as stainless steel, and machined to a smooth finish. Still larger venturi tubes may be constructed in smooth concrete, wood or steel plates, the throat being lined with finished metal. For working pressures up to 14 bar, the venturi tube may be made of cast iron, but for pressures above this the casing may be cast steel or a length of weldless steel tube, with a liner of gun metal or similar material.

The I.S.A. 1932 nozzle type of venturi tube. This tube is in effect a shortened venturi tube. The entrance cone is much shorter and has a curved profile. The inlet pressure tap is located at the mouth of the inlet cone, and the low pressure tap in the plane of minimum section as shown in *Figure 3.41*.

Advantages of the venturi tube. The main advantage is that the overall loss of pressure is less than for nozzles and orifice plates. This is important in some installations where the available head of pressure is small. The loss is usually from 10 to 20 per cent of the differential pressure, and decreases as the size of the venturi throat is increased. At high speeds, however, the total pressure recovery may not be obtained until the fluid has flowed a considerable distance beyond the end of the venturi tube. A venturi tube is useful for measuring the flow of slurries and suspensions of solids in liquids. A piezometer ring should not be used in such applications.

Disadvantages. The main disadvantages are (1) its high initial cost; (2) it cannot be easily installed in an existing pipe line because of its length; and (3) once the tube is manufactured and installed it is impossible to change the range of the flow installation except by modifying the differential pressure-measuring instrument or replacing the venturi tube.

Precautions in installation and use. The entrance portion of the tube must have the same diameter as the adjacent pipe and must be installed concentric with it. No jointing material may be allowed to project inside the pipe, nor may any projection raised by the method of manufacture be permitted. The diameter of the pressure holes at the entrance should not exceed one-tenth of the diameter of the entrance section. The diameter of the pressure holes at the throat should not exceed one-tenth of the throat diameter. No damage, incrustation or corrosion must be allowed within the tube, for this will influence the coefficient of discharge and so upset the calibration of the installation. As with the other primary elements a minimum length of straight run of pipe is required upstream of the venturi.

I.S.A. 1932 flow nozzle. This is in effect a very short venturi tube. The entrance cone is bell-shaped and there is no exit cone. Its shape is shown in *Figure 3.42.* It is usually made of gun metal, stainless steel, or Monel metal. The flange is held between pipe flanges, and the pressure tappings may take the form of annular rings with slots opening into the main at each side of the flange of the nozzle, or of single holes (corner taps) drilled through the flanges of the main close to the nozzle flange. The diameter of the pressure holes or the width of the pressure slot must not exceed $0.03D$ where d/D is equal to or less than 0.67, and $0.02D$ where d/D is greater than 0.67. The minimum permissible size is determined only by the possibility of accidental blockage.

A nozzle may be used in pipelines of internal diameter less than 50 mm provided the throat is not less than 11.4 mm and not greater than $0.742D$. It may also be used at the inlet of a pipe of diameter not less than 50 mm receiving from a large space, or in the outlet of a similar pipe discharging into such a space. It may be used for metering liquids, gases or vapours and for critical flow measurement. When used for measuring compressible fluids the numerical value of the pressure difference in mm w.g. must not exceed 1493 times the numerical value of the absolute upstream pressure in N/m^2a except in critical flow metering where the numerical value of the pressure difference in mm w.g. must be at least 3528 times the absolute upstream pressure in N/m^2a.

It is not suitable for metering viscous liquids.

It is considerably cheaper than a standard venturi tube and may be installed in an existing main without great difficulty.

Owing to the smooth entrance cone, fluids flow more easily through a nozzle than through an orifice so that a smaller value of *m* may be used for a given rate of flow. The nozzle is, therefore, used in high-velocity mains, such as steam distribution mains, where orifice plates of the approved ratios would produce extremely high differential pressures.

Its main disadvantage is that owing to its having no exit cone it produces

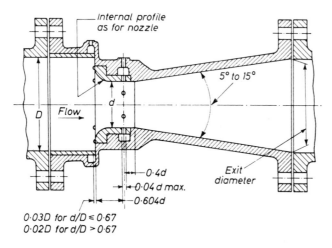

Figure 3.41   Venturi nozzle. (By courtesy, B.S. Code 1042: Part 1: 1964)

a large overall loss of pressure, although this loss is slightly less than that produced by an orifice plate. The value of this loss is indicated in *Figure 3.35*.

(c) Pitot venturi tube

This device combines the pitot and venturi effects and will, when placed in a fluid stream, produce a differential pressure which is much larger than that produced by a pitot static tube under similar conditions. Like the pitot static tube it is essentially a velocity measuring device and it is used in the same way.

Figure 3.43 shows a cross-sectional drawing of the Taylor pitot venturi tube mounted through a flange fitting. The element consists of two concentric venturi tubes arranged so that their openings lie in the same plane, and the inner venturi tube ends at the throat section of the outer venturi tube. The pressure at the throat of the inner venturi tube is measured by means of a slit which communicates with the differential pressure-measuring instrument by means of the inner support tube. The high pressure is measured by means of the impact hole drilled in the front face of the outer support tube. This hole communicates with the differential pressure-measuring instrument by means of the space between the inner and outer tubes; and the tee piece at the bottom of the support tube.

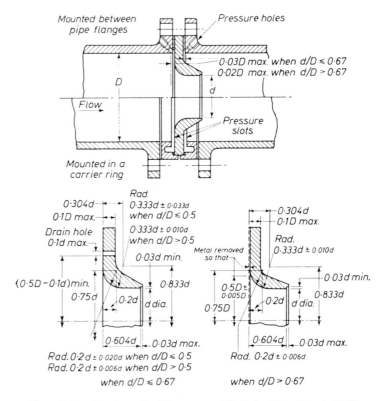

Figure 3.42   Nozzle profile. (By courtesy, B.S. Code 1042: Part 1: 1963)

Fluid passes through both the inner venturi tube and the space between the two venturi tubes. The pressure at the throat of the outer venturi tube will be less than the pressure at its front opening, and this reduced pressure acts at the end of the exit cone of the inner venturi tube. The pressure at the exit cone of the inner venturi tube is, therefore, less than at its entrance so that the fluid passing through it will be accelerated. The difference between the velocity of the fluid at the entrance and at the throat of the inner venturi tube will, therefore, be considerably more than the difference of velocity in the same venturi tube if it were placed as a unit in the main. As the increase of velocity in the venturi tube is larger, there will be, consequently, a greater decrease in pressure in this inner venturi tube than in a simple venturi tube. The pressure at the throat of the inner venturi tube is, therefore, less than the static pressure within the main. The pressure measured at the high pressure hole is, on the other hand, the impact pressure which is larger than the static pressure within the main. The instrument measures the difference between this impact pressure and the reduced pressure at the throat of the inner venturi tube.

A further multiplying effect on the differential pressure is produced by means of the calibrating ring. This ring produces a divergent cone in the fluid stream, causing a fall of static pressure at the exit of the outer venturi tube, thus accelerating the fluid through the tube and creating a still greater differential pressure. By moving the calibrating ring towards the downstream end of the tube its effect upon the differential pressure can be increased. Its position is, therefore, fixed when the element is calibrated, and should not be changed afterwards. Owing to the three-fold multiplying effect, the differential pressure produced by the pitot venturi tube is about 10 times larger than that produced by a simple pitot static tube.

(d) The Dall flow tube

The device is a modified venturi tube which, contrary to what would be expected from theoretical considerations, gives a higher differential pressure but a lower head loss than the conventional venturi tube.

Figure 3.44 shows a cross-section of a typical Dall flow tube. Fluid flowing through the tube first strikes the dam at *a*, flows through a short steep inlet cone to a cylindrical section on each side of the throat slot. Thus the fluid encounters sharp edges at *b* and *c*, traverses the open throat slot, passes two more sharp edges at *d* and *e*, through the short recovery cone having an included angle of 15° and finally undergoes a sudden enlargement to the pipe diameter at *f*. The whole device is about 2 pipe diameters long and has no smooth curves as in the venturi tube or nozzle.

The differential pressure produced by the Dall flow tube, which is the pressure between a tapping immediately in front of the dam and the pressure at the throat, is much higher than that of a corresponding nozzle or venturi tube. *Figure 3.45* shows the ratio of the differential pressure produced by a Dall tube to that produced by a venturi tube of the same diameter ratio.

Contrary to what would be expected, the overall pressure loss is less than that of the conventional venturi tube as can be seen from *Figure 3.46*. The head loss figures are those of the primary element when it is followed by sufficient straight pipe to provide full recovery. This is about 5 pipe diameters. The typical calibration of a Dall flow tube is more like an orifice than that of a venturi tube. *Figure 3.47 (a)* shows the discharge coefficients of a 203·2 mm × 149·1 mm Dall flow tube ($\beta = 0.732$), while *Figure 3.47 (b)* shows the value of the discharge coefficients at various values of the diameter ratio.

As with other primary devices the Dall tube is affected by disturbances upstream and in general requires a greater length of straight pipe upstream than a venturi tube. In addition the discharge coefficient for the Dall tube becomes variable below a Reynolds number considerably higher than that at which a venturi tube coefficient starts to vary. As with other primary elements having pressure tappings it is not suitable for measuring the flow of fluids containing solids which could settle out in the throat slot.

The discharge coefficient is not generally affected by radiusing of the sharp edges. Slight rounding of the sharp edge *a* reduces the coefficient by 0·4 per cent while slight rounding of the edge at *c*, or the edges downstream of the throat produce no further change in calibration. Rounding the edge *a* to a 1·6 mm radius reduces the discharge coefficient by 1 per cent. When used for measuring gases the expansibility factors may be taken to be identical with those for the sharp-edged concentric orifice.

A Dall orifice, which is a shortened version of the Dall tube, is also available which is 0·3 pipe diameters long and can be bolted between pipe flanges. For general use Dall units are cast in gun-metal but on 450 mm and larger sizes high grade cast iron is used. When required to protect the tube from corrosion it may be Lithcote lined.

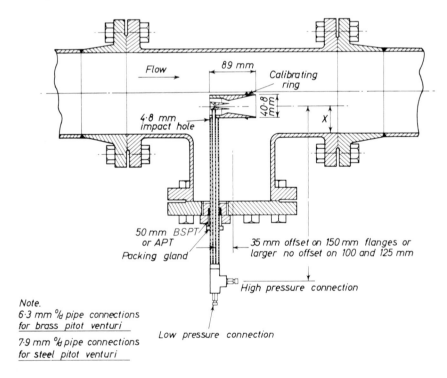

Figure 3.43   Pitot venturi tube. (By courtesy, Taylor Instrument Companies Ltd.)

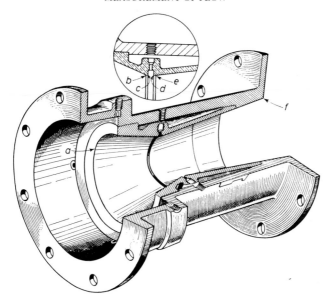

Figure 3.44   Dall flow tube. (By courtesy, Kent Instruments Ltd.)

(e) Orifice plate

The oldest and most common form of pressure differential device is the orifice plate. It was first used by the Romans in Caesar's time as a means of controlling the amount of water delivered into the various channels of water supply systems. It was used much later to meter the flow of water from one large tank into another, but it is only recently that it has been systematically studied as a device for metering fluids in pipes.

In its simplest form it consists of a thin sheet of metal, having in it a square edged circular hole which is arranged to be concentric with the pipe. In the metering of dirty fluids or fluids containing solids, the hole is placed so that its lower edge coincides with the inside bottom of the pipe. This allows the solids to flow through without obstruction. In this case, the pressure taps were placed at a point diametrically opposite to the point where the hole coincides with the

Figure 3.45
Ratio of Dall flow tube differential
to Herschel-type venturi-type differential

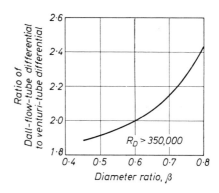

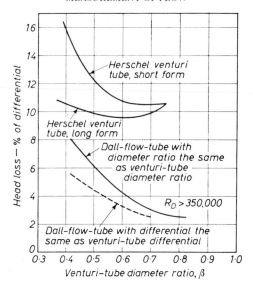

Figure 3.46   Head-loss comparison

pipe, although in the latest practice it is found that this is not necessary, and
they are placed to the side as air entrained in the liquid readily enters impulse
lines at the top of the pipe. The plate is inserted into the main between adjacent
flanges, the outside diameter of the plate being turned to fit within the flange
bolts as shown in *Figure 3.48*.

An orifice place may be used for measuring the flow of gases, liquids or
vapours but for compressible fluids, the numerical value of the pressure drop
in mm w.g. must not exceed 1493 times the numerical value of the absolute
upstream pressure in N/m^2a. The device is not suitable for measuring the flow
of viscous fluids or for critical flow metering. When used with corner or D and
$D/2$ tappings it may be used in all pipelines having an internal diameter of
at least 25 mm provided the orifice diameter is not less than 6·4 mm. When used

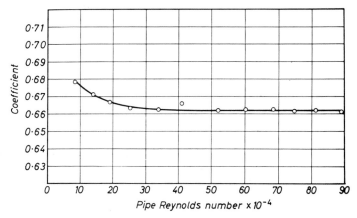

Figure 3.47(a)   Calibration of 203·2 × 149·1 mm Dall flow tube

with flange tappings the minimum internal pipe diameter is 50 mm and the minimum orifice diameter 5 mm. In general it is suitable for measurements for area ratios from 0 to 0·5 and Reynolds numbers of 10 000 upwards for pipe diameters of 25 mm to 50 mm, and area ratios from 0 to 0·7 and Reynolds numbers of 20 000 upwards for pipe diameters of 50 mm and above.

The upstream edge of the orifice must be square and free from burrs or wire edges. The upstream face must also be flat over the whole face and smooth over a circle concentric with the orifice over a diameter not less than $1\frac{1}{2}$ times the diameter of the orifice. The sharp edged form of the orifice plate is chosen because it is the most easily reproduced form, and so the cheapest to produce as no elaborate checking gauges are required. The edge thickness of the orifice must not be greater than $0·1d$ where d/D is less than or equal to 0·2, or $0·02D$ where d/D is greater than 0·2. If the thickness of the plate exceeds this the downstream edge is bevelled as shown in *Figure 3.48*. In general this is not necessary for a thin plate can usually be used as it need only be strong enough to withstand the differential pressure it produces, and not the actual static pressure within the main.

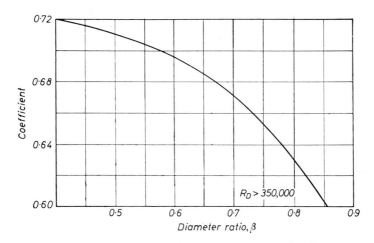

Figure 3.47(b)   Dall flow tube coefficient versus diameter ratio. (By courtesy, Kent Instruments Ltd.)

The plate is usually made of stainless steel, Monel metal or gun metal, although for the measurement of corrosive gases and liquids, glass, plastics, ebonite and other non-metallic materials have been used. It is vital that the material should not corrode in the fluid being metered, otherwise the edge of the orifice will become rounded to a sufficient extent to interfere with the character of the flow and the accuracy of the measurement.

Position of taps. As has already been stated, the area of the fluid stream continues to contract after the stream has left the orifice, and it has a minimum diameter at the vena contracta. The pressure of the fluid will therefore continue to fall after leaving the orifice; the pressure distribution curve is as shown in *Figure 3.33*.

There is a slight fall in pressure in the approach section owing to pipe friction and viscosity of the fluid, and the static pressure is at a minimum about

1 pipe diameter before the orifice plate. The pressure of the fluid then rises comparatively sharply, near the face of the orifice plate. There is then a sudden fall of pressure as the fluid passes through the orifice, but the minimum pressure is not attained until the vena contracta is reached. Beyond the vena contracta there is a rapid increase in the static pressure. Owing to friction and dissipation of energy in turbulence, the maximum downstream pressure is always smaller than the upstream pressure. The pressure loss so caused depends upon the differential pressure and increases as the orifice ratio m decreases for a given rate of flow.

The position of the vena contracta, however, depends upon the velocity of the fluid through the orifice, and upon the ratio m. With decreasing velocity and consequent decrease of the Reynolds number, the vena contracta moves towards the orifice plate. When m is small, the velocity component of the stream towards the centre of the stream will be large, and the jet contraction is a maximum. As m increases, the degree of contraction decreases, and the plane of the vena contracta moves towards the orifice.

The differential pressure obtained with an orifice plate will depend upon the position of the pressure taps. These taps should be located so that:

(1) They are in the same position relative to the plane of the orifice for all pipe sizes.

(2) The tap is located at a position for which the slope of the pressure distribution curve (*Figure 3.33*) is least, so that slight errors in tap location will have less effect on the value of the observed pressure. Positions of least slope are located at the vena contracta, and at 8 pipe diameters downstream of an orifice plate. At 25·4 mm downstream, the position chosen for flange taps, the slope is small except when the value of m is high. A rapidly changing pressure is a sign of very turbulent flow, and the readings of pressure taken in such locations are not as reliable as those taken where the rate of change of pressure is small.

(3) The tap location in the installation is identical with that used in evaluation of the coefficients on which the calculation is based.

The choice of tap location is, therefore, restricted to the standard positions for which adequate experimental values of the flow coefficients are available.

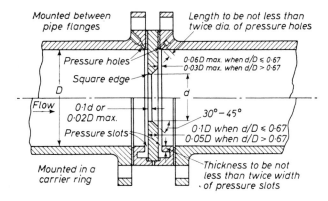

Figure 3.48   Orifice plate with corner tappings. (By courtesy, B.S. Code 1042: Part 1: 1964)

The most common tap locations used in practice are mentioned below.

1. *Flange taps*. These are shown in *Figure 3.49*. Taps are situated 25·4 mm from the upstream and the downstream faces of the orifice plate. Where the pipe diameter is equal to or less than 100 mm the orifice size must be between 0·1 and 0·7D, but for larger pipes the upper limit is increased to 0·748D. They are used universally in American practice and in the oil industry, probably the largest industrial user of the differential-pressure type of meter in this country. They are easier to manufacture than the corner taps, and do not suffer from the troubles associated with corner taps and plate taps when measuring dirty fluids, or when metering in a vertical main. They are incorporated in an integral unit with the orifice plate, and can, thus, be located accurately under workshop conditions by the manufacturer, and are less likely to suffer from faulty location than taps which have to be located by the user.

2. *Corner taps*. Details of these taps are shown in *Figure 3.48*. The pressure holes open into the corner formed by the pipe wall and the orifice plate. This is done by drilling a small hole, or a number of such holes distributed around the circumference of the pipe. A single radial hole is used in the vast majority of installations, and this is connected directly to the pressure pipe. If a number of holes are used, these open out into an annular chamber or piezometer ring, which is connected to the pressure pipe. The hole or holes open exactly into the corner. Alternatively, a circumferential slit may be provided between the end of the pipe and the orifice plate, as shown in the lower half of the figure. This slit leads to a recess in the flange where the pressure connection is made.

The diameter of the pressure hole or width of the pressure slot should not exceed 0·06D where d/D is equal to or less than 0·67, and 0·03D where d/D is greater than 0·67. The diameter of the orifice should not be less than 6·4 mm and not greater than 0·707D where the pipe diameter is less than 50 mm or not greater than 0·837D where the pipe diameter is 50 mm or greater.

3. D and $\dfrac{D}{2}$ taps are located at a distance of $D \pm \dfrac{D}{10}$ from the upstream face of the plate, and $\dfrac{D}{2} \pm \dfrac{D}{20}$ from the downstream face of the plate and are placed as shown in *Figure 3.45*. They should not interfere with the smooth wall of the pipe, or have a diameter of more than $\dfrac{D}{10}$. These taps will be in the position of minimum static pressure on the upstream side of the orifice plate, and at the position of the vena contracta for certain rates of flow and certain orifice ratios so that the differential pressure developed will not be so susceptible to errors due to faulty location. The limitations on the size of the orifice are the same as for corner tappings.

4. Pipe taps are located $2\frac{1}{2}D$ upstream and $8D$ downstream of the orifice plate, and are used for flow-metering fuel gas. These taps require a greater length of straight pipe runs than flange taps, as the necessary length must be measured from the taps and not from the orifice plate.

5. For Integral or Plate taps two radial holes are drilled into the orifice plate. One hole is then opened on to the upstream face of the plate to measure the upstream pressure, by a small hole drilled in the face of the plate. The second hole is opened into the stream on the downstream side of the plate. The main advantage of these taps is that the orifice plate and pressure connections are

all in one unit, but this advantage is very much outweighed by the difficulty experienced in cleaning out the taps when they become blocked.

Advantages. Because of its simple form the orifice plate is easy to manufacture accurately. It is therefore cheap in comparison with the other pressure-differential-producing devices. Its wide use has led to a great deal of information becoming available about its calibration constants.

It can be fitted with a carrier ring containing the pressure taps and inserted in the space between adjacent pipe flanges. If the maximum flow is changed a great deal, then the orifice plate may be replaced by another having a different orifice ratio, and the remainder of the installation used to measure the fluid flow on a new scale.

With D and $D/2$ taps it can be fitted in an existing main where there is no room between the pipe flanges for the carrier ring. In certain applications, where there is solid matter in the measured fluid the orifice may be made to one side of the plate, or the restricting area limited to a segment of a circle only. The bottom of the pipe may therefore be left completely free from obstruction, so that solid matter can move freely by, and not collect and interfere with the orifice. The orifice may be made in the form of a variable area gate, so that for a wide range of flows the differential pressure produced may be kept within the limits set by the pressure-measuring portion of the instrument.

Disadvantages. This type of pressure-differential device produces a loss of about 50 per cent of the differential pressure, and cannot be used where the pressure head producing the flow is already very low. Modern differential pressure meters can, however, be produced with differential ranges as low as 250 mm w.g. for liquids and 25 mm w.g. for gases, so this disadvantage may often be overcome.

When the orifice area ratio becomes greater than 0·55, inaccuracies due to irregularities in the pipe make it impractical to use an orifice plate for precise measurement. A flow nozzle gives a lower differential than an orifice plate for the same throat size, which makes it possible to use a smaller throat for the same amount of flow, thus reducing these inaccuracies. In its simple form, the orifice plate cannot be used with fluid containing extraneous matter which would settle against the faces of the plate. In such cases a venturi tube or a flow nozzle should be used.

Precautions in installing and in use. The actual diameter of a commercial pipe may vary considerably from the nominal diameter. Thus, for large values of the orifice ratio, this may result in a large error in the measurement. It is, therefore, essential to measure the pipe into which the plate is being inserted.

During installation, and in use, nothing must be allowed to damage the sharp upstream edge of the orifice. For this reason, the orifice plate should never be cleaned with emery cloth, owing to the danger of rounding the sharp edge of the orifice. The upstream face should be smooth for a diameter of at least twice that of the bore, and be flat to within 2 per cent of the pipe diameter, when the plate is clamped between the flanges.

When the plate is in use, no extraneous matter must be allowed to collect on or near the orifice plate. Where gases containing tar and other similar substances have to be metered, arrangements are usually made to remove their deposits with a blast of high-pressure steam. In some cases, the orifice plates are constructed of fire brick, so that the main may be burnt out periodically.

When the plate is being installed, gaskets, usually 1·6 mm thick and graphited on the side next to the plate, are used between the plate and the flanges. For superheated steam or other high temperature installations special gaskets are used. The gaskets must never extend into the pipe and should be cut to size when they are being installed. When the flange bolts are being tightened care must be taken to see that a uniform pressure is maintained all around the plate, so that the danger of buckling the plate is eliminated. Great care must also be taken to see that the orifice is concentric with the pipe, particularly when it has a high orifice ratio.

Care must also be taken with the pressure holes. Pressure connection holes in the pipe are usually 12·7 mm diameter for 100 mm and larger pipes, 9·5 mm for 80 mm pipes and 6·4 mm for 50 mm pipes. After drilling, the inside edge of the hole should be rounded off slightly with a reamer or file to be sure that no burrs exist.

Bleed holes should always be provided in orifice plates. In liquid flow measurement the hole should be at the top to allow gas, which would otherwise be trapped by the plate, to pass on with the liquid. In gas and steam flow measurement the hole should be at the bottom to allow liquids to pass; 1·6 mm diameter holes are used for most line sizes.

The measured diameter d_m of the throat of the primary element should be corrected to allow for the additional orifice or throat area represented by the bleed hole of diameter d_h as shown in the following equations:

$$d = d_m \left[1 + 0·55 \left(\frac{d_h}{d_m} \right)^2 \right] \text{ for square edged orifice plates} \qquad (3.38)$$

$$d = d_m \left[1 + 0·40 \left(\frac{d_h}{d_m} \right)^2 \right] \text{ for nozzles} \qquad (3.39)$$

(f) Centrifugal type of meter

A moving body will continue to move at a constant speed in the same straight line unless a force causes it to change its motion.

In order to make a body move in a circle, it is necessary to apply a force towards the centre of the circle. This force is necessary because the direction of motion of the body is constantly changing, i.e. the body has no acceleration towards the centre of the circle. To bring about this acceleration a force must be applied towards the centre of the circle.

If a fluid is flowing through a right-angled bend in a pipe and the bend is in the form of a smooth arc of a circle, then, owing to the tendency of the fluid to continue to move in a straight line, the pressure of the fluid on the pipe on the outside of the bend will be greater than the pressure on the inside. The difference of pressure developed will depend upon the density of the fluid, and upon its velocity. It has been found by experiment that the pressure difference bears a reasonably constant relationship to the mass-rate of flow of the fluid. This relationship may be expressed by the equation

$$w = CA \sqrt{\rho (p_1 - p_2)} \qquad (3.40)$$

where A = Area in m² of the channel cross-section in which the pressure holes are placed. p_1 = pressure (N/m²) at hole in outer circumference of the bend. p_2 = pressure (N/m²) at hole in inner circumference of the bend. w = mass rate of flow in kg per sec. ρ = density of fluid (kg/m³) at the plane of the pressure holes. C = Discharge coefficient, which must be found experimentally for each installation. The difference of pressure is measured by means of two taps which are located in the inner and outer circumferential walls of the bend, on the centre lines of the two legs of the bend, and in the same radial plane, as shown in *Figure 3.51*.

Conditions of use of primary elements

The fluid to be measured should be a continuous single-phase fluid or behave in the same way as such a fluid. Single-phase fluids include gases, vapours, liquids and solutions of liquids, or solids, in liquids. The fluid should not change phase while flowing through the device, i.e. a liquid should not freeze or vaporise, and a vapour should not condense. The minimum permissible diameter of the pipe upstream of a primary element is governed by the type of device and the roughness of the internal surface of the pipe and is indicated in Table 3.1.

Swirls or eddies upstream of the primary element caused by partially closed valves, by regulators, or by combinations of elbows in different planes, will lead to inaccuracy in the flowmeter reading. The flowmeter will usually read low, and the size of the error due to various disturbing influences at various distances in front of the orifice in steam flow and liquid flow measurement is shown in *Figure 3.52*.

In order that a measurement may be made to the accuracy specified in B.S. 1042: Part 1:1964, the pipeline immediately upstream of the primary element should be straight and of constant internal diameter for a distance not less than that indicated in Table 3.2, under the heading (a). In addition the

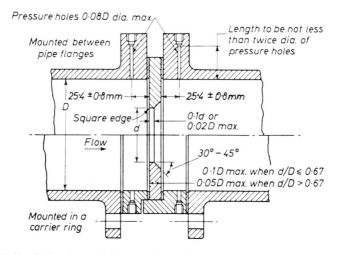

Figure 3.49   Orifice plate with flange tappings. (By courtesy, B.S. Code 1042: Part 1: 1964)

Table 3.1 MINIMUM INTERNAL DIAMETER OF UPSTREAM PIPELINE (B.S. 1042: Part 1:1964)

Type of pipe and internal surface	Minimum internal diameter of upstream pipe, mm					
	Square-edge orifice plate			Conical-entrance or quarter-circle orifice plate	Nozzle or venturi nozzle	Venturi tube
	Corner tappings	D and D/2 tappings	Flange tappings			
Brass, copper, lead, glass, plastics	25	25	50	25	50	50
Steel						
Not rusty						
Cold-drawn	25	25	50	25	50	50
Seamfree	25	25	75	50	50	50
Welded	25	25	100	75	50	50
Slightly rusty	25	25	125	100	50	200
Rusty	50	50	250	200	50	*
Slightly encrusted	200	200	350	*	100	*
Bitumenised						
New	25	25	50	25	50	50
Used	25	25	125	75	50	125
Galvanised	25	25	125	75	50	125
Cast iron						
Not rusty	50	50	250	200	50	50
Rusty	200	200	350	—	50	—
Bitumenised	25	25	125	100	50	50

*Not permissible.

pipeline upstream of the first fitting should be of constant diameter for a distance indicated under the heading (*b*).

The distances given in the table are for primary elements having multiple tappings, and are measured from the upstream tapping. Where a single tapping is used these distances should be doubled or the tolerance on the accuracy of the measurement increased by 0·5 per cent.

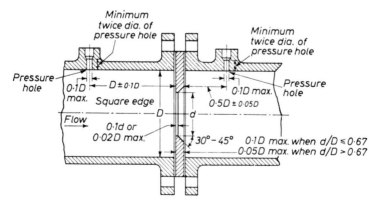

Figure 3.50   *Orifice plate with D and D/2 tappings. (By courtesy, B.S. Code 1042: Part 1:1964)*

Where the upstream line includes fittings producing asymmetrical disturbances and swirling motion, the minimum distance can be reduced by the use of flow straighteners as indicated in Table 3.2, but not in any other cases. Where there is a very strong source of swirling motion such as a tangential or off-set pipe entry upstream of the primary element, flow straighteners may not be effective unless they cause a substantial pressure loss many times greater than that of the primary element. A more effective arrangement consists of two or three perforated plates spaced one pipe diameter apart. The perforations should have a diameter of half the pipe diameter and be uniformly distributed. A straight length of 10 pipe diameters should exist between the last plate and the primary element.

With steam and liquids, the type of flow straightener used is shown in *Figure 3.53*. It is made from a single piece of stainless steel, folded and welded.

The tubular type shown in *Figure 3.54* is made from carbon steel tubes welded together and is used in gas flow measurement. Both types should be securely fixed in the pipe. Spot welding is often used. Since a damaged vane will upset rather than improve meter accuracy, vanes should be accessible for periodic inspection if there is any danger of their being damaged.

Straightening vanes will not improve the accuracy of the reading if the primary element is installed immediately after an enlargement or a contraction of the pipe. Sudden reduction of the pipe diameter without any taper piece will cause the meter to read low. The internal diameter of the downstream

Table 3.2 MINIMUM LENGTHS OF STRAIGHT PIPELINE UPSTREAM OF DEVICE
(B.S. 1042: Part 1:1964)

	Minimum number of pipe diameters								
	(a) Minimum length of straight pipe immediately upstream of device								(b) Minimum length between first upstream fitting and next upstream fitting
Diameter ratio d/D less than:	*0·22*	*0·32*	*0·45*	*0·55*	*0·63*	*0·7*	*0·77*	*0·84*	
Area ratio in less than:	*0·05*	*0·1*	*0·2*	*0·3*	*0·4*	*0·5*	*0·6*	*0·7*	
Fittings producing symmetrical disturbances. Case A. Reducer (reducing not more than 0·5D over a length of 3D. Enlarger (enlarging not more than 2D over a length of 1·5D). Any device having an area ratio not less than 0·3.	16	16	18	20	23	26	29	33	13
Case B. Gate valve fully open (for ¾ closed see Case H)	12	12	12	13	16	20	27	38	10

TABLE 3.2 (*continued*)

Case C. Globe valve fully open (for ¾ closed see Case J)	18	18	20	23	27	32	40	49	16
Case D. Reducer (any reduction including from a large space)	25	25	25	25	25	26	29	33	13
Fittings producing asymmetrical disturbances in one plane									
Case E. Single bend up to 90°, Y-junction, T-junction (flow in either but not both branches).	10	10	13	16	22	29	41	56	15
Case F. Two or more bends in the same plane, single bend of more than 90° swan neck and loops	14	15	18	22	28	36	46	57	18
Fittings producing asymmetrical disturbances and swirling motion									
Case G†. Two or more bends, elbows, loops or Y-junctions in different planes, T-junction with flow in both branches.	34	35	38	44	52	63	76	89	32
Case H†. Gate valve up to ¾ closed‡ (for fully open see Case B).	40	40	40	41	46	52	60	70	26
Case J†. Globe valve up to ¾ closed‡ (for fully open see Case C).	12	14	19	26	36	60	80	100	30
Other fittings									
Case K. All other fittings (provided there is no swirling motion).	100	100	100	100	100	100	100	100	50

†If swirling motion is eliminated by a flow straightener installed downstream of these fittings they may be treated as Cases F, B and C respectively.

‡The valve is regarded as three quarters closed when the area of the opening is one quarter of that when fully open.

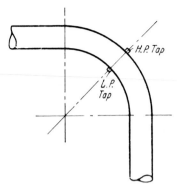

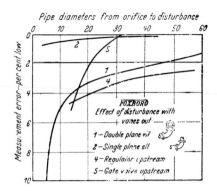

Pipe diameters from orifice to disturbance

Figure 3.51
Centrifugal type meter. Tap location

Figure 3.52
Effect of disturbance on orifice meters.
(By courtesy, Foxboro-Yoxall Ltd.)

pipe should be the same as the upstream pipe. It should be straight and free
from fittings for a distance of 5 diameters, where the area ratio is equal to or
less than 0·4, or 7 pipe diameters for an area ratio greater than 0·4.

Flowmeter pipe installations. For any flowmeter to measure accurately, it is
essential that the pressure differential at the meter end of the pressure lines
should be a true reproduction of the differential pressure set up by the primary
element. The fluid in both pipes must, therefore, be identical, and contain
nothing which would obstruct the pressure reading, or set up a false differential.
The two pressure pipes must be maintained at the same temperature so that
the density of the fluid they contain is the same in both. It is essential to give all
pressure lines a slope of at least 1 in 12. In the case of liquid filled lines, this
causes bubbles of trapped gas to rise to the highest point in the line, where it is
collected. From time to time the vent at the highest point is opened and the gas
allowed to escape, so preventing 'air locks' in the lines.

In the same way any condensed vapour or other liquid, together with any
solid which may be present, may be removed from gas pressure lines by collecting it in a moisture sump at the lowest point in the pipe lines. This sump should
be drained from time to time. In this way liquid, which being much denser
than gas would cause a large false differential, is removed from the lines, and
errors avoided.

When a flowmeter is used on steam mains, a condensing chamber should
be fitted to each tapping. Steam entering these chambers is converted into
water, and the condensing chambers are arranged so that there is a constant
and equal level of water in each under all conditions. The area of horizontal
cross-section of the chambers must be large in comparison with the area of
cross-section of the meter limbs, so that the level of liquid in the chamber does
not change appreciably when the level of mercury in the meter falls.

When the liquids being measured are such that they would corrode the
lines or instrument, vaporise in the pressure lines, or block them, liquid seal
chambers or a purge system must be used, as described in the section on pressure measurement.

Examples of typical pipe layouts are shown in *Figure 3.55 (a–j)*. *Figure
3.55 (k and l)* shows a 3-valve and 5-valve manifold. These are frequently

Figure 3.53
Flow straighteners, one-piece type.
(By courtesy, Foxboro-Yoxall Ltd.)

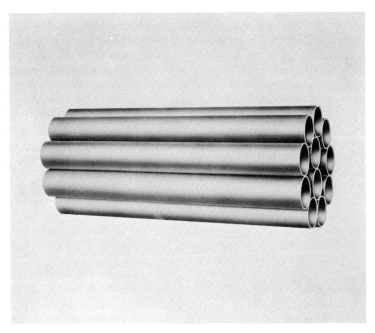

Figure 3.54
Flow straightener, tubular type.
(By courtesy, Foxboro-Yoxall Ltd.)

used in the types of measurement indicated. It is important to see that the pressure line runs for flowmeters are as simple as possible, with the minimum number of fittings which give rise to potential leaks. Right-angled elbows, tees, etc. should be avoided where possible; smooth bends are far preferable.

3.4.1.2 SECONDARY ELEMENTS

Methods of measuring pressure differential and converting it into measurement of flow. The methods of measuring differential pressure have been dealt with in the section on pressure measurement. It will, therefore, be sufficient in this section merely to make some observations on the application of these methods to the measurement of flow.

The differential-pressure measuring portion of an instrument must have a suitable range for the particular operation. It must be capable of withstanding the static pressure of the fluid in the main to which it is applied, and also be capable of retaining its accuracy in the adverse conditions which may be found at the point at which it is installed. As the pressure differential produced may be small, particularly in the measurement of gas flow, the measuring-element must be sensitive to small changes of differential pressure.

The differential pressure produced by any primary element is proportional to the square of the velocity through the throat, so that a choice must be made between having a simple pressure measuring device and a scale which opens out as the flow increases; or having a more complex system, which corrects for the square law and so uses a linear scale. Both methods are in general use, although the simpler measuring device is probably the more common. Compensation for the square law may be achieved in several ways. If a simple U tube is used as a secondary element, one limb may be curved so that equal movements of liquid in the curved limb correspond to equal increments in increase of flow, as shown in *Figure 1.18*. A well type of manometer may be used, and one chamber shaped as described in the pressure measurement section (page 36) and illustrated in *Figure 1.18*.

If a parallel-sided well is used, a displacer may be fixed inside it so as to produce the same variation in cross-section as that in the shaped well. The shaped displacer is more frequently used than the shaped well as it can be turned from the solid and is much easier to manufacture.

With the inverted-bell type of secondary element, correction for the square law may be achieved by shaping the bell so that the variation in the up-thrust due to the displaced liquid compensates for the square law relation. This is the principle of the 'Le Doux' Bell, and is used in the steam flow measuring mechanism shown in section in *Figure 3.56*. It will be noticed that the bell is so shaped internally that as it rises, equal increments of lift cause increasing volumes of the bell to leave the mercury. The guide link is provided to prevent the bell from touching the sides of the mercury chamber. As an additional safeguard against overloads, the bell clamp is arranged to seal off the low pressure connection from the measuring chamber when the high pressure becomes excessive. This prevents mercury being blown out into the low pressure lines and being lost, and so upsetting the calibration.

In the air flow meter shown in section in *Figure 3.57* a similar principle is used, but instead of the bell being shaped, a shaped displacer is used in a

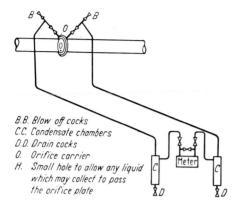

B.B. Blow off cocks
C.C. Condensate chambers
D.D. Drain cocks
O. Orifice carrier
H. Small hole to allow any liquid
   which may collect to pass
   the orifice plate

(a) Clean non-corrosive gases—meter below orifice.

(b) Clean non-corrosive gases—meter above orifice.

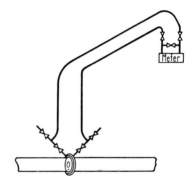

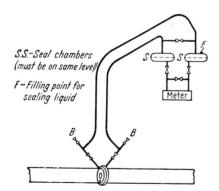

S.S.-Seal chambers
(must be on same level)

F-Filling point for
   sealing liquid

(c) Installation for metering corrosive gas—meter above the orifice. Seals added to prevent corrosive gas from entering the meter body.

Figure 3.55(a–l)   Typical pipe layouts

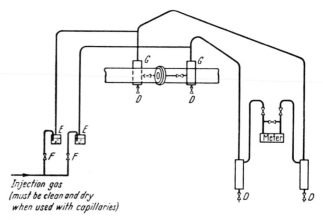

Injection gas
(must be clean and dry
when used with capillaries)

E.E.-Sight bubblers, capillaries, constant differential gas purge valves, or
    rotameter type of restrictor
F.F.-Fine adjustment valves used with sight bubblers only
G.G.-Gas capacity vessels

(d) Installation used for metering dirty or contaminated low or high
    pressure gas or liquid (when used with liquid injection the
    bubblers are replaced by capillaries, catchpots are omitted, and
    capacity vessels inverted and installed at the highest point of
    the main).

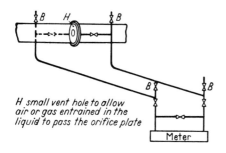

H small vent hole to allow
air or gas entrained in the
liquid to pass the orifice plate

(e) Installation for clean liquids—
    meter below orifice.

(f) Installation for clean liquids—
    meter above orifice (necessary
    pressure must be available to
    maintain the liquid head to the
    meter).

(V.V.–Vapours traps)

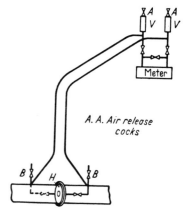

A. A. Air release
cocks

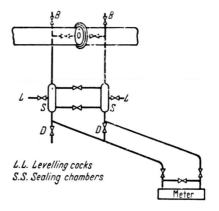

(g) Installation for dirty liquids requiring seals—meter below orifice. Metered liquid less dense than sealing liquid.

L.L. Levelling cocks
S.S. Sealing chambers

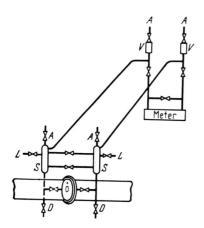

(h) Installation for liquids requiring seals—meter above orifice. Metered liquid more dense than sealing liquid.

(i) Installation for steam or other condensable vapour—meter below orifice.

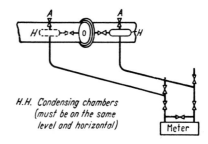

H.H. Condensing chambers
(must be on the same
level and horizontal)

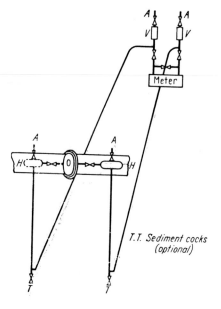

(j) Installation for steam. Pressure of steam in the line must be sufficient to force condensate up to the level of meter which is above the orifice.

T.T. Sediment cocks (optional)

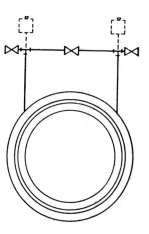

(k) 3-Valve manifold used generally for liquids and low-pressure gases. Vapour traps required when meter is above orifice plate.

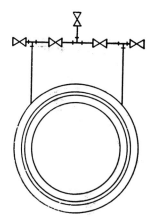

(l) 5-Valve manifold used generally for hazardous or high-pressure gases—permits the venting of either side of meter to atmosphere.

separate mercury chamber. The displacer is carried on a forked arm so that it rises with the bell. This type of instrument is used for measuring the steam-flow/air-flow ratio in boilers, and for this reason has linear charts. It is used to increase the efficiency of the firing of boilers by maintaining a constant steam-flow/air-flow ratio. The optimum value of this ratio is determined by flue gas analysis.

In the ring-balance type of secondary element, square law correction may be achieved by suspending the control weight on a flexible strip which passes over a cam mounted on the container. By suitably shaping the cam, it is possible

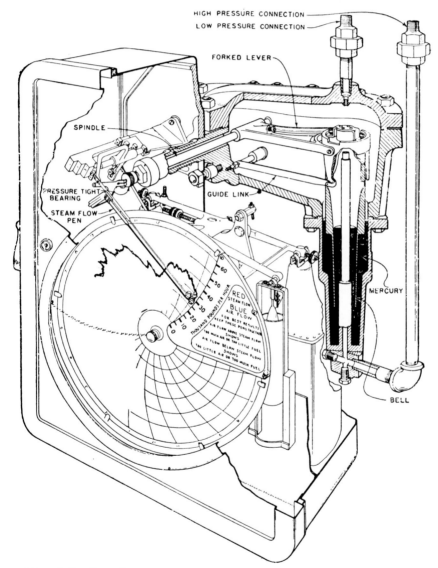

Figure 3.56 Steam-flow meter mechanism. (By courtesy, Bailey Meters & Controls Ltd.)

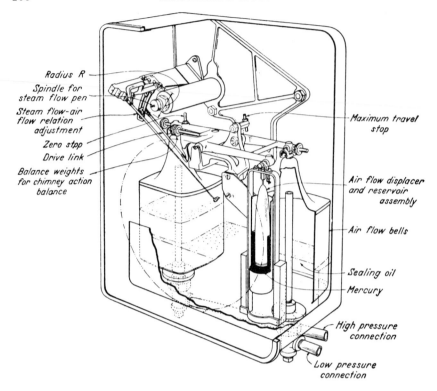

Radius R
Spindle for
steam flow pen
Steam flow-air
flow relation
adjustment
Zero stop
Drive link
Balance weights
for chimney action
balance

Maximum travel
stop

Air flow displacer
and reservoir
assembly

Air flow bells

Sealing oil
Mercury

High pressure
connection

Low pressure
connection

Figure 3.57   Air-flow meter mechanism. (By courtesy, Bailey Meters & Controls Ltd.)

to vary the controlling torque, and so to obtain a linear scale over the whole range.

If the measuring element is not compensated, a cam may be incorporated in the recording or indicating mechanism to correct for the square law.

The compensated type of secondary element has a great advantage when employed with an integrator, for it is rather difficult to integrate the indications of an instrument which is not corrected.

Where a transmitter giving an air signal is used on flow measurement the square root of the signal may be extracted by the instrument shown in *Figure 3.58.* It is based on the cosine relationship between: (*a*) increments of horizontal beam movement, and (*b*) corresponding angles of the vertical beam movement. Within the limited movement permitted, the square root relationship is virtually identical with the cosine relationship. The instrument operates in the following manner:

Motion of the measuring element is directly proportional to differential pressure. This motion is transmitted mechanically to the drive link (1) which moves the horizontal beam (2). Resulting change of vane-nozzle relationship causes air pressure change in booster bellows (3), thus actuating booster relay (4). Pneumatic signal (5) produced by booster relay is also applied to restoring bellows (6) to reposition vertical beam (7). As the vertical beam is repositioned, it restores original vane-nozzle relationship and rebalances transmitter.

Because of the transmitter's design, the vertical beam must move a distance proportional to the square root of the distance moved by the horizontal beam. In this manner, a pneumatic signal corresponding to rate of flow (a function of the square root of differential pressure) is transmitted.

Models suitable for inputs of 0·2–1 bar g or 0·2–1·8 bar g are available, having outputs which may be 0·2–1 bar g or 0·2–1·8 bar g. The air supply is at 2 bar g and the air consumption is 0·003 m³ at stp at balance on dead-end service. The transmitter is calibrated at the factory to give a transmitted signal which is accurate to within ±0·5 per cent of the signal range span.

It is because of this square-law relationship that the usual practice is to design the orifice, or other primary element, to give a suitable differential pressure at the rate of flow usually measured. The scale of the instrument is then calibrated by applying to the secondary element a series of differential pressures equivalent to those produced by the primary element when metering the appropriate rates of flow. The values of these pressures are calculated from the flow equations by substituting the values of the factors appropriate to the installation. This procedure may also be followed if it is required to check the calibration of any secondary element. The differential pressure equivalent to a flow of $\frac{1}{4}$, $\frac{1}{2}$, $\frac{3}{4}$, full scale, as measured on a simple mercury or water manometer, is applied to the measuring element, and its indication compared with the correct indication. If the error is small, it may be corrected by making the appropriate adjustment. If the error is large, the instrument should be carefully

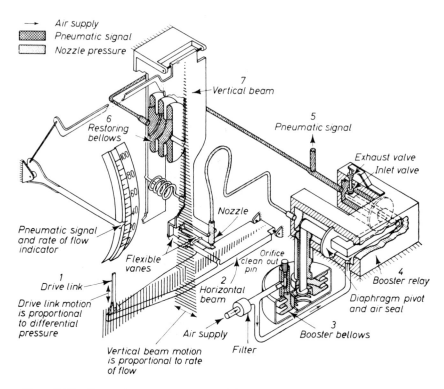

Figure 3.58 Pneumatic square root extractor. (By courtesy, Bailey Meters & Controls Ltd.)

overhauled and the source of the error found. In many cases it is sufficient to check the indication at one point on the scale, usually at about two-thirds of the full scale.

The instrument will not give very accurate indications at the lower end of the scale, whether it is compensated or not. There is a tendency to regard compensated instruments as being more accurate at the lower end of the scale than the non-compensated type, because the scale divisions are larger, but this is not so. The differential pressure at small flows will be small in comparison with that at full scale flows whatever the type of instrument, so that small inaccuracies in the measurement of the differential pressure will produce serious errors in the indication of flow. The differential pressure produced by 1/20th of the maximum flow will be only 1/400th of the differential pressure produced at the maximum flow, so that errors of indication are more likely to occur at this part of the range than higher up the scale. Instruments are usually designed so that they normally measure flows of about two-thirds full scale.

Range of application of types of secondary elements. The tendency at present is for secondary elements to be limited to a few types. The well type or the bellows type is used for high static pressures and larger pressure differentials, particularly for metering liquids with uncompensated indicators. For the compensated instruments, the shaped bell is used for the higher pressure differentials, while the parallel-sided bell with the shaped displacer, is used for the medium and lower ranges. For the lowest ranges, the thin-walled bell or bellows type are used. In some cases the ring balance types are still used, particularly in gas measurement. For the measurement of process liquids, the force-balance type of differential meter is being increasingly used and is described in volume 3.

High static pressure types of ring balance elements are usually made of heavy gauge tube with mercury as a sealing liquid, and are still used for measuring the flow of steam, water, or high pressure gas, for pressures up to 70 bar and pressure differentials up to 2·5 m water gauge.

The low pressure type of ring balance elements are made of light gauge sheet metal, or of a plastic material, with water or oil as the filling medium, and are used in the measurement of air or gas at low pressures of the order of 2 bar or less, with differential pressures up to 100 mm water gauge. The range may be changed in both the high and low pressure type of ring balance by changing the control weight.

Installation and maintenance. In all cases, the maker's instructions concerning the installation and maintenance must be carefully observed. As types of flowmeters are so varied, it is impossible to indicate all the points to be observed in the installation and maintenance of all types, so that the remarks will be limited to the common causes of faults which may be found in many types of installation.

All U tubes must be mounted with their limbs vertical, and this is true of all instruments containing liquids. Loss of liquid by leakage or by evaporation will affect the zero point of many instruments. In instruments containing a shaped well, displacer or bell, it will also influence the square-law compensation. The zero point will also be influenced by contamination of the liquid by condensation in the limbs or the incursion of dirt and other impurities. Deposition of dirt on the walls of the U tube can also affect the reading by influencing the adhesive forces between the liquid and the U tube. If the contamination of the walls of the tube is considerable it will alter the reading of the instrument by altering the cross-sectional area of the tube. Dirt in the range chamber of a recorder will influence the zero point and the calibration. Dirt on a bell or float will also affect the zero point by changing the weight of the bell or float.

Gas trapped in the liquid in the limb of a tube will increase the level of the liquid on the side of the trapped gas, as the density of the gas is considerably less than that of the liquid. One of the commonest faults in flow measurement is the presence of leaks in the lines between instrument and the differential-producing device. The instrument reading may be high or low, depending upon whether the leak increases or decreases the differential pressure. A leak at the meter end of the lines has more effect on the reading than a leak further away from the instrument, owing to the pressure drop produced in the line by the flow towards the leak. A leak of air or gas can usually be heard if it is large, but if it is small it can be detected by painting the lines over with soap solution, when the presence of bubbles will indicate the location of the leak.

Choked connecting pipes are not common, but choking does occasionally occur at the orifice nipples, which are cleaned by blowing or poking through.

Steam installations with cooling chambers are also affected by chokes in orifice nipples. Steam continuously condensing in the cooling chambers causes them to become hot, but they may become excessively hot because of a leaking line or by-pass valve. Gas or air trapped in a liquid-filled connecting pipe causes the instrument to become sluggish, because the gas expands when the pressure falls and contracts when the pressure rises, and so acts as a cushion between primary and secondary elements of the instrument. It also changes the reading because a vertical length of pipe filled with gas will produce less pressure than a corresponding length filled with liquid. Further, the trapped gas may move from an inclined to a horizontal portion of the line as the mercury moves in the meter thus giving rise to a variable head correction at varying parts of the scale. For these reasons gas release cocks are always fitted at the highest points of an installation for venting away the trapped gas.

Liquid in gas-filled, or air-filled, pipes will also affect the instrument reading by the obstruction it causes to the flow of gas, and by the fact that a short vertical column of liquid will produce the same pressure effect as a long column of gas. It may also cause the reading to become erratic because of gas bubbling through the liquid at the lowest point in the line. For these reasons, catch pots are fitted at the lowest point in a gas-filled installation to collect the liquid which is drained off from time to time.

The pen action in a recorder may be sluggish owing to excessive pen pressure on the chart, bending or tightness of the pen spindle, or tightness in the linkage.

Testing. Low pressure gas flowmeters can be checked by disconnecting the lines at the orifice plate, and measuring the pressure differential by means of another instrument such as a U tube manometer. It is very important that the standard against which the meter is checked should be of known reliability and accuracy. A commercial differential meter may have an accuracy of ± 0.25 per cent and the standard used must be more accurate than this. The reading obtained in this way may then be compared with that indicated by the meter. This method cannot, however, be used in many installations, such as those used for the metering of high pressure or medium pressure gas, water, liquor or steam, when the following tests should be used.

1. Isolate the meter and open the by-pass. The meter should now read zero. If it does not, the error may be due to loss of indicating liquid, which should be made up, or renewed if necessary. The pointer may be adjusted if the error is small and a zero adjustment is provided. Zeroing of meters by adjustment of the pointer is not permissible under any circumstances if the instrument contains compensators in the form of displacers or shaped chambers. Error in recorders may also be due to the mechanism sticking owing to excessive pen pressure, which can be checked by pressing the chart plate away from the pen tip. The pen should just touch the chart. Light tapping of the instrument will detect stickiness in the gland linkage at the float chamber, and friction at any of the bearings or links. The magnitude of the error due to backlash is best found by taking 'up and down' readings when calibrating the instrument.

2. Open isolating valves and close the by-pass. The meter should commence to read. Note the speed of rising from zero.

3. Repeat 1 and 2 at least twice, and, except in the case of steam meters, the same reading should be obtained. Steam meters will probably take longer to reach the maximum reading with each successive test. If the meter is slow, or does not rise when the isolating valves are opened, a partial or complete choke is indicated. This is most likely to be found at the orifice plate nipples. Proceed as follows:

    (a) Gas flowmeters. Isolate the meter, open blow-offs at the orifice plate. A good blow will indicate that the orifice shanks are clear. Now open the catch pot drain valves to blow through the lines. If the lines are clear vent gas from the instrument through each line in turn. Examine the lines for kinks where liquid locks may occur.

    (b) Liquid flowmeters. Isolate the meter, isolate at the orifice plate, open the orifice plate blow-off valves. Any blow-out will indicate the presence of gas locks in the lines. Now open the orifice isolation valves to blow through the shanks. If clear, close the orifice plate blow-offs and vent liquid from the instrument blow-off through each line in turn to prove them clear and to remove trapped air or gas. A good stream is required to ensure the line is free from air.

    (c) Steam flowmeters. Proceed as for liquid flowmeters, but, to prove the lines are clear and free from gas-locks, pump water through each line in turn from the instrument to the orifice blow-off. A good, clear, air-free stream proves that the lines are clear and free from gas-locks.

4. Isolate at the orifice plate with the meter valves in the running position when the reading should be stationary. If the reading is not stationary, it indicates the presence of a leak. It may be in the meter by-pass, and it may be located by repeating the test with the meter isolated. If the reading is still not stationary, the leak is on the instrument side of the isolating valves.

The above tests merely check the functioning of the instrument and give no information about its calibration, which must be checked by other means. The state of the orifice plate or other differential-producing device is vital, and the device should be checked from time to time to ensure that it is clean and free from wear or damage.

If straightening vanes are fitted, they also should be checked periodically if there is any danger of their becoming damaged.

When seal chambers are used, the equalising valve at the plant side of the seal chamber must always be opened before the equalising valves at the meter, otherwise sealing liquid will be moved from one side of the meter owing to the movement of the mercury.

3.4.2 Constant area: constant head methods

From the formulae developed at the beginning of this section, it can be seen that, for a particular installation, the flow through the primary element will be constant if the area of the element and the differential pressure across it are kept constant. This principle is used in the funnel meter, the flow prover and the 'constant flow niveau'.

The funnel meter and the flow prover have been used as working standards for testing large-capacity displacement-type gas meters *in situ*. When used in this way they are connected to the outlet of the displacement meter, and discharge into the air.

The funnel meter consists of a series of orifices in a plate in the end of a tank, or in the large end of a funnel. The orifices discharge into the air. The differential pressure across the meter is held constant for any particular test and the rate of flow is controlled by the number of holes that are open, the rest of the holes being closed by rubber plugs. Thus, in any particular test, both the differential pressure and the area are kept constant. The flow characteristics of the meter are assumed to be the same whatever the number of holes open. The general equation for orifice meters can be used to calculate the rate of flow, but it is more usual to calculate the pressure differential to be used from an empirical formula involving the specific gravity and pressure of the gas.

The flow prover is an improved funnel meter. It consists of a pair of orifice-meter flanges fitted with inlet and outlet sections between 8 and 10 pipe diameters long. Between these flanges a number of interchangeable orifices of different sizes may be fitted. The pressure differential across the orifice is measured by a manometer. Each orifice is calibrated, under a differential pressure which is calculated from an empirical formula, with a gas of known density, and the nominal rate of flow established. The meter may then be used to check other gas flowmeter installations.

The niveau in its more usual form is a constant area, variable head meter, and in its usual form it is shown in *Figure 3.59*. It consists of an orifice in the bottom or side of a vessel, and the flow varies with the height of the liquid above the orifice plate. The height of the liquid above the orifice is measured by a sight glass, float, or air-operated depth gauge. The usual equations for orifice meters apply in this case.

In the constant-flow niveau, the installation is modified so as to maintain a constant head of liquid above the orifice, and so maintain a constant flow. This may be arranged by fitting an adjustable overflow, and feeding the instrument with sufficient liquid to maintain a small flow over the overflow.

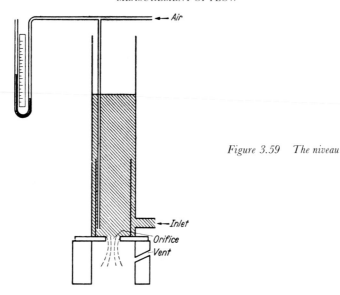

Figure 3.59   *The niveau*

The flow through the orifice will then remain constant as long as the orifice does not become worn or dirty.

3.4.3   Variable area: constant head methods

From the formula $(Q = A_2 E \sqrt{2gh})$ it can be seen that for a particular installation, if the pressure differential is maintained constant and the flow pattern does not change, the rate of flow through an orifice will depend upon the area A_2 of the orifice, and upon the velocity of approach factor E. Thus, if the pressure differential across an orifice is maintained constant by manually, or automatically, adjusting its area, then the area of the orifice will be an indication of the rate of flow.

In the meter is arranged as explained later, so that the rate of flow is independent of the velocity-of-approach factor, then the rate of flow is directly proportional to the area of the orifice. This is a great advantage, for the indication does not involve any square law, which is the source of several difficulties with the constant-area, variable-head meters.

In the earlier forms, variable area meters had to be calibrated for different rates of flow as the discharge coefficient varied with the rate of flow and the Reynolds number. The calibration of a float type of area meter also depended upon the density of the metered fluid. A meter has now been designed in which the accuracy is almost independent of the rate of flow and the viscosity of the fluid it meters. The accuracy may also be made independent of small variations in the density of the metered fluids.

3.4.3.1   GATE TYPE OF AREA METER

In this form of the meter, the area of the orifice may be varied by raising or lowering a gate, either manually, or by an automatically controlled electric motor. The gate is moved so as to maintain a constant pressure drop across the orifice. This pressure drop is measured by two taps in the main, one on each side of the gate, as shown in *Figure 3.60*. The position of the gate is indicated by a scale. As the rate of flow through the orifice increases, the area of the orifice is increased. If all other factors in the flow equation 3·16, except the area A_2 of the orifice, are kept constant, then the flow through the orifice will depend upon the product

$$A_2 E \quad \text{or} \quad A_2 \left(\frac{1}{\sqrt{1 - \left(\dfrac{A_2}{A_1} \right)^2}} \right)$$

As A_2 increases, $\left(\dfrac{A_2}{A_1} \right)^2$ increases, $\left\{ 1 - \left(\dfrac{A_2}{A_1} \right)^2 \right\}$ decreases,

$$\therefore \quad \frac{1}{\sqrt{\left\{ 1 - \left(\dfrac{A_2}{A_2} \right)^2 \right\}}} \text{ increases}$$

∴ The product $A_2 E$ increases at a greater rate than A_2, i.e. the relationship between the flow and A_2 is not strictly linear, but the flow increases more rapidly than the area A_2. Thus, decreasing increments of increased area will be required for equal increments of rate of flow. If the vertical movement of the gate is to be directly proportional to the rate of flow, the width of the opening A_2 must decrease towards the top, as shown in *Figure 3.60*.

The flow through the meter may be made to depend directly upon the area of the orifice, if, instead of the normal static pressure being measured at the upstream tapping, the impact pressure is measured. In order to do this, the upstream tap is made in the form of a tube with its open end facing directly upstream as shown in *Figure 3.61*. It is, in effect, a pitot tube.

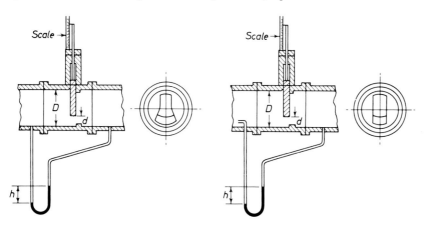

Figure 3.60   Gate type area meter (From 'Flow Meters' by courtesy, American Society of Mechanical Engineers)

Figure 3.61   Gate type area meter corrected for velocity of approach. (From 'Flow Meters' by courtesy, American Society of Mechanical Engineers)

The difference in pressure as measured by the usual upstream and downstream taps is given by equation 3.8, where h is the amount the pressure at up stream tap is greater than that at the downstream tap.

$$h = \frac{V_2^2}{2g} - \frac{V_1^2}{2g} \qquad (3.40)$$

The difference in pressure h_1 between the impact pressure, as measured by the pitot tube, and the normal upstream pressure, as measured by a tap in the pipe, is given by equation 3.34.

$$h_1 = \frac{V_1^2}{2g}$$

where h_1 is the amount the impact pressure is greater than the normal upstream static pressure. Thus, the difference between the impact pressure and the pressure measured at the downstream tap will be h_2 where:

$$h_2 = h + h_1$$

$$= \frac{V_2^2}{2g} - \frac{V_1^2}{2g} + \frac{V_1^2}{2g} = \frac{V_2^2}{2g} \qquad (3.41)$$

Therefore the velocity V_2 through the section A_2 is given by: $V_2 = \sqrt{2gh_2}$, where h_2 is the difference between the impact pressure and the downstream pressure. The normal flow equations for the type of installation shown in *Figure 3.61*, will be the same as for other orifices, but the velocity of approach factor is 1: i.e. the flow through the orifice is directly proportional to the area A_2. The opening at the gate may therefore be made rectangular, and the vertical movement will be directly proportional to the rate of flow.

3.4.3.2   ORIFICE AND PLUG TYPE

This type of meter, shown in *Figure 3.62*, consists of a circular orifice, into which fits a tapered plug of such a form that the area of the annular space

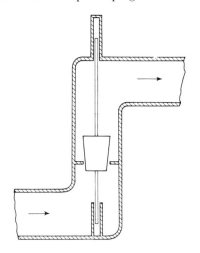

Figure 3.62   Orifice and plug meter. (From 'Flow Meters', by courtesy, American Society of Mechanical Engineers)

between plug and orifice is proportional to the lift of the plug. When a fluid flows past the plug, the plug will rise, and the amount it rises will be a measure of the rate of flow.

Suppose the fluid which is being metered has a density of ρ_e kg/m³ and that the plug has a density of ρ_p kg/m³ and a volume v_p m³. There will be a force on the plug owing to the tendency of the fluid to adhere to the surfaces of the plug parallel to the stream, and owing to the viscosity of the fluid; but in most cases this force will be small and can be neglected. The plug is, therefore, in equilibrium under the action of three forces; its weight, the upthrust of the fluid, and the difference between the forces on its upper and lower surfaces. The pressure on its lower surface will be the impact pressure, p_1 N/m², of the fluid flowing through the annular space, while the pressure p_2 N/m² acting on its upper surface, will be the downstream static pressure. If A_e m² is the effective area of the plug at right angles to the stream, this difference of pressure will produce a force $(p_1 - p_2) A_e$ N in the direction of the stream

∴ Weight of plug = (Pressure difference × effective area) + upthrust on plug.

$$v_p \rho_p g = (p_1 - p_2) A_e + v_p \rho_e g$$

or

$$(p_1 - p_2) A_e = v_p \rho_p g - v_p \rho_e g = v_p (\rho_p - \rho_e) g$$

$$p_1 - p_2 = \frac{v_p}{A_e} (\rho_p - \rho_e) g \tag{3.42}$$

But velocity of the fluid through the annular space is given by equation 3.41.

$$V_2^2 = 2gh_2$$

and

$$h_2 \rho_e g = p_1 - p_2 \tag{3.43}$$

$$\therefore V_2^2 = 2g \frac{(p_1 - p_2)}{\rho_e g} = 2g \frac{v_p}{A_e} \frac{(\rho_p - \rho_e)}{\rho_e} \tag{3.44}$$

Quantity flowing through the annular space is then

$$Q = CA_2 V_2 = CA_2 \sqrt{2g \frac{v_p}{A_e} \frac{(\rho_p - \rho_e)}{\rho_e}} \text{ m}^3/\text{s} \tag{3.45}$$

or, in mass units

$$w = CA_2 \sqrt{2g \frac{v_p}{A_e} (\rho_p - \rho_e) \rho_e} \text{ kg/s} \tag{3.46}$$

where C is the discharge coefficient for the installation. In the simple orifice-and-plug type of primary element, the effective area of the plug is not constant but changes as the area of the annular space increases. The flow, therefore, is not directly proportional to the area of the annular space. The contour of the plug, however, can be determined by experiment, so that the quantity rate of flow is directly proportional to the lift of the plug.

A modified form of this meter is the cone-and-disc type of meter. The primary element is a piston or disc moving in a conical chamber so that the force due to the differential pressure between the two sides of the disc balances the weight of the disc. By suitably shaping the chamber, the flow can be made proportional to the lift of the disc. In the form of this meter shown in *Figure 3.63*, the flow is downwards. The truncated bronze cone 5 is fixed vertically inside the cast iron meter body 4, and water passed down through the cone from the top, moving a flat bronze disc 20 with it. The disc is guided vertically by the stem 19 to which it is attached. When at the top of its travel, the disc exactly fits the small end of the cone. From the top end of the stem, a phosphor-bronze wire 25 passes through a packing gland 26, to the dry chamber above, which contains the recording gear. This wire is connected to a carriage 24 holding a pencil. The carriage is suspended from a flexible phosphor-bronze cord which passes over a pulley 22 and carries a balance weight. The counter-weight is a certain amount heavier than the disc, and this difference in weight is counterbalanced by the difference of pressure on the two sides of the disc, owing to the flow of water through the meter. The disc will move downwards as the flow increases, so that the annular area increases to keep the difference in pressure across the disc constant, and equal to the difference between the counterweight and disc weight, divided by the effective area of the disc.

The movement of the gold-pointed pencil is recorded on a metallic paper chart carried on a cylindrical drum 8, which is rotated at a uniform rate by a spring-driven clock. The metallic paper type of chart is used because it is less susceptible to the effects of moisture than ordinary paper.

This type of meter is used to measure the flow of water into district main water pipes, and, in particular, to measure the flow at night, in order to detect leaks.

Usually the meter is housed above ground on the water authority estate, in a weatherproof metal cabinet; but a portable form is also available.

The meter is protected against damage by excessive flows, by the fact that at very large flows, the disc comes into contact with the bottom cover disc guide pin 7.

3.4.3.3   VARIABLE AREA FLOWMETERS

Although the principle of the float type of variable-area flowmeter has been known for over one hundred years, the development of this type of instrument had to await the appearance of the carbon mandril method of producing heavy-walled, tapered glass tubing of reasonably uniform wall-section and reasonably uniform taper, before the commercial manufacture of large numbers became practicable. Until precision was attained in producing the tapered glass tubing, each meter had to be calibrated over the whole of its scale. In the last 30 years a great deal of work has been done on the rotameter, resulting in far-reaching improvements. The principle is exactly the same as that of the orifice-and-plug type of instrument, and the equations developed for that type of instrument apply equally to this instrument. It consists of a long graduated tube having a uniform taper, usually arranged with the smaller section at the bottom, and the axis of the tube vertical. A 'float', which in its simplest form has the shape of a plumb bob, moves freely within the tube, being maintained centrally on a single central guide or a number of bead guides moulded in the tube.

As the rate of flow through the instrument increases, the float rises in the tube increasing the area of the annular space, and keeping the differential

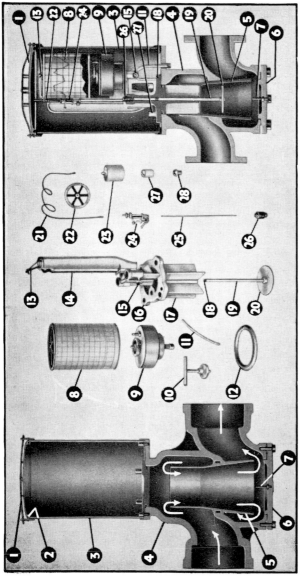

Figure A. Sectional view through socket meter with internal mechanism removed. (Body fitted with blank flange.)

Figure B. Sectional view of assembled meter. (Meter shown with flanged connections.)

1.—Meter box cover
2.—Rubber joint ring
3.—Meter Box
4.—Cast Iron Body
5.—Brass Gauging Cone
6.—Bottom Cover
7.—Bottom Cover Disc Guide Pin
8.—Chart drum
9.—Clockwork
10.—Clock winding key
11.—Drain tube
12.—Hydraulic packing ring (India rubber)
13.—Hinged drum pivot
14.—Standard to carry pulley and guide the pencil frame
15.—Screw for fastening the clock
16.—Clock bracket forming cover to the meter body
17.—Guide blades
18.—Disc cap
19.—Disc stem
20.—Brass disc
21.—Flexible phosphor bronze cord
22.—Balance cord pulley
23.—Balance weight
24.—Pencil carriage
25.—Hard drawn phosphor bronze wire
26.—Stuffing box and gland
27.—Cap nuts
28.—Relief plug

Figure 3.63 Cone and disc meter. (By courtesy, Palatine Engineering Co. Ltd.)

pressure across the float at a fixed value. The discharge coefficient C required in equation 3.45 is defined in the same way as for other differential pressure flowmeters, and will depend upon the flow pattern within the tube. The flow pattern will, to a large extent, depend upon the viscosity of the fluid and the rate of flow within the tube, and for the simple plumb bob float, the discharge coefficient will increase rapidly with increasing Reynolds number up to 7000, and will then remain constant.

The change in the flow pattern brought about by increased velocity is shown in *Figure 3.64a*. In order to reduce the tendency of the liquid to carry the float along with it because of viscous forces, the float is given the form shown in *Figure 3.64b*.

The effect of this is to produce a float which gives the instrument a constant coefficient of discharge for all Reynolds numbers above 300.

Still further improvement is achieved by giving the float the form shown in *Figure 3.64c*, when the flow pattern and discharge coefficient are constant for all flows having Reynolds numbers above 40. The instrument is, therefore, capable of giving a constant calibration for a very large range of flows, and for a large range of viscosities of metered fluid. This makes it possible to produce a series of instruments having a wide variety of ranges without the necessity of individual calibration, as the discharge coefficient for this type of flow remains constant at 0·61. When this float is used, the body of the float, which gives it its weight, is outside the flowing liquid.

Another advantage of the rotameter is that it can be arranged to give an indication in weight units which is independent of small changes in the specific gravity of the metered fluid. This is extremely useful in metering liquids like petrol, when specific gravity may vary through a limited range.

It will be seen from equation 3.46 that the flow measured in mass units depends upon $\sqrt{(\rho_p - \rho_e)\,\rho_e}$

Now if ρ_p is made equal to $2\rho_e$ then $(\rho_p - \rho_e) = \rho_e$

Thus for small changes in ρ_e the product $(\rho_p - \rho_e)\,\rho_e$

will remain substantially the same. Increase in ρ_e will be compensated by a similar decrease in $(\rho_p - \rho_e)$. Thus, in metering petrol of which the specific gravity is about 0·72, if the float is made of a solid whose specific gravity is 1·44 the product will be reasonably constant for values of ρ_e from 9·70 to 0·74 as shown below.

ρ_e	$(\rho_p - \rho_p)\,\rho_e$
0·70	$0·74 \times 0·70 = 0·5180$
0·72	$0·72 \times 0·72 = 0·5184$
0·74	$0·70 \times 0·74 = 0·5180$

By choosing the float material from a wide range of possible materials, floats suitable for metering a large number of commercial liquids can be made.

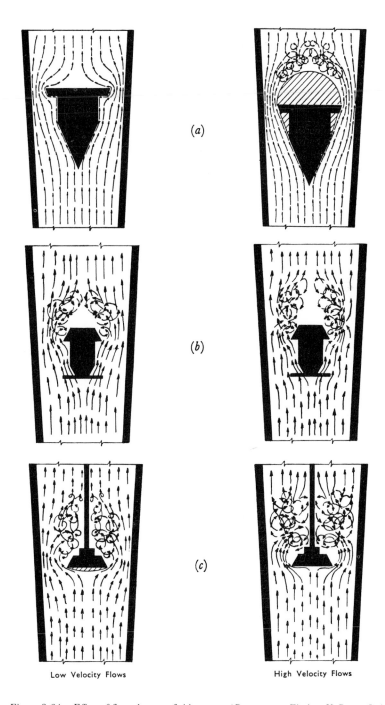

(a)

(b)

(c)

Low Velocity Flows   High Velocity Flows

Figure 3.64   Effect of float shape on fluid stream. (By courtesy, Fischer & Porter Ltd.)

230

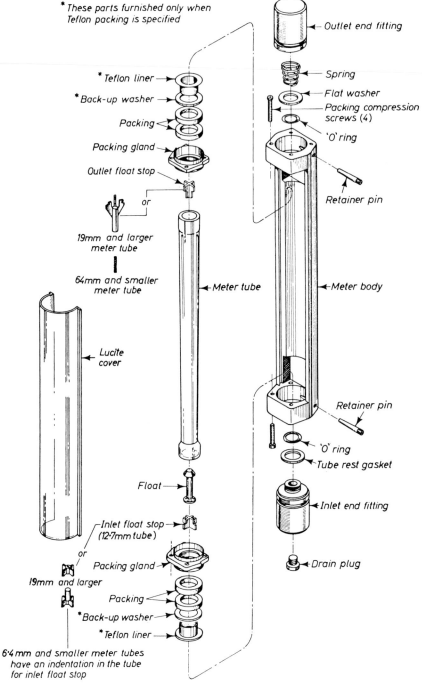

* These parts furnished only when Teflon packing is specified

Outlet end fitting

* Teflon liner
* Back-up washer
Packing
Packing gland
Outlet float stop

Spring
Flat washer
Packing compression screws (4)
'O' ring

or

19mm and larger meter tube

64mm and smaller meter tube

Retainer pin

Meter tube

Meter body

Lucite cover

Float

Inlet float stop (12·7mm tube)

or

19mm and larger

6·4 mm and smaller meter tubes have an indentation in the tube for inlet float stop

Retainer pin

'O' ring
Tube rest gasket

Inlet end fitting

Drain plug

Packing gland
Packing
* Back-up washer
* Teflon liner

Figure 3.65   Variable area meter (exploded view). (By courtesy, Fischer & Porter Ltd.)

To make the quantity indication independent of small changes in the specific gravity of the metered fluid, the fluid must flow downwards, so that the term $(\rho_p - \rho_e)$ becomes $(\rho_e - \rho_p)$; the remainder of the equation remaining the same. If the float is made of a very light material, then

$$\frac{(\rho_e - \rho_p)}{\rho_e}$$

remains reasonably constant for small changes in the fluid density.

Construction of the instrument and its installation. Details of the construction of a rotameter are shown in the exploded view given in *Figure 3.65*. The actual material used in the construction of the instrument depends to a very large extent on the nature of the fluid which it is used to meter. The instrument is extremely useful in metering concentrated acids, chlorine, in either the liquid or gaseous form, and other corrosive liquids and gases. For this purpose, the tube and float are made of glass, and the flow can be measured by reading directly the position of the float in the tube.

Where the liquid being metered is opaque, or the circumstances are such that it is inadvisable to use glass, the tube and the float are made of non-magnetic stainless steel or other suitable material. In this case, it is necessary to show the position of the float on the outside of the tube.

This is achieved by having an extension to the float which carries a magnet as shown in *Figure 3.66*. The position of the internal magnet is picked up by a follower magnet which transmits the position to the pointer.

The accuracy of a rotameter is normally ±2 per cent of the full scale flow

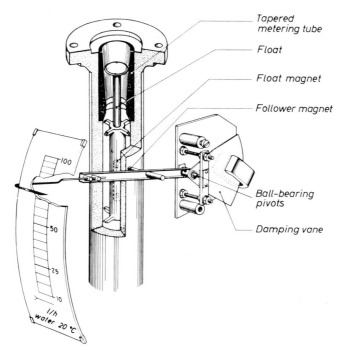

Figure 3.66   Magnetic rotameter. (By courtesy, G.E.C.-Elliot Process Instruments Co. Ltd.)

over a 10:1 range of flows, but other rangeabilities are available, and in-struments having an accuracy of $\pm 2\%$ of the indicated flow are available. A wide range of tube sizes is available, and a large variety of pressure ratings. The flow range depends upon the tube size, the material of the float, the permissible pressure drop and the nature of the metered fluid. A wide range of float materials is available to give the necessary flow ranges, and corrosion and wear resistance in a wide variety of applications. Among the materials used are aluminium, brass, ceramic materials, Fluon, glass, Hastelloy, Korannite, nickel, Monel, Polyethylene, phosphor-bronze, sapphire, stain-less steel, titanium, and tantalum.

Glass tube rotameters having flow ranges of from 30–450 ml/h up to 1·36–13·6 m³/h are readily available. The corresponding flow ranges for air would be approximately 0·001–0·012 m³/h up to 160–1600 m³/h. Glass tube rotameters withstand higher pressures in the smaller sizes than in the larger sizes. When measuring liquids, the smaller sizes may be safely used at pressures up to 32 bar, but the larger sizes are limited to 8 bar. When measuring gases the pressures should be very much lower. In order to protect personnel from hazards associated with glass under pressure, armoured glass protection boxes should be provided on glass rotameters carrying gas at pressure. In all cases the maker's advice should be followed.

Metal tube rotameters, on the other hand, are available for working pressures up to 2000 bar, temperatures up to 300° C, and may have capacities as large as 410 m³/h.

The rotameter is mounted with the measuring tube vertical, because inclining it from the vertical will reduce the effective weight of the float and so introduce errors. As long as the angle of inclination is small, these errors will not be serious, for an inclination of up to 10° to either side of the vertical produces an error of only 1 per cent of the scale indication.

Maintenance and fault finding. The rotameter requires very little servicing as there is so little that can go wrong. Deposits on the float or taper-tube can be seen when the taper-tube is made of glass. In the armoured or metal taper-tube type, the deposits will be seen in the sight glass. These deposits are readily removed by flushing the rotameter with a suitable washing liquid. The rotameter may be tested for sensitivity by noticing its response to small changes of flow brought about by adjusting the process flow control valve. Periodically, the rotameter should be removed from the pipe line, dismantled, cleaned, and examined for corrosion, wear, etc. The weight of the float should be checked to see that it has not been reduced by the abrasive action of the fluid flowing past.

3.4.4   Variable head, variable area, meters

3.4.4.1   THE GILFLO PRIMARY ELEMENT

In order to increase the rangeability of a differential pressure type flowmeter, a meter has been developed in which the differential pressure is measured across an orifice whose area also increases with flow. In this way it is possible to measure flows down to 1/1000 part of the meter maximum with an accuracy better than ± 1 per cent of the meter range when the meter is used in its semi-logarithmic mode. Meters are also available with a linear relationship where the differential pressure produced is a direct analogue of the flowrate from zero to full scale giving a measurable range of 100:1 of the meter's maximum.

A cut-away illustration of the meter is shown in *Figure 3.67*. In the rest position the orifice is closed. When gas or liquid is flowing through the meter the difference between the pressures acting on the upstream and downstream faces of the orifice plate will tend to displace the orifice plate until the differential pressure is balanced by the force produced by the bellows. This moves the orifice relative to the fixed central control member increasing the

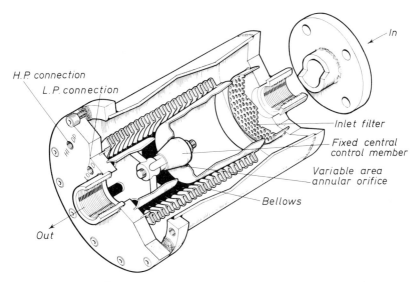

Figure 3.67   Gilflow primary element. (By courtesy, Gervase Instruments Ltd.)

annular area through which the liquid flows. Thus, increase of flow has two effects, increasing the area and increasing the differential pressure, and for any design of instrument a unique relationship will exist between these three parameters producing a stable differential pressure to flow relationship. This relationship is purely a function of the meter design.

From the equation 3.18,

$$Q = A_2 E \sqrt{\frac{2g\Delta p}{\rho}} \ \mathrm{m^3/s}$$

it will be seen that the differential pressure Δp across the orifice will be proportional to Q^2.

The area of the orifice, however, will increase with flow because as the flow increases the bellows is compressed along its axis so increasing A_2. The actual relationship between the flowrate Q and Δp will depend on the shape of the fixed control member, or plug, and the springrate of the bellows, both of which determine the free orifice area. By suitable choice of these factors, the relationship between Δp and flow can be made either linear or $\propto \Delta p^{\frac{3}{2}}$. The $\Delta p^{\frac{3}{2}}$ relationship to flow enables the wide range form of the instrument to be produced such that, if it is accepted that with an orifice plate the differential pressures at 10 per cent of maximum gives a rangeability of 3:1 on flow, then the Gilflo linear primary element will give 10:1 and the wide-range element 30:1. However, the overruling advantage of these devices is that at 1 per cent of maximum differential pressure, the linear primary element will measure 1 per cent of meter maximum flow and the wide range element will measure 0·1 per cent of meter maximum flow.

Gilflo primary elements are available from 2 mm size, having a maximum flowrate of 24 litres/h, up to 600 mm size with a maximum flowrate of 3000 m³/h. In all sizes the differential pressure at maximum flow is 250 mmHg.

3.4.4.2 OPEN CHANNEL METERS

In this type of meter (described in B.S. 3280 Part 4A 1964, where the value of the discharge coefficient will be found) which measures the flow of liquids, usually water, both the head and the area of the stream vary, but the area of the stream will depend upon the head. The meters are of two forms.

3.4.4.2(a) The Weir, which is merely a dam over which the liquid is allowed to flow, the depth of liquid over the sill of the weir being a measure of the rate of flow.

3.4.4.2(b) Hydraulic Flumes, an example being the venturi flume, in which the channel is given the same form in the horizontal plane as a section of a venturi tube, while the bottom of the channel is given a gentle slope up to the throat.

3.4.4.2(a) WEIRS

Weirs may have a variety of forms, and are classified according to the shape of the notch or opening.

The simplest is the rectangular notch, or in certain cases the square notch.

The V or triangular notch, which is a V-shaped notch with the apex downwards, is used to measure rates of flow which may become very small. Owing to the shape of the notch, the head is greater at small rates of flow with this type of notch than it would be for the rectangular notch.

Notches of other forms, which may be trapezoidal, parabolic, or of other forms, are designed so that they have a constant discharge coefficient, or a head which is directly proportional to the rate of flow.

The velocity of the liquid increases as it passes over the weir because the centre of gravity* of the liquid falls. Liquid which was originally at the level of the surface above the weir can be regarded as having fallen to the level of the centre of pressure† of the issuing stream. The head of liquid ˙producing

---

*Centre of gravity. The weight of a body is made up of the weights of the individual particles which make up the body. These weights make up a system of vertical forces. This system of forces may be replaced by a single force called the 'total weight of the body'. This single force produces the same effect in all respects as the system of forces. There is a point (not necessarily in the body) through which this single force acts for all positions of the body. This point is called the centre of gravity of the body.

†Centre of pressure. In the same way as the weight of a body is made up of a system of forces due to the weight of the individual particles, the pressure due to a liquid on a restraining boundary is made up of a similar system of forces. These forces are at every point perpendicular to the boundary.

the flow is therefore equal to the vertical distance from the centre of pressure of the issuing stream to the level of the surface of the liquid upstream.

If the height of the centre of pressure above the sill can be regarded as being a constant fraction of the height of the surface of the liquid above the sill of the weir, then the height of the surface above the sill will give a measure of the differential pressure producing the flow. If single particles are considered, some will have fallen a distance greater than the average, but this is compensated for by the fact that others have fallen a smaller distance.

The term 'head of a weir' is usually taken to mean the same as the depth of the weir, and is measured by the height of the liquid above the level of the sill of the weir just upstream of where it begins to curve over the weir, and is denoted by H usually expressed in metres.

Rectangular Notch

Consider the flow over the weir in exactly the same way as the flow through other primary differential-pressure elements. If the cross-section of the stream approaching the weir is large in comparison with the area of the stream over the weir, then the velocity V_1 at section 1 upstream can be neglected in comparison with the velocity V_2 over the weir, and in equation 3.9, $V_1 = 0$, and the equation becomes:

$$V_2^2 = 2gh, \quad \text{or} \quad V_2 = \sqrt{2gh}$$

The quantity of liquid flowing over the weir will be given by:

$$Q = A_2 V_2$$

But the area of the stream is $B\,H$, where H is the depth over the weir, and B the breadth of the weir, and h is a definite fraction of H.

By calculus it can be shown that, for a rectangular notch,

$$Q = \frac{2}{3} B H \sqrt{2g\,H} \tag{3.47}$$

$$= \frac{2}{3} B \sqrt{2g\,H^3} \ \mathrm{m^3/s} \tag{3.48}$$

The actual flow over the weir is less than that given by equation 3.47 for the following reasons:

1. The area of the stream is not BH, but something less, for the stream contracts at both the top and bottom as it flows over the weir as shown in

---

† *Centre of pressure (continued)*

Figure 3.68   Centre of pressure

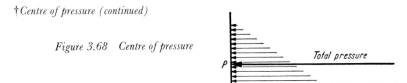

Total pressure

P

The side of a tank is subjected to a system of horizontal forces, the size of the force increasing with the depth as shown in *Figure 3.68*.

This system of forces may be replaced by a single force which produces the same effect as the system, called the 'Total Pressure', and the point through which this force acts is called the 'centre of pressure', P.

Figure 3.69 making the depth at the weir less than *H*.

2. Owing to friction between the liquid and the sides of the channel, the velocity at the sides of the channel will be less than that at the middle. This effect may be reduced by making the notch narrower than the width of the stream as shown in *Figure 3.70*. This, however, produces side contraction of the stream. $(B_1 - B)$ should, therefore, be at least equal to $4H$, when the side contraction is equal to $0 \cdot 1H$ on both sides, so that the effective width becomes $(B - 0 \cdot 2H)$.

When it is required to suppress side contraction and make the measurement more reliable, plates may be fitted as shown in *Figure 3.71* so as to make the stream move parallel to the plates as it approaches the weir.

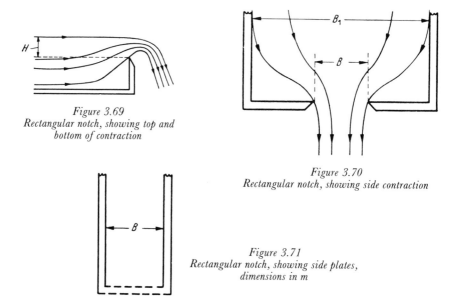

Figure 3.69
Rectangular notch, showing top and
bottom of contraction

Figure 3.70
Rectangular notch, showing side contraction

Figure 3.71
Rectangular notch, showing side plates,
dimensions in m

To allow for the difference between the actual rate of flow and the theoretical rate of flow, the discharge coefficient *C*, defined as before, is introduced, and equation 3.48 becomes:

$$Q = \frac{2}{3} CB \sqrt{2gH^3} \text{ m}^3/\text{s} \tag{3.49}$$

The value of *C* will vary with *H* and will be influenced by the following factors, which must remain constant in any installation if its accuracy is to be maintained; (*a*) the relative sharpness of the upstream edge of the weir crest; (*b*) the width of the weir sill.

Both these factors influence the bottom contraction and influence *C*, so the weir sill should be inspected from time to time to see that it is free from damage.

In developing the above equations it was assumed that the velocity of the liquid upstream of the weir could be neglected. As the rate of flow increases, this is no longer possible and a velocity of approach factor must be introduced. This will influence the value of *C*, and as the velocity of approach increases it will cause the observed head to become less than the true or total head so that a correcting factor must be introduced.

Triangular notch

If the angle of the triangular notch is θ, as shown in *Figure 3.72,*

$$B/2/H = \tan (\theta/2) \qquad B = 2H \tan (\theta/2)$$

The position of the centre of pressure of the issuing stream will now be at a different height above the bottom of the notch from what it was for the rectangular notch. It can be shown by calculus that the numerical factor involved in the equation is now

$$\frac{4}{15}$$

Substituting this factor and the new value of A_2 in equation 3.49

$$Q = \frac{4}{15} C B \sqrt{2g\,H^3} \text{ m}^3/\text{s}$$

$$= \frac{4}{15} C \; 2H \tan\frac{\theta}{2}\sqrt{2g\,H^3}$$

$$= \frac{8}{15} C \qquad \tan\frac{\theta}{2}\sqrt{2g\,H^5} \qquad\qquad (3.50)$$

Experiments have shown that θ should have a value between $35°$ and $120°$ for satisfactory operation of this type of installation.

While the cross-section of the stream from a triangular weir remains geometrically similar for all values of H, the value of C is influenced by H. C varies from 0·57 to 0·64, and takes into account the contraction of the stream.

If the velocity of approach is not negligible the value of H must be suitably corrected as in the case of the rectangular weir.

Installation and Operation of Weirs

1. Upstream of a weir there should be a wide, deep, and straight channel of uniform cross-section, long enough to ensure that the velocity distribution in the stream is uniform. This approach channel may be made shorter if baffle plates are placed across it at the inlet end to break up currents in the stream.

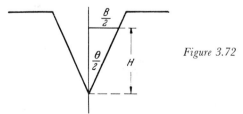

Figure 3.72

2. Where litter is likely to be brought down by the stream, a screen should be placed across the approach channel to prevent the litter reaching the weir. This screen should be cleaned as often as it is found necessary.
3. The upstream edge of the notch should be maintained square or sharp edged according to the type of installation.
4. The weir crest should be level from end to end.
5. The channel end wall on which the notch plate is mounted should be cut away so that the stream may fall freely and not adhere to the wall. To ensure this, a vent may be arranged in the side wall of the channel so that the space under the falling water is open to the atmosphere.
6. Neither the bed, nor the sides of the channel, downstream from the weir should be nearer the weir than 150 mm and the water level downstream should be at least 75 mm below the weir sill.
7. The head H may be measured by measuring the height of the level of the stream above the level of the weir sill, sufficiently far back from the weir to ensure the surface is unaffected by the flow. This measurement is usually made at a distance of at least 6 H upstream of the weir. It may be made by any appropriate method for liquids described in the section on level measurement, such as the Hook gauge, float operated mechanisms, or air purge systems. It is often more convenient to measure the level of the liquid in a well alongside the channel at the appropriate distance above the notch. This well is connected to the weir chamber by a small pipe, or opening, near the bottom. Liquid will rise in the well to the same height as in the weir chamber and will be practically undisturbed by currents in the stream.

3.4.4.2(b)   HYDRAULIC FLUMES

Where the rate of fall of a stream is so slight that there is very little head available for operating a measuring device, or where the stream carries a large quantity of silt or litter, a flume is often much more satisfactory than a weir. Several flumes have been designed, but the only one mentioned here is the Venturi flume. This may have more than one form, but where it is flat bottomed and of the form shown in *Figure 3.73* the volume rate of flow is given by the equation:

$$Q = C \, B \, h_2 \sqrt{\frac{2g \, (h_1 - h_2)}{1 - \left(\dfrac{Bh_2}{B_1 h_1}\right)^2}} \quad \text{m}^3/\text{s}$$

$$(3.51)$$

where: B_1 = Width of channel (m).
$\quad\quad B$ = Width of the throat (m).
$\quad\quad h_1$ = depth of water measured immediately upstream of the entrance to the converging section (m).
$\quad\quad h_2$ = minimum depth of water in the throat (m).
$\quad\quad C$ = Discharge coefficient whose value will depend upon the particular outline of the channel and the pattern of the flow. Tests on a model of the flume may be used to determine the coefficient, provided that the flow in the model and in the full-sized flume are dynamically similar.

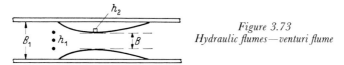

Figure 3.73
Hydraulic flumes—venturi flume

The depths of water h_1 and h_2 are measured as in the case of the weir by measuring the level in wells at the side of the main channel. These wells are connected to the channel by small pipes opening into the channel near, or at, the bottom.

As in the case of the closed venturi tube, a certain minimum uninterrupted length of channel is required before the venturi is reached, in order that the stream may be free from waves or vortices.

By carefully designing the flume, it is possible to simplify the actual instrument required to indicate the flow. If the channel is designed in such a manner that the depth in the exit channel at all rates of flow is less than a certain percentage of the depth in the entrance channel, the flume will function as a free-discharge outlet. Under these conditions, the upstream depth is independent of the downstream conditions, and the depth of water in the throat will maintain itself at a certain critical value, at which the energy of the water is at the minimum, whatever the rate of flow. When this is so, the quantity of water flowing through the channel is a function of the upstream depth h_1 only, and may be expressed by the equation:

$$Q = k \, h_1^{3/2}$$

where k is a constant for a particular installation and can be found.

It is now necessary to measure h_1 only, and this may be done by means of a float in a well, connected to the upstream portion of the channel. This float operates an indicating, recording, and integrating instrument.

The channel is usually constructed of concrete, the surface on the inside of the channel being made smooth to reduce the friction between water and channel. Flumes of this kind are used largely for measuring flow of water or sewerage, and may be made in a very large variety of sizes, to measure anything from the flow in a small stream to that of a large river.

3.4.5   Target flow meters

As the primary device and responsive element are in a single unit, the differential pressure fluid connections required in most head meters are eliminated so that target flow meters may be used in applications where other flow meters were not successful. These include the flow measurement of high viscosity liquids, hot asphalt, tars, synthetic dopes, oils and slurries at pressures up to 100 bar, at Reynolds numbers as low as 2000. *Figure 3.74* shows the meter and working principle.

The liquid impinging on the target will be brought to rest so that the pressure increases by $V^2/2g$ in terms of head of liquid so that the force F on the target will be

$$F = \frac{K\gamma V_1^2 A_t}{2} \text{ N} \qquad (3.52)$$

Figure 3.74   Target flow transmitter (a) working principle; (b) the 18 series pneumatic target flow transmitter. (By courtesy, Foxboro-Yoxall Ltd.)

where γ is the mass per unit volume in kg/m³; A_t is the area of the target in m³; K is a constant; V_1 is the velocity in m/s of the liquid through the annular ring between target and pipe;

If the pipe diameter is D m and the target diameter d m then area A of the annular space equals

$$\pi \frac{(D^2 - d^2)}{4} \text{ m}^2$$

Volume flow rate

$$Q = AV_1 = \pi \frac{(D^2 - d^2)}{4} \sqrt{\frac{8F}{K\gamma\pi d^2}} = C \frac{(D^2 - d^2)}{d} \sqrt{\frac{F}{\gamma}} \text{ m}^3/\text{s} \qquad (3.53)$$

where C is a new constant including the numerical factors.
Mass flow rate

$$W = Q\gamma = C \frac{D^2 - d^2}{d} \sqrt{F\gamma} \text{ kg/s} \qquad (3.54)$$

The force F is balanced through the force bar by the air pressure in the bellows so that a 0·2 to 1 bar signal proportional to the square root of the flow is obtained. The circular square edged target is attached to the force bar so that it is exactly concentric with the pipe forming an annular orifice. The lower section of the meter is made of mild or 316 stainless steel according to the application and a vent connection provided in the upper section. This connection can be used for back flushing, or purging, whenever the flowing fluid may tend to harden or settle out in the area of the force bar. A leakproof Elgiloy seal isolates the process fluid from the topwork. A silicone oil-filled damping device is installed in the transmitter mechanism and is factory set.

A calibrated range rod and setting wheel for range change provides up to a 2 to 1 change in the maximum flow rate for a given target. For greater changes the target size is changed.

The ranges vary from 0–52·7 to 0–123 litres/min for the 19 mm size at temperatures up to 400° C to from 0–682 to 0–2273 litres/min for the 100 mm size at temperatures up to 260° C. Meters are also available for gas flow measurement.

The instrument requires the same length of straight pipe upstream and downstream as an orifice plate type instrument having a large d/D ratio.

The overall accuracy of the meter is ±0.5 per cent, the repeatability within 0.1 per cent, and the sensitivity less than 0.1 per cent of the full span.

A maximum zero shift of 2 per cent occurs if the meter is calibrated in the horizontal position but it is mounted in a vertical position. This error can be calibrated out at the time of installation.

3.4.6 Integration

The flow records produced by a flowmeter will vary in type with the kind of meter used and the nature of the chart. The method used to integrate the flow will, therefore, depend upon the type of record. Probably the simplest way of finding the total quantity which has flowed through a rate-of-flow meter during the course of a day of 24 hours, if the flow has not varied much, is to examine the chart at the end of the day. The value of the rate of flow is then read at hourly intervals on the chart, and these twenty-four readings added. This will give an accuracy which will depend upon the amount the chart reading has fluctuated between the hourly readings. If greater accuracy is required, the chart may be read at half-hour intervals, and the sum divided by two. This is because the rate, measured in m^3/h, say, has been multiplied by a time interval of half an hour. Still greater accuracy will be obtained by taking readings of the chart at more frequent intervals, finding the total, and multiplying by the appropriate fraction. The value obtained for the total quantity will probably require multiplying by a correcting factor, if the temperature or pressure at which the measurement is made is different from that at which the meter was calibrated. If the pressure of a metered gas varies continuously it is usual to record the pressure at the same time as the flow. The correcting factor will then be obtained from the pressure record.

If there is a considerable fluctuation in the reading, the average reading for each hour may be found, and the sum of the hourly averages obtained. This average may be found on a strip chart by taking a piece of transparent plastic such as Celastoid or Perspex, about the size of a foot rule, with a hair line down the middle, along its length. This is then placed on the strip chart to be integrated, and arranged so that the area between the hair line and the chart record above the hair line, is equal to the corresponding area below the hair line. This method may be used for linear or square root charts if the flow scale is taken into account.

Another method, which may be used to find the hourly average for either circular or strip charts, is to divide the chart in the region of the record into rectangles or areas representing equal quantities of flow, by drawing lines parallel to the rate of flow and time axes. These areas may then be counted, and the position of an average line chosen so that there is as great a flow represented by the area above the line as is represented by the area below the line.

3.4.6.1 PLANIMETER METHOD

A planimeter is an instrument used for finding the area of a plane figure.

Chart planimeters are made in a variety of forms for linear and square root charts which may be either rectangular or circular.

The planimeter is used to find the area between the record line and the zero line. When this area is found, and multiplied by a factor which is determined by the scale of the chart, the total flow is obtained.

In one form of planimeter, the counter is set to zero, the chart carefully positioned, and the stylex (D) moved so as to trace along the record line of the chart. The stylex must start and finish on the zero line of the chart. To do this, the stylex is moved from the zero line along the hour line to the beginning of the chart record, along the chart record, and then down an hour line to the zero line, at the end of the tracing process. For integrating square root charts, a square root attachment with its own stylex is used. This attachment stylex is made to follow the record line, and drives the main stylex and counter.

Planimeters for circular charts, of the form shown in *Figure 3.75*, are available for both square root and uniform charts. The centre of the chart is placed on the planimeter hub (F) and the chart rotated and its centre moved so that the goose neck pointer follows the chart record. At the start the reading on the planimeter wheel is noted, and the chart is rotated through one complete revolution by means of a rubber tipped pencil held in one hand. With the other hand, the concave chart hub is moved back and forth in its slot so as to keep the record under the goose-neck point. The initial reading is then subtracted from the final reading, and the result is the planimeter reading.

Planimeters are also available with a tracing pen in place of the goose-neck pointer. This superimposes its own line, in ink of a different colour, directly on the original chart record line thus providing a record of the accuracy of the planimetering operation.

3.4.6.2   INTEGRATORS—MECHANICAL

A mechanical integrator may work in one of two ways.

1.   Planimeter type

The index, which may be of the counting train or cyclometer type, may be rotated continuously, at a rate which depends upon the rate of flow shown by the indicator, so that the total rotation is a measure of the total quantity of flow. This will produce the same effect as finding the area under the rate-of-flow time curve by dividing it into a very large number of very narrow strips, and should theoretically give the true summation of the product of rate-of-flow × time. A cam which is rotated by the flow indicator is so shaped that the motion of the follower is proportional to the rate of flow. The follower in turn moves a carriage whose displacement from its zero position is proportional to the rate of flow. This carriage carries the integrating mechanism which consists of a counting train driven by the integrating wheel. The edge of the integrating wheel is in contact with the rough face of a disc which, driven by a clock, is rotating at a constant speed. This clock may be driven by spring, weight, or electricity. The plane of rotation of the wheel is at right angles to the plane of rotation of the integrating wheel. The linear velocity of any point on

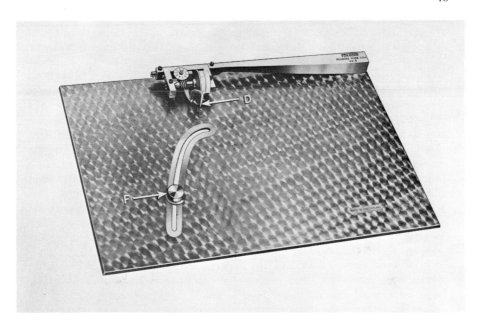

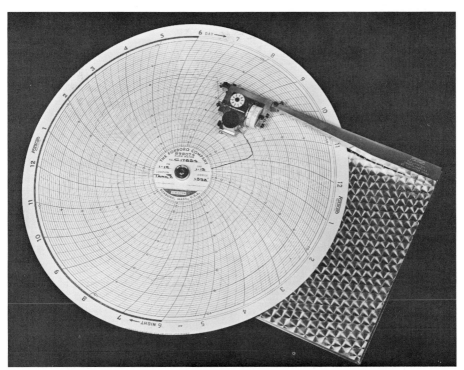

Figure 3.75   Square-root planimeter. (By courtesy, Foxboro-Yoxall Ltd.)

the face of a rotating disc will be equal to the product of angular velocity of the disc and the distance of the point from the centre. If, therefore, the integrating wheel is moved radially across the face of the disc, the rate at which it rotates will depend upon its distance from the centre of the disc. The integrating wheel is arranged to be over the centre of the disc when there is no flow, and its distance from the centre is arranged to be proportional to the rate of flow, at all other flows. Since the disc rotates at a constant speed, the speed of rotation of the integrating wheel will be a measure of the rate of flow, and the total rotation will be a measure of the total flow. This total rotation is indicated by the counting train which can be arranged to show the total flow in the appropriate units.

In order to reduce the friction between the integrating wheel and the rotating disc as the wheel traverses the disc, a series of small wheels are sometimes mounted in slots around the circumference of the wheel, the axes of these wheels being tangential to the main wheel. As the wheel moves across the face of the disc these wheels rotate, but this does not affect the driving force between the wheel and disc.

In order to reduce the possibility of large errors due to the integration of incorrect values of flow near the zero flow, integrators are often arranged to cease integrating at flows below 10 per cent of the maximum flow.

On flowmeters depending upon the measurement of differential pressure this corresponds with

$$\frac{100}{10^2} = 1 \text{ per cent}$$

of the differential pressure. The suppression of this initial portion of the range is easily achieved in the first type of integrator described, by recessing the surface of the disc near the centre. In this way, the wheel is not driven in the initial portion of the range. In other instruments a feeler is used which prevents rotation of the integrator wheel as the rate of flow approaches zero.

2.   *Escapement type*

Another method, which will give a reasonably good summation if the rate of flow is not fluctuating too rapidly, is to drive the integrator index at a constant speed, but the fraction of the total time for which it is driven is made proportional to the flow indication at a certain point in the time interval. In effect the instrument produces the same result as that which would be obtained by using the mid-ordinate rule.

Using the mid-ordinate rule, the area $ABCD$ in *Figure 3.76*

$$= (f_1 \times 1) + (f_2 \times 1) + (f_2 \times 1) \text{ m}^3,$$

where f_1, f_2 etc., are the rates of flow in m³ per minute at the middle of the minute intervals. An instrument can be arranged, so that at the middle of the first minute interval, the integrator is driven at a fixed speed for a time which depends upon the value of the flow at the middle of the interval. When the integrator index has been turned the required amount, it stops until the middle of the second minute interval, when it is again driven for a time which depends upon the value of the flow at the middle of this interval. If the fluctuation in

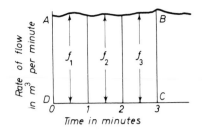

Figure 3.76   Integration—mid-ordinate rule

the rate of flow is not too rapid, the instrument will, over a period of time, indicate the total flow with a reasonable degree of accuracy, for the amount the true average rate of flow is greater than that which the integrator registers in one time interval will tend to cancel the amount the integrator over-registers in another time interval.

It will be seen that, the shorter time interval, the nearer will be the value of the indication at the middle of the time interval to the average value, and the more accurate will be the integration. Shortening the time interval has the same effect as dividing the area into a larger number of strips when using the mid-ordinate rule.

Many methods have been adopted to achieve intermittent rotation for a time proportional to the rate of flow, but only one mechanical method will be described here.

The principal parts of the integrator are shown in *Figure 3.77*. The heart-shaped cam has a uniform angular rise, and is driven at a constant speed of two revolutions per minute by a synchronous motor. By means of a friction clutch between the cam and the escape wheel, the motor also drives the escape wheel and the integrating counter at a constant speed, when the pawl is not engaged. When the pawl engages with the escape wheel, the integrator counter is held stationary, but the friction clutch slips and allows the cam to rotate.

The roller arm is pivoted near its left end to the meter flow arm, so that the position of the pivot *A* varies only with changes in the rate of flow. The right end of the roller arm moves up and down under the action of the rotating cam

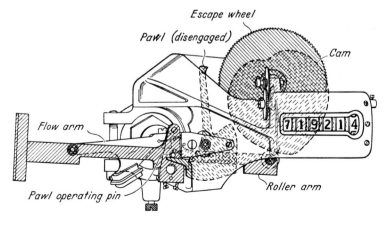

Figure 3.77   Integrators, escapement type. (By courtesy, Bailey Meters & Controls Ltd.)

so that the pawl-operating pin moves up and down also, causing the pawl to engage with, or disengage from, the escape wheel.

Figure 3.78 illustrates the action of instrument. Positions 1 and 2 show the flow recorder at zero rate of flow. In position 1 the cam and roller are at maximum throw, but the integrator counter is still stationary, because the path of the pawl-operating pin does not come sufficiently low to disengage the pawl.

Positions 3 and 4 also show the cam and roller in their extreme positions but the flow recorder is now at 50 per cent of maximum flow. This increased rate

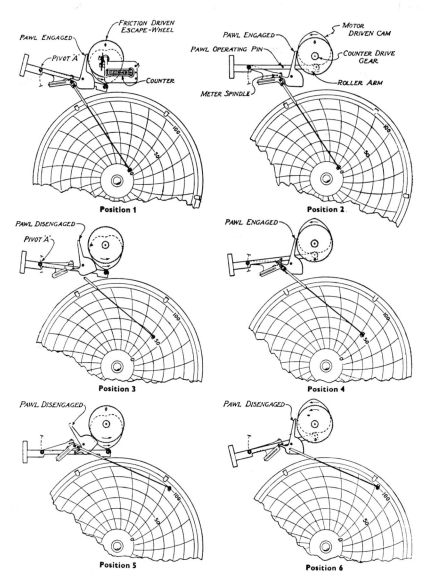

Figure 3.78   Integrators, escapement type. (By courtesy, Bailey Meters & Controls Ltd.)

of flow has resulted in the lowering of pivot A, thus lowering the path travelled by the operating pin. Consequently during 180 degrees rotation of the cam, the pawl remains engaged with the escape wheel, but during the remaining 180 degrees rotation, it is kept disengaged by the pin. Under these conditions the integrator counter runs for half the time.

Positions 5 and 6 show the cam and roller in similar positions when the flow has increased to 100 per cent. In this case, the pivot A is so low that the pawl operating pin keeps the pawl disengaged during the full rotation, so that the integrator counter runs continuously.

In this way, it is arranged that the counter turns at a constant speed for a fraction of the time, this fraction being the same as the fraction of the maximum flow which the measured flow represents.

In this type of instrument, the escapement wheel and clutch assembly should not be taken apart. Handling will destroy the special treatment of the friction pads. The clutch spindle bearings are made of self-lubricating bronze, and have sufficient oil to last the life of the integrator. They, therefore, do not require oiling, and should not be oiled because of the danger of oil contaminating and destroying the friction pads. If the integrator registers when no flow exists, both the zero adjustment of the pointer and that of the integrator should be checked. The zero of the integrator is usually checked at 10 per cent of the scale because it is difficult to calibrate the integrator with the pointer at zero. The calibration at other points on the scale may be checked by counting the number of teeth the pawl allows to pass before it re-engages.

3.4.6.3 INTEGRATORS—ELECTRICAL

Electrical integrators may be of the continuous rotation or the intermittent rotation types.

In the continuous rotation type the construction is similar to that of the induction type of watt-hour meter. By means of the specially designed resistance unit already shown in *Figure 1.22* (page 41), the resistance of the measuring circuit is arranged to be inversely proportional to the rate of flow of fluid through the primary element. The current produced by the measuring unit is therefore directly proportional to the rate of flow. This varying current operates the indicator, recorder, and integrator, which are connected to the mains transformer and measuring unit as shown in *Figure 3.79*.

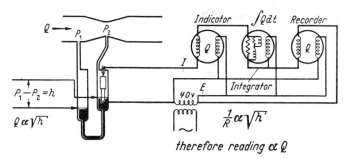

Figure 3.79   Integrators, electrical—Electroflo, circuit diagram. (By courtesy, G.E.C.-Elliot Process Instruments Ltd.)

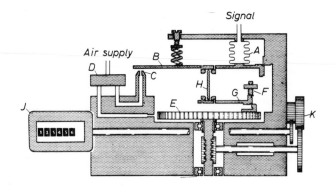

Figure 3.80(a)   Pneumatic integrator—principle of operation. (By courtesy, Foxboro-Yoxall Ltd.)

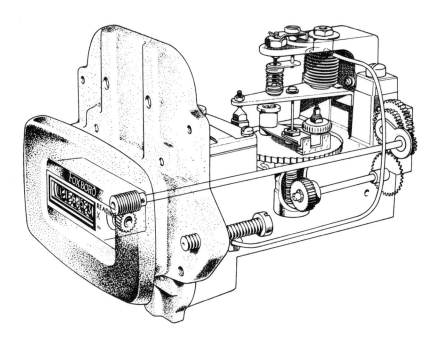

Figure 3.80(b)   Pneumatic integrator—Cutaway view of the instrument. (By courtesy, Foxboro-Yoxall Ltd.)

The essential portion of the integrator consists of an aluminium disc pivoted at its centre. It is arranged that the current through the instrument governs the driving force, and the applied voltage governs the value of the controlling force on the disc. The rate of rotation of the disc is, therefore, a measure of the quotient

$$\frac{current}{voltage}$$

Since, $$current = \frac{voltage}{resistance}$$

this quotient is equal to

$$\frac{1}{resistance}$$

and is, therefore, proportional to the rate of flow. The total rotation of the disc in any period of time is, therefore, a measure of the product, rate-of-flow × time; i.e. a measure of the total quantity which has flowed through the primary element in the given time. Since a change in the mains voltage will result in a proportional change in the current, the quotient

$$\frac{current}{voltage}$$

will be unaffected by variations of the mains voltage. The rate of rotation of the disc, therefore, depends only upon the rate of flow, and is independent of variation in the mains voltage. The disc in turn drives a counting train which indicates the total quantity which has flowed.

In an intermittent type of electrical integrator recently introduced, an electronic circuit is used to detect the pen position twelve times every minute. The electronic detector operates a relay unit and so controls the motor which rotates the counting assembly. The counter unit is driven forward a distance proportional to the rate of flow for each cycle of operations, so that its total rotation is an indication of the total quantity which has flowed through the meter.

3.4.6.4 PNEUMATIC INTEGERS

The principle of operation of this type of flow integrator is illustrated in *Figure 3.80(a)* and the actual instrument is shown in *Figure 3.80*(b). The instrument is designed to provide a continuous integration of the flow measurement from any flow transmitter giving a 0·2–1 bar air signal. It is powered by clean dry air or gas at 1·4 bar g and operates in the following manner:

A 0·2–1 bar signal, proportional to 0–100 per cent of differential pressure from a flow transmitter, is applied to the integrator receiver bellows (A). The force exerted by the bellows positions a force bar (B), which acts as a flapper, in relation to a nozzle (C). With an increase in differential pressure the flapper approaches the nozzle. The resulting increase in nozzle back pressure regulates —through air relay (D)—the jet which drives turbine rotor (E).

Weight (F) is mounted on the flexure-pivoted bell crank (G). As the rotor revolves, the centrifugal force of the weight (F) is applied to force bar (B), through thrust pin (H), to balance the force exerted by the bellows (A). Therefore a condition of force balance is continuously maintained.

The centrifugal force is proportional to the square of the rotor speed. This force balances the signal pressure which is proportional to the square of the flow. Therefore, rotor speed is directly proportional to flow, and, as the counter (J) is connected to the turbine rotor through gear train (K), integrator count is also directly proportional to the flow.

The receiver bellows is manufactured in bronze and the force bar in stainless steel. The zero spring is made of 'Ni Span' alloy which has a very low temperature coefficient thus giving very stable zero setting. The air-driven aluminium turbine rotor is carried in pre-lubricated sealed-for-life ball bearings. The force feedback system consists of a bell crank, a tungsten-carbide alloy thrust pin, and synthetic sapphire thrust pivots. The gears are polished nylon, and easily accessible standard range change gears permit change of unit of measurement as required. The integrated flow is shown on a six-digit counter. For intermittent flows, as in a batch process, a model with an integral brake assembly, bringing the turbine rotor from full speed to stop in less than one second, is available. The brake greatly reduces errors due to inertia of the rotor.

4

MEASUREMENT OF TEMPERATURE

It is very important to distinguish between temperature and heat. Temperature may be defined as 'degree of heat' but heat is usually taken to mean 'quantity of heat'. A bucketful of warm water will melt more ice than a teacupful of boiling water. The bucketful must, therefore, contain a greater quantity of heat than the teacupful, but its temperature is lower than that of the boiling water, a fact which is readily apparent if a finger is dipped into both vessels.

A thermometer measures the temperature of a body. The quantity of heat the body contains will depend upon its temperature, but it also depends upon the mass of the body and the nature of the material from which the body is made. A body, A, is said to be hotter than a body, B, if heat flows from A to B when the bodies are placed in contact. Heat will flow from a cupful of boiling water to a bucketful of warm water although the bucketful contains the greater amount of heat. Temperature is 'heat level'. This idea may be made clearer if one thinks in terms of two mountain lakes. One lake, A, is at a height of 2,000 m above sea level, and contains 1 million m^3 of water; the other lake, B, is at a height of 1000 m above sea level, but contains 5 million m^3 of water. Water will flow from lake A to lake B because A is at the greater height. If, however, the water is used to produce electrical energy, more energy will be available from lake B than from lake A, because lake B has the greater potential energy by virtue of its much larger mass.

Galileo was probably the first person to measure temperature. This he did towards the end of the sixteenth century. His thermometer depended upon the expansion of air. Some form of scale was attached to the thermometer, for he mentions degrees of heat in his records. Many types of thermometers have been made since, but, before temperatures measured on different thermometers could be compared, it was necessary to have some recognised fixed points. At least two fixed points are required which are constant in temperature and can be easily reproduced.

The two fixed points chosen were:

(1) The lower fixed point, or ice-point, which is the temperature of ice, prepared from distilled water, when melting under a pressure of 760 mm of mercury. The pressure of the atmosphere does not have a great influence on the melting point, but the ice should be in the form of fine shavings and mixed with ice-cold water, for the ice-point is the temperature at which water and ice can exist together.

(2) The upper fixed point, or steam-point, which is the temperature of steam from pure distilled water boiling under a pressure of 760 mm of mercury

in latitude 45°. As the boiling point of water varies a great deal with the applied pressure, it is very important to note the pressure at which the water is boiling. The boiling point, t_p, of water at a pressure p mm of mercury is given by the formula:

$$t_p = 100 + 0.0367 \, (p-760) - 0.000023 \, (p-760)^2 \qquad (4.1)$$

A small change in latitude has no great influence. The temperature interval between the ice-point and the steam-point is known as the 'fundamental interval'.

4.1   SCALES OF TEMPERATURE

In order to graduate a thermometer between these fixed points, the temperature interval between the points is divided into a number of equal parts. On the Celsius scale the interval is divided into 100 parts, and the melting point of ice is made 0° Celsius.

Absolute thermodynamic or Kelvin scale

The earlier scales of temperature depended upon the change with temperature of some property, such as size, of a substance. Such scales depended upon the nature of the substance selected. About the middle of the nineteenth century, Lord Kelvin defined a scale of temperature in terms of the mechanical work which may be obtained from a reversible heat engine working between two temperatures, and which, therefore, does not depend upon the properties of a particular substance. Kelvin divided the interval between the ice and steam points into 100 degrees, so that one Kelvin degree represented the same temperature interval as one Celsius degree.

It has also been established that an ideal gas obeys the gas law (see page 178) $PV = RT$, where T is the temperature on the Absolute or Kelvin Scale. Thus, the behaviour of an ideal gas forms a basis of temperature measurement on the Absolute Scale. Unfortunately, the ideal gas does not exist, but the so-called permanent gases, such as hydrogen, nitrogen, oxygen and helium, obey the law very closely, provided the pressure is not too great. For other gases and for the permanent gases at greater pressures, a known correction may be applied to allow for the departure of the behaviour of the gas from that of an ideal gas. By observing the change of pressure of a given mass of gas at constant volume, or the change of volume of the gas at constant pressure, it is possible to measure temperatures on the Absolute Scale.

The constant-volume gas-thermometer is simpler in form, and is easier to use than the constant-pressure gas-thermometer. It is, therefore, the form of gas-thermometer which is most frequently used. Nitrogen has been found to be the most suitable gas to use for temperature measurement between 500° and 1500° C, while at temperatures below 500° C hydrogen is used. For very low temperatures, helium at a low pressure is used.

International practical temperature scale of 1968 (I.P.T. S.-68)

The gas-thermometer, which is the final standard of reference, is, unfortunately, rather complex and cumbersome, and entirely unsuitable for industrial use.

In addition, it is not capable of such a high degree of reproducibility of reading as is required for many purposes. On the other hand, there are available temperature measuring instruments capable of a very high degree of reproducibility. Use of these instruments enables temperatures to be reproduced to a very high degree of accuracy, although the actual value of the temperature on the thermodynamic scale is not known with the same degree of accuracy. In order to take advantage of the fact that temperature scales may be reproduced to a much higher degree of accuracy than they can be defined, an International Practical Temperature scale was adopted in 1929 and revised in 1948. The latest revision of the scale was in 1968 (IPTS68) and this is the scale used in this work. The 1948 scale is still used in many places in industry, but as can be seen from Table 5.7 in the appendices which gives the differences between the temperatures on the two scales, the differences in many cases are within the accuracy with which industrial temperature measurements are made. Thus the errors which arise from using the calibration tables based on the 1948 scale may be neglected for most practical purposes. For work of high accuracy a correction based on Table 5.7 may be made where necessary.

The International Practical Temperature scale is based on a number of defining fixed points each of which has been subject to reliable gas-thermometer or radiation pyrometer observations and these are linked by interpolation using instruments which have the highest degree of reproducibility. In this way the International Practical Temperature scale is conveniently and accurately reproducible and provides means for identifying any temperature within much narrower limits than is possible on the thermodynamic scale.

The defining fixed points are established by realising specified equilibrium states between phases of pure substances. These equilibrium states and the values assigned to them are given in Table 4.1.

The scale distinguishes between the International Practical Kelvin Temperature with the symbol T_{68} and the International Practical Celsius Temperature with the symbol t_{68} the relationship between T_{68} and t_{68} is

$$t_{68} = T_{68} - 273 \cdot 15 \text{ K} \tag{4.2}$$

The size of the degree is the same on both scales being $1/273 \cdot 16$ of the temperature interval between absolute zero and the triple point of water $(0 \cdot 01° \text{ C})$. Thus, the interval between the ice point $0° \text{ C}$ and the boiling point of water $100° \text{ C}$ is still 100 Celsius degrees. Temperatures are expressed in Kelvin below $273 \cdot 15 \text{ K } (0° \text{ C})$ and $° \text{ Celsius above } 0° \text{ C}$.

Temperatures between and above the fixed points given in Table 4.1 can be interpolated as follows.

(1) From $13 \cdot 81 \text{ K}$ to $630 \cdot 74° \text{ C}$, the standard instrument is the platinum resistance thermometer. The thermometer resistor must be strain free, annealed pure platinum having a resistance ratio $W(T_{68})$ defined by

$$W(T_{68}) = \frac{R(T_{68})}{R(273 \cdot 15 \text{ K})} \tag{4.3}$$

where R is the resistance, which must not be less than $1 \cdot 392\,50$ at $T_{68} = 373 \cdot 15 \text{ K}$, i.e. the resistance ratio

$$\frac{R(100^\circ \text{ C})}{R(0^\circ \text{ C})}$$

is greater than the 1·392 0 of the 1948 scale so the platinum must be purer.

Below 0° C the resistance temperature relationship of the thermometer is found from a reference function and specified deviation equations. From 0° C to 630·74° C two polynomial equations provide the resistance tempera-

Table 4.1   DEFINING FIXED POINTS OF THE IPTS-68[1]

Equilibrium state	Assigned value of International Practical temperature	
	T_{68}	t_{68}
Equilibrium between the solid, liquid and vapour phases of equilibrium hydrogen (triple point of equilibrium hydrogen).	13·81 K	−259·34° C
Equilibrium between the liquid and vapour phases of equilibrium hydrogen at a pressure of 33 330·6 N/m² (25/76 standard atmosphere).	17·042 K	−256–108° C
Equilibrium between the liquid and vapour phases of equilibrium hydrogen (boiling point of equilibrium hydrogen).	20·28 K	−252·87° C
Equilibrium between the liquid and vapour phases of neon (boiling point of neon).	27·102 K	−246·048° C
Equilibrium between the solid, liquid and vapour phases of oxygen (triple point of oxygen).	54·361 K	−218·789° C
Equilibrium between the liquid and vapour phases of oxygen (boiling point of oxygen).	90·188 K	−182·962° C
Equilibrium between the solid, liquid and vapour phases of water (triple point of water)[3]	273·16 K	0·01° C
Equilibrium between the liquid and vapour phases of water (boiling point of water).[2][3]	373·15 K	100° C
Equilibrium between the solid and liquid phases of zinc (freezing point of zinc).	692·73 K	419·58° C
Equilibrium between the solid and liquid phases of silver (freezing point of silver).	1235·08 K	961·93° C
Equilibrium between the solid and liquid phases of gold (freezing point of gold).	1337·58 K	1064·43° C

(1)   Except for the triple points and one equilibrium hydrogen point (17·042 K) the assigned values of temperature are for equilibrium states at a pressure $p_0 = 1$ standard atmosphere (101.325 N/m²).

   In the realization of the fixed points small departures from the assigned temperatures will occur as a result of the differing immersion depths of thermometers or the failure to realize the required pressure exactly. If due allowance is made for these small temperature differences, they will not affect the accuracy of realization of the Scale.

(2)   The equilibrium state between the solid and liquid phases of tin (freezing point of tin) has the assigned value of $t_{68} = 231\cdot9681°C$ and may be used as an alternative to the boiling point of water.

(3)   The water used should have the isotopic composition of ocean water.

ture relationship. This will be discussed further in the section on resistance thermometers.

(2) From 630·74° C to 1 064·43° C the standard instrument is the platinum 10 per cent rhodium/platinum thermocouple, the electromotive force-temperature relationship of which is represented by a quadratic equation and is discussed in the appropriate section.

(3) Above 1 337·58 K (1 064·43° C) the scale is defined by Planck law of radiation with 1 337·58 K as the reference temperature and the constant c_2 has a value 0·014 388 metre Kelvin. This will be discussed in the section on radiation thermometers.

In addition to the defining fixed points the temperatures corresponding to secondary points are given. These points, particularly the melting or freezing points of metals, form convenient workshop calibration points for temperature measuring devices (see Table 4.2).

Table 4.2

Equilibrium state	International Practical Temperature	
	T_{68}	t_{68}
Triple point of normal hydrogen	13·956 K	−259·194° C
Sublimation point of carbon dioxide	194·674 K	−74·476° C
Freezing point of mercury	234·288 K	−38·862° C
Equilibrium between ice and air saturated water (ice point)	273·15 K	0° C
Freezing point of		
indium	429·784 K	156·634° C
bismuth	544·592 K	271·442° C
lead	600·652 K	327·502° C
antimony	903·87 K	630·74° C
aluminium	933·52 K	660·37° C
copper	1357·6 K	1084·5° C
nickel	1728 K	1455° C
palladium	1827 K	1554° C
platinum	2045 K	1772° C
rhodium	2236 K	1963° C
iridium	2720 K	2447° C
Melting point of tungsten	3660 K	3387° C

The full text of the English Version of IPTS 1968 is obtainable from H.M. Stationery Office.

In order to use any device as a thermometer for measuring temperature, it must be brought to the same temperature as the body whose temperature is required. This may be relatively simple when the substance whose tempera-ture is being measured is a liquid or gas, but it is much more difficult when the substance is a solid. The circumstances in which a thermometer is used should be carefully considered before it can be assumed that the reading on the thermometer is a true indication of the temperature of the tested body. Thus, if thermometers are used out-of-doors to measure air temperature, the reading of a thermometer on which the sun is shining will be higher than that of a thermometer in the shade, even if the temperature of the air surrounding both thermometers is the same. This is because the thermometer in the direct sunlight will be receiving radiant heat from the sun, while that in the shade will not. The radiant heat will be absorbed by the thermometer, raising its temperature above that of the surrounding air, so that its reading will be too high. For this reason, thermometers used for meterological work are always protected from direct sunlight by a screen.

On the other hand, a thermometer placed in a stream of hot gas will indicate a temperature which is lower than that of the gas if the gas is retained by walls whose temperature is lower than that of the gas. This is because the thermometer will be radiating more heat to the walls than it receives from them. Thus, although the thermometer is receiving heat from the gas, tending to bring its temperature up to that of the gas, its temperature will always be a little lower than that of the gas. The rate at which the thermometer absorbs heat from the surrounding gas will depend upon the nature of the thermometer and of the gas, so that these also will influence the thermometer reading.

When a thermometer is used to find the temperature of a body, sufficient time must be allowed for the thermometer to reach the same temperature as the body. If the temperature of a hot body is changing continuously, then the thermometer will rarely indicate the true temperature of the hot body. This is because the temperature of the body has changed before the thermometer has attained the same temperature.

The time taken for the thermometer to reach the temperature of the hot body can be reduced by making the thermal capacity of the thermometer as small as possible, and ensuring that the contact between the hot body and the thermometer is as close as possible. Close thermal contact between thermometer and tested body is particularly difficult to achieve when a refractory sheath is being used to protect the thermometer. This is because the thermal conductivity of the sheath is low. A sheath of mild steel ($0 \cdot 1$ per cent C) would transmit 330 times as much heat as a silica sheath of identical dimensions, and a similar copper sheath would transmit 3000 times as much.

In order that the transfer of heat to or from a temperature sensor along its stem and along the thermowell protecting it, may be kept to a permissible level, the depth of immersion of the thermowell should be as large as possible. The actual depth of immersion required will depend upon the temperature gradient along the thermowell and the thermal conductivity and the actual cross-sectional area of the thermometer stem and of the thermowell. Where high accuracy is required a stem and thermowell conduction calculation may be carried out and the error reduced to an acceptable level or a correction made. In industrial applications where a mild or stainless steel thermowell having, say, an external diameter of 24 mm and a wall thickness of 5 mm is used, the immersion depth should be at least 150 mm in liquid temperature measurement, and at least 250 mm in gas measurement. When the temperature element is installed on a tank this depth of immersion can be attained without difficulty but in pipework this is not always easy as the tip of the thermowell should not be immersed more than two-thirds of the pipe diameter.

Where the pipe diameter is insufficient, the depth of immersion may be achieved by installing a pipe branch with its axis at 30° to the main pipe axis, or the thermowell connection installed on a bend with the thermowell along the centre line of the pipe facing upstream. If the pipe diameter is small so that the thermowell would occupy an excessive cross-sectional area of the pipe, an enlarged T may be installed on the bend and the thermowell housed in this, again lying along the centre of one branch of the bend facing into the on-coming stream. Lagging the pipework and branch in the vicinity of the temperature sensor will also reduce errors owing to conduction as this will reduce the temperature gradient along the thermowell.

When measuring the mean temperature of a fluid in a duct, the centre of the duct is not the position of mean temperature owing to the velocity variation in the duct. When a single sensor is used, a position 0·72 radii from the centre will give a more representative temperature when the fluid is fully turbulent, and 0·58 radii from the centre in the case of laminar flow.

When a temperature sensor is used in a location where there is severe vibration or very high flow rates it may be necessary to protect it by means of a long tapered boss or a thermowell which goes right through the pipe and is rigidly fixed to the opposite side of the pipe or duct. High velocity fluids may also cause friction heating of the sensor or of the walls of the pipe so that the sensor should be mounted midway between the wall and the centre of the pipe. As the heating effect is proportional to the square of the fluid velocity, a correction will only be necessary at very high velocities.

When surface mounting sensors are used they should make good thermal contact with the surface and be firmly anchored by cementing, welding or clamping to withstand any vibration or shocks. The leads should also be securely fastened to the surface to prevent lead breakage.

The fact that a thermometer takes a certain time to reach the temperature of the tested body must be taken into account in designing a temperature controller. The longer the time lag, the more difficult is the task of the con-troller which may cause the temperature to oscillate about the controlled temperature instead of settling down at that temperature.

The method of measuring temperature in any industrial process must be carefully chosen, so that the indication obtained for the same temperature is always the same. In other words, the results must be consistent. This is even more important than that the results should be absolutely accurate. If it is known that a certain indication on an instrument represents the correct temperature for the process and that the instrument is consistent in its indica-tion, then the process conditions can be kept constant. If, however, it is required to compare the actual temperature used, with that used in a similar process and indicated on another instrument, then it is important to know, not only that the indication is consistent, but also the value of the indication represented on a definite scale of temperature.

Unless the thermometer and the tested body are in close thermal contact it is necessary to know the relationship between their temperatures.

There must be no chemical action between a thermometer and the sub-stance which surrounds it. No vapour must be allowed to condense on the thermometer, for this would give up heat to the thermometer and warm it. No liquid must be allowed to evaporate from the thermometer, for this would take heat from the thermometer and cause it to be cooled.

4.2 INSTRUMENTS FOR MEASURING TEMPERATURE

The instruments for measuring temperature described in this section have been classified in the first place according to the nature of the change produced in the testing body by the change of temperature. This section is, therefore, divided up as follows:

Expansion thermometers.
Change-of-state thermometers.
Electrical methods of determining temperature.
Radiation and optical pyrometry.

Expansion of solids

When a solid is heated, it increases in volume. It increases in length, breadth and thickness. The increase in length of any side of a solid will depend upon the original length, l_0, the rise in temperature, t, and the coefficient of linear expansion, α.

The coefficient of linear expansion may be defined as the increase in length per unit length when the temperature is raised 1° C. Thus, if the temperature of a rod of length, l_0, is raised from 0° C to t° C, then the new length, l_t, will be given by:

$$l_t = l_0 + l_0\alpha t = l_0(1 + \alpha t) \qquad (4.4)$$

The value of the coefficient of expansion varies from substance to substance, and the coefficients of linear expansion of some common materials are given in the appendix Table 5.3.

It can be easily shown that the coefficient of superficial expansion (i.e. coefficient of increase in area) is twice the coefficient of linear expansion. The coefficient of cubical expansion (i.e. coefficient of increase in volume) is three times the coefficient of linear expansion.

Frequently, allowance has to be made in engineering practice for the fact that substances expand. In laying railway lines, allowance must be made for the differences in lengths of the lines in summer and winter; while bridges are supported on rollers and gaps are left to allow for expansion.

If great accuracy is required when measuring lengths with a scale made of metal, allowance should be made for the increase in length of the scale when its temperature is greater than that at which it was calibrated. Owing to the expansion of the scale, a length which was originally l_1 at the temperature, t_1, at which the scale was calibrated, will have increased to l_2 where

$$l_2 = l_1\{1 + \alpha(t_2 - t_1)\} \qquad (4.5)$$

t_2 is the temperature at which the measurement is made, and α the coefficient of expansion of the metal of the scale.

A 1 mm division on the scale will therefore now measure

$$1 + \alpha(t_2 - t_1) \text{ mm}$$

An actual length l_2 mm will therefore measure

$$\frac{l_2}{1 + \alpha(t_2 - t_1)} \text{ mm}$$

The length will therefore appear to be smaller than it actually is. To make this error negligibly small, secondary standards of length are made of invar, a nickel steel alloy whose linear coefficient of expansion at ordinary temperatures is very nearly zero (1.5×10^{-6} for commercial invar).

Expansion of liquids and gases

In dealing with the expansion of liquids and gases it is necessary to consider the volume expansion, or cubical expansion. Both liquids and gases have to be held by a container, which will also expand, so that the apparent expansion of the liquid or gas will be less than the true or absolute expansion. Actually, the true coefficient of expansion of a liquid is equal to the coefficient of apparent expansion plus the coefficient of cubical expansion of the containing vessel. Usually the expansion of a gas is so much greater than that of the containing vessel that the expansion of the vessel may be neglected in comparison with that of the gas.

The coefficient of expansion of a liquid may be defined in two ways. First, there is the zero coefficient of expansion, which is the increase in volume per degree rise in temperature, divided by the volume at $0°$ C, so that volume, v_t, at temperature t is given by:

$$v_t = v_o(1 + \beta t) \qquad (4.6)$$

where v_o is the volume at $0°$ C and β is the coefficient of cubical expansion.

There is also the mean coefficient of expansion between two temperatures. This is the ratio of the increase in volume per degree rise of temperature, to the original volume.

i.e.
$$\beta = \frac{v_{t_2} - v_{t_1}}{v_{t_1}(t_2 - t_1)} \qquad (4.7)$$

where v_{t_1} is the volume at temperature, t_1, and v_{t_2} is the volume at temperature, t_2.

This definition is useful in the case of liquids that do not expand uniformly, e.g. water.

Correction of a manometer reading for temperature changes

When a very high degree of accuracy is required and a column of liquid is being used to measure pressures, allowance must be made for the change in the density of the liquid, and for change in the length of the scale when the temperature changes.

A mass, m, of liquid, which has a volume, v_o, at a temperature, $0°$ C, will occupy a volume, v_t, at a temperature, $t°$ C, given by

$$v_t = v_o(1 + \beta t) \qquad (4.6)$$

The density of the liquid is therefore reduced from

$$\frac{m}{v_o} \quad \text{to} \quad \frac{m}{v_o(1 + \beta t)}$$

or
$$\rho_t = \frac{\rho_o}{1 + \beta t} \qquad (4.8)$$

where ρ_o is the density at a temperature, $0°$ C, and ρ_t is the density at $t°$ C. Thus, a column of liquid, length h, which represented a pressure, $hg\rho_o$, at $0°$ C will represent a pressure of

$$hg\rho t = \frac{hg\rho_o}{1 + \beta t}$$

at a temperature, $t°$ C. If h is measured on a scale which was correct at a temperature, t_1, and the coefficient of linear expansion of the scale is α, then the true pressure is

$$hg\rho_o \frac{\{1 + \alpha(t - t_1)\}}{1 + \beta t}$$

4.3   EXPANSION THERMOMETERS

Expansion thermometers may be classified according to the nature of the substance which expands, and are therefore described under three headings.

Expansion of solids:
    Solid rod thermostats
    Bimetallic strip thermometers.
Expansion of liquids:
    Liquids-in-glass thermometers
    Liquids-in-metal thermometers.
Expansion of gases:
    Gas thermometers.

4.3.1   Expansion of solids

4.3.1.1   SOLID ROD THERMOSTATS

A temperature-controlling device may be designed using the principle that some metals expand more than others for the same range of temperature. An example of such a device is the thermostat used with water heaters and shown in *Figure 4.1*.

The temperature-sensitive portion of the instrument consists of an invar rod encased in a concentric brass tube. The lower end of the invar rod is hard-soldered to the containing tube. When this combination, or stem, is heated, the brass rod ($\alpha = 19 \times 10^{-6}$) will expand more than the invar rod ($\alpha = 1.5 \times 10^{-6}$) so that the position of the free end of the rod relative to the end of the tube will change. This change in relative position is used to operate a sensitive microgap switch. The temperature change or 'stem differential' required to change the position of the switch from on to off depends upon the length of the stem. As the stem length increases, the differential temperature gets less. The portion of the switch relative to the invar rod can be adjusted by means of a small knob, so that the supply of electricity to the heater is cut off when the temperature of the thermostat has reached a certain predetermined temperature.

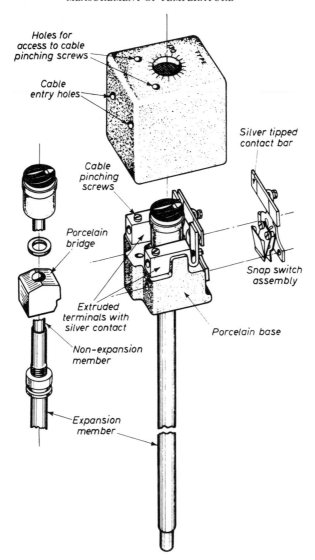

Holes for
access to cable
pinching screws

Cable
entry holes

Cable
pinching
screws

Porcelain
bridge

Extruded
terminals with
silver contact

Non-expansion
member

Expansion
member

Silver tipped
contact bar

Snap switch
assembly

Porcelain base

Figure 4.1   Rod type thermostat. (By courtesy, Robert Maclaren & Co. Ltd.)

An exactly similar thermostat is used to control the temperature of domestic electric ovens, but the sensitive element consists of an aluminium-bronze tube with an inner nickel-iron rod. Again, the tube expands more than the rod for a given rise in temperature, and cuts off the electricity supply to the heaters at a preset temperature.

In a similar way an invar rod in a brass tube may be arranged to reduce the supply of gas to an oven as the temperature rises, so that an oven heated by gas may be maintained at a preset temperature.

When the current required to operate an electrical heater is greater than 20 amps at 250 volts (A.C.), the thermostat-switch is usually used to operate

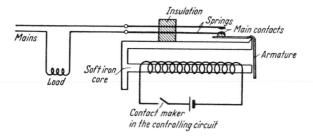

Figure 4.2   Electromagnetic type relay

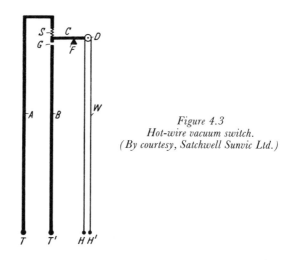

Figure 4.3
Hot-wire vacuum switch.
(By courtesy, Satchwell Sunvic Ltd.)

a relay which, in turn, switches on the heater. In this way, large heating loads may be controlled without damage to the contacts of the thermostat switch. The relay may be an electromagnetic type similar to that shown in *Figure 4.2* or a hot-wire switch which is shown diagrammatically in *Figure 4.3*.

Hot-wire vacuum switches

T and T' are terminals of the main circuit to be controlled. A and B are fixed conductors, while C is a lever pivoted so that it is free to rotate about the fulcrum, F. The spring, S, between A and C, tends to rotate the lever in an anti-clockwise direction and close the gap, G (seen open in the diagram). At the other end of the lever, C, is the bobbin, D, round which passes the resistance wire, W, rigidly fixed to the terminals H and H'. By tensioning this wire the spring, S, is compressed and the gap, G, opened, breaking the circuit between A and B. Now if a small current is passed through the wire, W, it is heated and expands, allowing the spring, S, to force the lever, C, down, so closing the gap, G, and completing the main circuit, $T\ T^1$.

The actual arrangement of the various parts of the switch is shown in *Figure 4.4*. The hot-wire winding consists of a number of turns of special steel wire wound between two steatite insulating bobbins. The temperature of the wire when the winding is energised is well below that at which creep is liable to occur so that it does not suffer from a permanent set even if it is frequently energised or kept energised for long periods. The energising current varies from about 25 mA in the smallest type to 60 mA in the largest. The contacts are made of tungsten, and when the switch is used on inductive loads a surge suppressor is connected across the contacts.

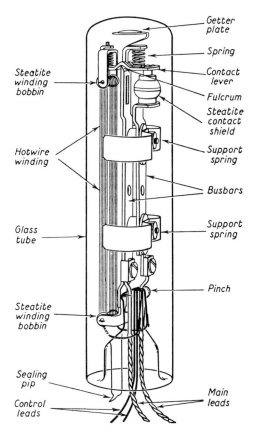

Figure 4.4   Sunvic hot-wire vacuum switch. (By courtesy, Satchwell Sunvic Ltd.)

The whole switch is enclosed in a glass tube which is evacuated and sealed off. After sealing, the tube is 'gettered'* to remove the remaining traces of gas and to assure the absorption of occluded gas which may be liberated during use.

The switch has a number of uses and *Figures 4.5* and *4.6* show two simple circuits in which the switch may be used. *Figure 4.5* shows the circuit arranged

*In order to produce very high vacua in valves and similar evacuated enclosures, chemical substances known as getters are used. The commonest is magnesium. After the valve has been exhausted by an oil pump, some magnesium on the anode is vaporised by electrical heating. The magnesium vapour condenses on the walls of the envelope forming a bright mirror and absorbing the residual gases.

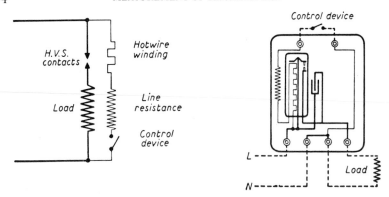

Figure 4.5   Circuit using hot-wire switch; main contacts are closed when the control device contacts close

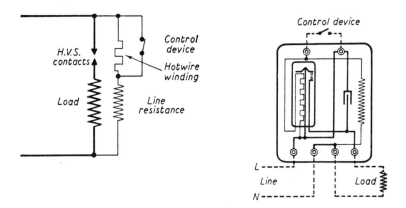

Figure 4.6   Alternative form of circuit; in this the main contacts close when the control device contacts open. (By courtesy, Satchwell Sunvic Ltd.)

so that the main contacts close when the control device contacts close, while, *Figure 4.6* shows the circuit arranged so that the contacts close when the control device contacts open. One of the main advantages of the hot-wire vacuum switch is that the hot wire is a non-inductive load, and requires only a small current for its operation. The switch can, therefore, be used in conjunction with very sensitive thermostats operating with very light contacts, without the troubles which the thermostat contacts give when handling the inductive load presented by a magnetic type of relay.

4.3.1.2   BIMETAL-STRIP THERMOMETERS

Bimetal strips consist of strips of two metals such as invar and brass welded together to form a cantilever as shown in *Figure 4.7a*. When heated, both metals expand, but the brass expands much more than the invar. The result is that the bottom of the strip becomes longer than the top, and the cantilever curls upwards as shown in *Figure 4.7b*.

Bimetal strips are often used in instruments to compensate for the effects due to changes in ambient temperature at the instrument. Such a use is

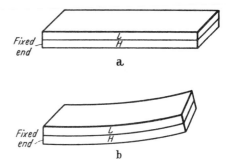

Figure 4.7 Simple bimetal strip
L metal, such as invar, having a low coefficient of expansion.
H metal, such as brass, having a high coefficient expansion.

illustrated in *Figure 4.8*. Instead of the free end of the Helical Bourdon Tube being connected directly to the pointer operating mechanism, it is connected to it through a bimetal strip. The length and arrangement of the bimetal strip are chosen so that the pointer movement due to the bimetal strip for a particular temperature change is exactly equal and opposite to that produced by the change in dimensions and elastic properties of the rest of the instrument.

Figure 4.8 Bimetal strip used for ambient temperature compensation.
(By courtesy, Taylor Instrument Companies (Europe) Ltd.)

Aneroid barometers and similar instruments are compensated by introducing a bimetal strip into the lever, as shown in *Figure 1.40*, page 59.

Watches are compensated for the effects of temperature changes by making the balance wheel in the form of two curved bimetal strips as shown in *Figure 4.9*. The metal with the greater coefficient of expansion is put on the outside so that when the temperature rises the strips curl, moving part of the mass of the rim towards the centre of the balance wheel and reducing its moment of inertia. This effect can be arranged to counter balance the effects due to the changes in elastic properties of the hair spring and main spring.

Many common temperature controlling devices utilize a bimetal strip.

An example is the energy regulator or the 'Simmerstat'. *Figure 4.10* shows the principle of its construction.

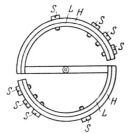

Figure 4.9   Temperature compensated balance wheel
L and H as *Figure 4.7*. S screws which provide mass whose distance from the centre of rotation may be adjusted in order to give correct compensation

Normally the contact D is in the position shown, completing the circuit through the heater winding, A, on the bimetal strip and through the load. The heat produced by the current through the heater winding warms the bimetal strip. The strip therefore curls, and in doing so moves the arm, B, which opens the switch, C, breaking the circuit at D, and interrupting the current both to the heater winding and to the load. The bimetal strip then cools and tends to return to its normal shape, allowing the circuits to be completed at D, and the cycle is repeated.

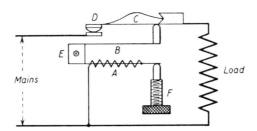

Figure 4.10   Simmerstat. (By courtesy, Satchwell Sunvic Ltd.)

The ratio of the time during which the contacts are made to the total time for the complete cycle determines the average input into the load. This ratio is controlled by the adjusting screw F, and is approximately proportional to its position. If the screw is slackened off so that the contacts remain closed even with the maximum deflection of the bimetal strip, the power will be permanently on. If, however, the screw is tightened up until the switch D is open when the bimetal strip is cold, then, the power will be permanently off. Between these two extreme positions of the screw, there is an infinite number of positions, each corresponding to a given input to the load. When the load exceeds 15 A at 230 V (A.C.), the energy regulator is used to control the current to the heating wire of a hot-wire vacuum switch. This in turn is used to make and break the circuit of the main load.

Helical Bimetal Strips

A long bimetal strip, consisting of an invar strip welded to a higher expansion nickel molybdenum alloy wound without a break into several compensated helices, arranged coaxially one within the other as shown in *Figure 4.11*, forms the temperature-sensitive element of an instrument which may be designed to measure, or control, temperature. This method of winding the strip, enables a length, sufficient to produce an appreciable movement of the free end, to be concentrated within a small space. It also makes it possible to keep the thermal capacity of the element and its stem at a reasonably low value, so the instrument will respond rapidly to small temperature changes.

The helices in the winding are so compensated that any tendency towards lateral displacement of the spindle in one helix is counteracted by an opposite tendency on the part of one or more of the other helices. Thus, the spindle of the instrument is fully floating, retaining its position at the centre of the scale without the help of bearings. The instrument is, therefore, not injured by mechanical shocks which would damage jewelled bearings.

This particular design also results in the angular rotation of the spindle being proportional to the change in temperature for a considerable temperature range. The instrument therefore has a linear temperature scale, and can be made to register temperatures below 300° C to within ±1 per cent of the scale range.

The construction of the actual instrument can be seen in the part-sectioned illustration shown in *Figure 4.12*.

If, instead of carrying a pointer, the spindle is arranged to carry a contact arm, which, when the sensitive element reaches a predetermined temperature, touches a fixed contact completing the circuit of a relay unit, then the instrument will operate a bell, or similar warning device, when the safe maximum temperature is reached.

Owing to its robust construction, this instrument is used on many industrial plants, and a slightly modified form is used in many homes and offices to indicate room temperature. It can be made for a large variety of temperature ranges and is used in many places where the more fragile mercury-in-glass thermometer was formerly used.

4.3.2 Expansion of liquids

4.3.2.1 LIQUID-IN-GLASS THERMOMETERS

The coefficient of cubical expansion of mercury is about eight times greater than that of glass. If, therefore, a glass container holding mercury is heated, the mercury will expand more than the container. At a high temperature, the mercury will occupy a greater fraction of the volume of the container than at a low temperature. If, then, the container is made in the form of a bulb with a capillary tube attached, it can be so arranged that the surface of the mercury is in the capillary tube, its position read, and the assembly used to indicate temperature. This is the principle of the mercury-in-glass thermometer.

The thermometer, therefore, consists simply of a stem of suitable glass tubing having a very small, but uniform, bore. At the bottom of this stem

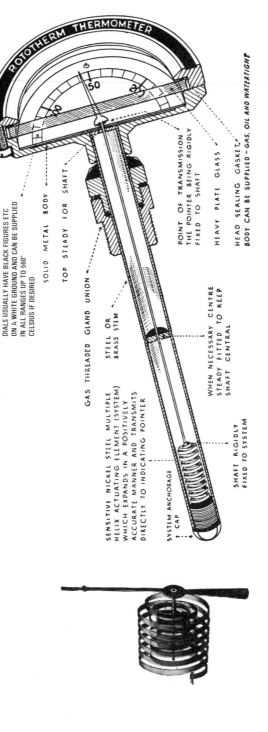

HEAVY DIE CAST BEZEL

DIALS USUALLY HAVE BLACK FIGURES ETC ON A WHITE GROUND AND CAN BE SUPPLIED IN ALL RANGES UP TO 560° CELSIUS IF DESIRED

SOLID METAL BODY

TOP STEADY FOR SHAFT

GAS THREADED GLAND UNION

STEEL OR BRASS STEM

POINT OF TRANSMISSION THE POINTER BEING RIGIDLY FIXED TO SHAFT

HEAVY PLATE GLASS

HEAD SEALING GASKET. BODY CAN BE SUPPLIED - GAS, OIL AND WATERTIGHT

WHEN NECESSARY CENTRE STEADY FITTED TO KEEP SHAFT CENTRAL

SENSITIVE NICKEL STEEL MULTIPLE HELIX ACTUATING ELEMENT (SYSTEM) WHICH EXPANDS IN A POSITIVELY ACCURATE MANNER AND TRANSMITS DIRECTLY TO INDICATING POINTER

SYSTEM ANCHORAGE CAP

SHAFT RIGIDLY FIXED TO SYSTEM

ROTOTHERM THERMOMETER

50

Figure 4.12 Rototherm thermometer. (By courtesy, British Rototherm Co. Ltd.)

*Figure 4.11
Helical bimetal strip.
(By courtesy, British Rototherm Co. Ltd.)*

there is a thin-walled glass bulb. The bulb may be cylindrical or spherical in shape, and has a capacity very many times larger than that of the bore of the stem. The bulb and bore are completely filled with mercury, and the open end of the bore sealed off either at a high temperature, or under vacuo, so that no air is included in the system. The thermometer is then calibrated by comparing it with a standard thermometer in a bath of liquid whose temperature is carefully controlled. The liquid is agitated rapidly to keep it all at the same temperature. The liquid used in the bath depends upon the temperature range of the thermometer, the most common liquids used being water, oil, alcohol and molten salt. For temperatures between 200° C and 600° C the contents of the salt bath consist of about equal quantities of sodium and potassium nitrates. When the standard thermometer and the thermometer to be calibrated have reached equilibrium with the bath at a definite temperature, the point on the glass of the thermometer opposite the top of the mercury meniscus is marked. The process is repeated for several temperatures. The intervals between these marks are then divided off by a dividing machine. In the case of industrial thermometers, the points obtained by calibration are transferred to a brass plate. The intervals are divided on the plate, which is then fixed with the tube into a suitable protecting case to complete the instrument.

The stem of the thermometer is usually shaped in such a way that it acts as a lens, magnifying the width of the mercury column. The mercury is usually viewed against a background of glass which has been enamelled white.

As a great deal of information about mercury-in-glass thermometers is readily available in almost any standard text-book on heat, remarks in this section will be confined to the industrial use of such thermometers.

Mercury-in-glass thermometers are available in three grades. The limits of error of grades A and B are specified by the National Physical Laboratory, and are given in the tables in the British Standards Code No. 1041 on 'Temperature Measurement'. Grade C is a commercial grade of thermometer, and no limits of accuracy are specified. Thermometers of this grade are, of course, cheaper than those of the other grades, and their price varies, to a certain degree, according to their accuracy. They should be compared from time to time during use with thermometers of known accuracy.

Whenever possible, thermometers should be calibrated, standardised, and used, immersed up to the reading, i.e. totally immersed, as this avoids errors which are due to the fact that the emergent column of mercury and the glass stem are at a lower temperature than that of the bulb, and are therefore not expanded by the same amount. Errors introduced in this way should be allowed for if accurate readings are required, particularly at high temperatures. Some thermometers are, however, calibrated for 'partial immersion', and should be used immersed to the specified depth.

When reading a thermometer an observer should keep his eye on the same level as the top of the mercury column. In this way errors due to parallax will be avoided.

Figure 4.13 shows the effect of observing the thermometer reading from the wrong position. When viewed from (a) the reading is too high. Taken from (b) the reading is correct, but from (c) it is too low.

A mercury-in-glass thermometer has a fairly large thermal capacity (i.e. it requires quite an appreciable amount of heat to change its temperature by

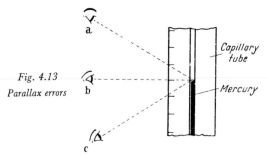

Fig. 4.13
Parallax errors

Figure 4.13   Parallax errors

one degree), and glass is not a very good conductor of heat. This type of thermometer will, therefore, have a definite thermal lag. In other words, it will require a definite time to reach the temperature of its surroundings. This time should be allowed for before any reading is taken. If there is any doubt as to whether the thermometer has reached equilibrium with a bath of liquid having a constant temperature, then readings should be taken at short intervals of time. When the reading remains constant the thermometer must be in equilibrium with the bath. If the temperature is varying rapidly the thermometer may never indicate the temperature accurately, particularly if the tested medium is a gas.

Glass thermometers used in industry are usually protected by metal sheaths. These sheaths may conduct heat to or from the neighbourhood of the thermometer bulb and cause the thermometer to read either high or low according to the actual conditions prevailing. A thermometer should, therefore, be calibrated, whenever possible, under the conditions in which it will be used, if accurate temperature readings are required. If, however, the main requirement is that the temperature indication be consistent for the same plant temperature, then an error introduced is not so important, so long as the conditions remain the same, and the error is constant.

Errors due to Ageing

It is often assumed that provided a mercury-in-glass thermometer is in good condition it will always give an accurate reading. This is not always so, particularly with cheap thermometers. A large error may be introduced by changes in the size of the bulb due to ageing. When glass is heated to a high temperature, as it is when a thermometer is made, it does not, on cooling, contract to its original volume immediately. Thus, for a long time after it has been made the bulb continues to contract very slowly so that the original zero mark is too low on the stem, and the thermometer reads high. This error continues to increase over a long period, and depends upon the type of glass used in the manufacture of the thermometer. In order to reduce to a minimum the error due to this cause, during manufacture thermometers are annealed by baking for several days at a temperature above that which they will be required to measure, and then cooled slowly over a period of several days.

Another error due to the same cause is the depression of the zero when a thermometer is cooled rapidly from a high temperature. When cooled, the glass of the thermometer bulb does not contract immediately to its original size so that the reading on the thermometer at low temperatures is too low, but returns to normal after a period of time. This period depends upon the nature of the glass from which the bulb is made.

High temperature thermometers

Mercury normally boils at 357° C at atmospheric pressure. In order to extend the range of a mercury-in-glass thermometer beyond this temperature, the top end of the thermometer bore is enlarged into a bulb having a capacity of about 20 times that of the bore of the stem. This bulb, together with the bore above the mercury, is then filled with nitrogen or carbon dioxide at a sufficiently high pressure to prevent the mercury boiling at the highest temperature at which the thermometer will be used. In order to extend the range to 550° C, a pressure of about 20 bar is required. In spite of the existence of this gas at high pressure, there is a tendency for mercury to vaporise from the top of the column at high temperatures and to condense on the cooler portions of the stem in the form of minute globules which will not join up again with the main bulk of the mercury. It is, therefore, inadvisable to expose a thermometer to high temperatures for prolonged periods.

At high temperatures the correction for the temperature of the emergent stem becomes particularly important, and the thermometer should, if possible, be immersed to the top of the mercury column. Where this is not possible, the thermometer should be immersed as far as conditions permit, and a correction made to the observed reading, for the emergent column. To do this, the average temperature of the emergent column should be found by means of a short thermometer placed in several positions near to the stem. The emergent column correction may then be found from the formula:

$$\text{Correction} = 0.0016(t_1 - t_2)n \text{ on Celsius Scale};$$

where, t_1, is the temperature of the thermometer bulb;

t_2, is the average temperature of the emergent column;

n, is the number of degrees exposed.

The numerical constant is the coefficient of apparent expansion of mercury in glass.

Adjustable mercury-in-glass electric-contact thermometer

A mercury-in-glass thermometer can form the basis of a simple two-step temperature controller which will control the temperature of an enclosure at any value between 40° C and 350° C.

Mercury is a good electrical conductor. By introducing into the bore of a thermometer two platinum contact wires, one fixed at the lower end of the scale and the other either fixed or adjustable from outside the thermometer stem, it is possible to arrange for an electrical circuit to be completed when a pre-determined temperature is reached. The current through the circuit is limited to about 25 mA, but this is ample to operate a relay which switches off the whole, or a portion, of the heating element of a small electrically-heated chamber. In this way the temperature of the chamber may be controlled to within $0 \cdot 1°$ C. Such chambers have a large variety of uses. One such use is maintaining at a constant temperature quartz crystals used to control the frequency of radio transmitters.

Use of liquids other than mercury

In certain industrial uses, particularly in industries where the escape of mercury from a broken bulb might cause considerable damage to the products, other liquids are used to fill the thermometer. These liquids are also used where the temperature range of the mercury-in-glass thermometer is not suitable. Details of the liquids used in thermometers are given below together with their range of usefulness.

Liquid	Temperature range °C
1. *Mercury*	−35 to +510
2. *Alcohol*	−80 to +70
3. *Toluene*	−80 to +100
4. *Pentane*	−200 to +30
5. *Creosote*	−5 to +200

4.3.2.2   LIQUID-IN-METAL THERMOMETERS

Mercury-in-steel thermometer

Two distinct disadvantages restrict the usefulness of liquid-in-glass thermometers in industry—glass is very fragile, and the position of the thermometer for accurate temperature measurement is not always the best position for reading the scale of the thermometer.

These difficulties are overcome in the mercury-in-steel thermometer shown in *Figure 4.14*. This type of thermometer works on exactly the same principle as the liquid-in-glass thermometer. The glass bulb is, however, replaced by a steel bulb and the glass capillary tube by one of stainless steel. As the liquid in the system is now no longer visible, a Bourdon tube is used to measure the change in its volume. The Bourdon tube, the bulb and the capillary tube are completely filled with mercury, usually at a high pressure. When suitably designed, the capillary tube may be of considerable length so that the indicator operated by the Bourdon tube may be some distance away from the bulb. In this case the instrument is described as being a 'distant reading' or 'transmitting' type.

When the temperature rises, the mercury in the bulb expands more than the bulb so that some mercury is driven through the capillary tube into the

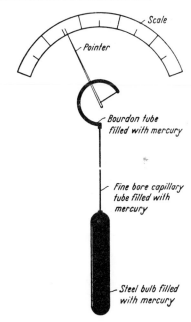

Figure 4.14 Mercury-in-steel thermometer

Bourdon tube. As the temperature continues to rise, increasing amounts of mercury will be driven into the Bourdon tube, causing it to uncurl. One end of the Bourdon tube is fixed, while the motion of the other end is communicated to the pointer or pen arm. As there is a large force available the Bourdon tube may be made robust and will give good pointer control and reliable readings.

The Bourdon tube may have a variety of forms, and the method of transmitting the motion to the pointer also varies. *Figure 4.15* shows one form of Bourdon tube in which the motion of the free end is transmitted to the pointer

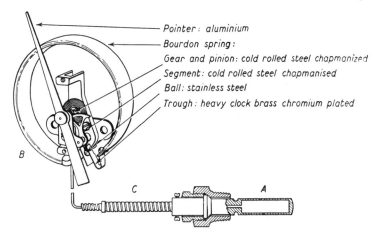

Figure 4.15 Construction of mercury-in-steel thermometer. (By courtesy, Foxboro-Yoxall Ltd.)

by means of a segment and pinion. The free end of the tube forms a trough in which a stainless ball at the end of the segment is free to move. The ball is held against the side of the trough by the tension in the hair spring. By using this form of construction lost motion* and angularity error† are avoided, and friction reduced to a minimum. Ambient-temperature compensation may be obtained by using a bimetallic strip, or by using twin Bourdon tubes in the manner described under the heading of capillary compensation.

Figure 4.16 shows a Bourdon tube having a different form, and a different method of transmitting the motion to the pen arm. This Bourdon tube is made of steel tube having an almost flat section. A continuous strip of the tubing is wound into two coils of several turns. The coils are arranged one behind the other so that the free end of each is at the centre, while the outer

Figure 4.16
Negretti and Zambra Bourdon tube
(By courtesy, Negretti and Zambra Ltd.)

turn of the coils is common to both, and can be seen in the illustration. One end of the continuous tube—the inner end of the back coil—is fixed and leads to the capillary tube; while the other end—the inner end of the front coil—is closed, and is attached to the pointer through a small bimetallic coil which forms a continuation of the Bourdon tube. This bimetallic coil compensates for changes brought about in the elastic properties of the Bourdon tube and in the volume of the mercury within the Bourdon tube owing to ambient temperature changes.

This particular formation of the tube causes the pointer to rotate truly about its axis without the help of bearings, but bearings are provided to keep the pointer steady in the presence of vibration. The friction at the bear-

---

*If there is any play in the various portions of the linkage which transmits the motion of the free end of the Bourdon tube to the pointer, it is possible for the free end of the Bourdon tube to move some short distance without there being a corresponding movement of the pointer. Such movement is called 'lost motion'.

†If the free end of the Bourdon tube is connected to the segment by means of a link which is pivoted at the lower end of the segment, then as the Bourdon tube uncurls, the angle between this link and the segment will change. Owing to this change in angle, equal movements of the free end of the Bourdon tube may not result in equal movement by the free end of the pointer. The errors introduced into the readings of an instrument due to this cause are called 'angularity errors'.

ings will, therefore, be very small as there is little load on them. As the end of the Bourdon rotates the pointer directly, there will be no backlash.

Thermometer bulbs

The thermometer bulb may have a large variety of forms, depending upon the use to which it is put. If the average temperature of a large enclosure is required, the bulb may take the form of a considerable length of tube of small diameter either arranged as a U or wound into a spiral. This form of bulb is very useful when the temperature of a gas is being measured, for it presents a large surface area to the gas and is therefore more responsive than the forms having a smaller surface area for the same cubic capacity.

In the more usual form, the bulb is cylindrical in shape and has a robust wall: the size of the cylinder depends upon many factors, such as the type of filling medium and the temperature range of the instrument, but in all cases, the ratio

$$\frac{\text{surface area}}{\text{volume}}$$

is kept at a maximum to reduce the time lag in the response of the thermometer.

The flange for attaching the bulb to the vessel in which it is placed also has a variety of forms depending upon whether the junction has to be gas-tight or not, and upon many other factors. *Figure 4.17* show some forms of bulbs.

The use of thermometer-wells, pockets or sheaths

Frequently the conditions on certain plants make it necessary to place the thermometer bulb in a protective pocket. Where the gas or liquid whose temperature is being measured is at a pressure other than atmospheric, the pocket will prevent the bulb being subjected to this pressure, and will enable the bulb to be changed without the necessity of shutting down the plant.

Where corrosive conditions exist, the material of the pocket is chosen so that it resists the corrosive action of the plant contents. In addition, the pocket

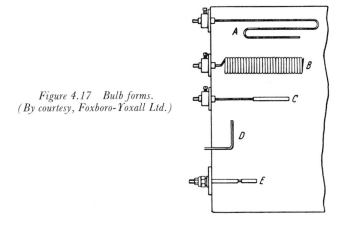

Figure 4.17   Bulb forms.
(By courtesy, Foxboro-Yoxall Ltd.)

itself may be coated on the outside to give greater protection. The well, or pocket, is coated with a lead sheath when sulphuric acid is present, while on other plants the well may be glass-coated. The use of refractory sheaths will be dealt with later in connection with electrical methods of measuring temperature.

The speed of response of a bulb in a separable pocket will be slower than that of an unprotected bulb.

This loss of speed may be reduced by keeping the clearance between bulb and pocket down to the absolute minimum and filling the space with oil, mercury, powdered metal or carbon. Mercury cannot be used except when the bulb and well are made of steel or an alloy which does not amalgamate with mercury. Oil evaporates or tends to break down into wax as it ages.

A satisfactory method of increasing the rate of heat transfer between the pocket and bulb is illustrated in *Figure 4.18*. In this method a very thin

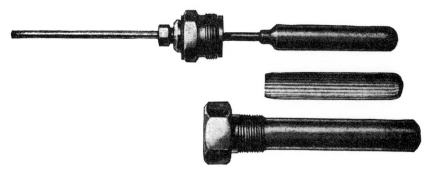

Figure 4.18   Taylor thermospeed separable-well tube systems. (By courtesy, Taylor Instrument Companies Ltd.)

corrugated aluminium sleeve is inserted between the bulb and pocket on one side. This forces the bulb over to the other side, ensuring metal-to-metal contact on this side, while on the other side, the sleeve itself, being made of aluminium which has a high thermal conductivity, provides a reasonable path for the heat. In addition the bulb should be placed well down in the pocket to reduce the possibility of errors due to heat conducted by the pocket to the outside with consequent reduction of the temperature at the bulb. For the same reason, the pocket itself should be well immersed into the fluid whose temperature is being measured.

Capillary tube and its compensation for ambient temperature

The capillary tube used in the mercury-in-steel thermometer is usually made from stainless steel, as mercury will amalgamate with other metals. Changes of temperature affect the capillary and the mercury it contains, and hence the thermometer reading; but if the capillary has a very small capacity, the error owing to changes in the ambient temperature will be negligible.

Where a capillary tube of an appreciable length is used, it is necessary to compensate for the effects brought about by changes in the temperature in the neighbourhood of the tube. This may be done in a number of ways. *Figure 4.19* illustrates a method which compensates not only for the changes of temperature of the capillary tube, but also for the changes of temperature within the instrument case. In order to achieve complete temperature compensation two thermal systems are used, which are identical in every respect except that one has a bulb and the other has not. The capillary tubes run alongside each other, and the Bourdon tubes are in close proximity within the same case. If the pointer is arranged to indicate the difference in movement between the free ends of the two Bourdon tubes, then it will be indicating an effect which is due to the temperature change in the bulb only. If compensation for case temperature only is required, then the capillary tube is omitted in the compensating system, but in this case the length of capillary tube used in the uncompensated system should not exceed about 8 m.

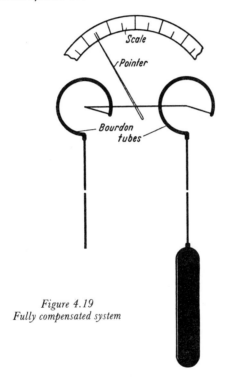

Figure 4.19
Fully compensated system

Another method of compensating for temperature changes in the capillary tube is to use a tube of comparatively large bore and to insert into the bore a wire made of invar, or other alloy with a very low coefficient of expansion. Mercury has a coefficient of cubical expansion about 6 times greater than that of stainless steel. If the expansion of the invar wire may be regarded as being negligibly small, and the wire is arranged to fill five-sixths of the volume of the capillary bore, then the increase in volume of the mercury, which fills the remaining one-sixth of the bore, will exactly compensate for the increase in volume of the containing capillary tube. This method requires the dimensions both of the bore of the capillary tube and of the diameter of the wire insert to be accurate to within very narrow limits for accurate compensation. The insert may not necessarily be continuous, but may take the form of short rods, in which case it is, however, difficult to eliminate all trapped gases.

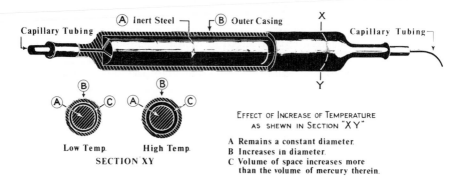

Figure 4.20   Ambient temperature compensation chambers. (By courtesy, Negretti and Zambra Ltd.)

Compensation for changes in the temperature of the capillary tube may also be achieved by introducing compensating chambers, of the form shown in *Figure 4.20*, at intervals along the length of the capillary tube. These chambers operate on exactly the same principle as the invar-wire-insert type of capillary tube, but the proportion of the chamber occupied by the invar is now arranged to compensate for the relative increase in volume of the mercury within the chamber and in the intervening length of capillary tube.

Other filling liquids

Admirable though mercury may be for thermometers in certain circumstances, it has its limitations, particularly at the lower end of the temperature scale. It is also very expensive to weld mercury systems in stainless steel. For these and other reasons, other liquids are used in place of mercury. Details of the liquids used in liquid-in-metal thermometers, with their usual temperature ranges, are given in Table 4.3.

When liquids other than mercury are used, the bulb and capillary tube need no longer be made of steel. The material of the bulb may, therefore, be

Table 4.3

Liquid	Temperature range °C
Mercury	$-39°$ to $+650°$
Xylene	$-40°$ to $+400°$
Alcohol	$-46°$ to $+150°$
Ether	$+20°$ to $+90°$
Other organic liquids	$-87°$ to $+260°$

chosen from a wide range of metals and alloys, and is selected to give the maximum resistance to any corrosive action which may be present where the bulb is to be used.

The capillary tube, too, may be made from a variety of materials, although copper and bronze are the most common. When capillary tubes are made from materials other than stainless steel, it may be necessary to protect them from corrosion or mechanical damage. This may be done by covering the tube with asbestos, and winding the whole in a heavy spiral of bronze. In cases where a bronze outer casing is likely to be damaged either by acid fumes or mechanically, it may be replaced by a stainless steel spiral which results in a much stronger but slightly less flexible construction. For use in damp places, or where the tube is liable to be attacked by acid fumes, the capillary and bronze spiral may be protected by a covering of moulded rubber, polyvinyl chloride, or rubber-covered woven-fabric hose. For use on chemical plants such as sulphuric-acid plants both the capillary tube and the bulb are protected by a covering of lead.

The construction of the liquid-in-metal thermometer is the same as that of the mercury-in-steel thermometer, and compensation for changes in ambient temperature may be achieved in the same ways.

Further facts about liquid-in-metal thermometers will be found in Table 4.5 showing the comparison of the various forms of non-electrical dial thermometers.

4.3.3 Expansion of gases; gas thermometers

It has already been stated (equation 3.25), that the volume of a gas increase with temperature, if the pressure is maintained constant; and the pressure increases with temperature, if the volume is maintained constant. If, therefore, a certain volume of inert gas is enclosed in a bulb, capillary and Bourdon tube, and most of the gas is in the bulb, then the pressure as indicated by the Bourdon tube, may be calibrated in terms of the temperature of the bulb. This is the principle of the gas-filled thermometer.

Since the pressure of a gas maintained at constant volume increases by $1/273$ of its pressure at $0°$ C for each degree rise in temperature, the scale will be linear provided the increase in volume of the Bourdon tube, as it uncurls, can be neglected in comparison with the total volume of gas.

An advantage of the gas-filled thermometer is that the gas in the bulb has a lower thermal capacity than a similar quantity of liquid, so that the response of the thermometer to temperature changes will be more rapid than that for a liquid-filled system with a bulb of the same size and shape.

The coefficient of cubical expansion of a gas is many times larger than that of a liquid or solid (air, 0·0037; mercury, 0·00018; stainless steel, 0·00003). It would therefore appear at first sight, that the bulb for a gas-filled system would be smaller than that for a liquid-filled system. The bulb must, however, have a cubical capacity many times larger than that of the capillary tube and Bourdon tube, if the effects of ambient temperature changes upon the system are to be negligible.

It is extremely difficult to get accurate ambient-temperature compensation in any other way. The change in dimensions of the capillary tube due to a temperature change is negligible in comparison with the expansion of the gas. Introducing an invar wire into the capillary bore would not be a solution to the problem, because the wire would occupy such a large proportion of the bore that extremely small variations in the dimensions of the bore or wire would be serious.

Placing an exactly similar capillary and Bourdon tube alongside that of the measuring system, and measuring the difference in the change of the two Bourdons does not give accurate compensation. This can be seen from the following example: Suppose the capillary and Bourdon tube have a capacity equal to 1/100th part of the capacity of the whole system. Let the ambient temperature rise by 10° C, while the bulb remains at the same temperature. In the compensating system the pressure will increase by 10/273 of the pressure at 0° C. In the measuring system this temporary increase in pressure in the capillary tube will soon be reduced by gas flowing into the bulb from the capillary tube until the pressures in the bulb and capillary are the same. Thus, the increase in pressure in the measuring system will only be about one-hundredth of the pressure increase in the compensating system.

More accurate compensation would be obtained by having a compensating system which also included a bulb which is maintained at ambient temperature, but this again would not give the completely accurate compensation.

Further facts about gas expansion thermometers will be found in Table 4.5, in which certain forms of dial thermometers are compared.

4.4   CHANGE-OF-STATE THERMOMETERS

4.4.1   Vapour-pressure thermometers

Suppose a container is partially filled with liquid as shown in *Figure 4.21*, the space above the liquid being completely evacuated. The molecules of liquid will be in motion, and will be moving in an entirely random manner. From time to time, molecules having a vertical component of velocity will reach the surface of the liquid. If this vertical component of velocity is great enough, a molecule will be able to leave the liquid in spite of the fact that other molecules are attracting it back into the liquid. Thus, after a time a number of liquid molecules will occupy the space above the liquid. These molecules, too, will be in a state of random motion, and from time to time molecules will leave the vapour and pass back into the liquid.

At first, the rate at which molecules are returning to the liquid will be less than the rate at which they leave, and the vapour above the liquid is said to be 'unsaturated'. Eventually, however, as the number of molecules in the vapour state increases, the rate at which the molecules leave the liquid will be exactly equal to the rate at which they return, and the quantity—and, therefore, the pressure—of the vapour in the space will remain constant. When this is so, the vapour is said to be 'saturated'.

If the temperature of the container is raised, the velocity of the molecules of liquid will be increased. The number of molecules now having sufficient energy to leave the liquid will be increased, and a greater number will leave the liquid. The rate at which molecules leave the liquid will now be greater than the rate at which they return so that the vapour pressure in the space will increase. It will continue to increase until the saturated vapour pressure for the new temperature is reached, when molecules of vapour will again be returning to the liquid at the same rate as that at which liquid molecules leave.

If instead of having a fixed top, the container has a movable piston, then, if the volume of the space is increased without the temperature changing, the vapour pressure will temporarily fall as the same number of vapour molecules are occupying a larger space. Now, however, the rate at which the molecules return to the liquid will be reduced, and will not be as great as the rate at which they leave the liquid. The pressure of the vapour will therefore increase, until the rate at which molecules return to the liquid again balances the rate at which they leave. The vapour pressure returns, therefore, to the 'saturated' vapour pressure for the particular temperature.

In the same way, if the volume of the space is reduced, molecules will leave the vapour at a greater rate than that at which they leave the liquid, so that the vapour pressure, which was temporarily increased, will fall until it is again the saturated vapour pressure of the liquid at the particular temperature.

Thus, provided there is always liquid and vapour present, the saturated vapour pressure of the liquid depends only upon its temperature, and is independent of the size of the container.

If a thermometer system similar to that described for gas-expansion thermometers is arranged so that the system contains both liquid and vapour, and the interface between liquid and vapour is in the bulb, i.e. at the temperature whose value is required, then the vapour pressure as measured by the Bourdon tube will give an indication of the temperature. This indication will be completely independent of the volume of the bulb, the capillary and the Bourdon tube and therefore independent of expansion due to ambient-temperature changes.

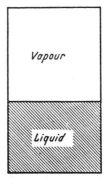

Figure 4.21   Vapour pressure

The form of the saturated-vapour-pressure graph has already been shown in *Figure 3.34* (page 182). The form of the saturated-vapour-graphs of other liquids is similar. It will be seen from the graph that the vapour pressure does not increase in a linear manner with temperature, but the rate of increase of pressure increases with temperature. Thus the instrument will have a scale on which the size of the divisions increases with temperature.

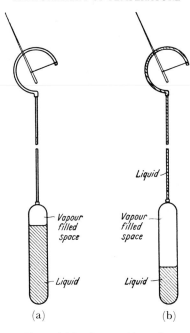

Figure 4.22   Cross-ambient effect

Form of the instrument

Many liquids are used for vapour-pressure-actuated thermometers. The
liquid is chosen so as to give the required temperature range, and so that the
usual operating temperature comes within the widely spaced graduations of
the instrument. In some forms of the instrument, a system of levers is arranged
to give a linear portion to the scale over a limited portion of its range. By
suitable choice of filling liquid, a wide variety of ranges is available, but the
range for any particular filling liquid is limited. The choice of material for
bulb construction is also very wide. Metals—such as copper, steel, Monel
metal, tantalum, etc.—may be used, and may be plated with nickel, chromium or silver where necessary.

In the actual instrument, shown diagrammatically in *Figure 4.22*, a quantity
of liquid partially fills the bulb. The surface of the liquid in the bulb should be
at the temperature which is being measured. The method by which the vapour
pressure developed in the bulb is transmitted to the Bourdon tube will depend
upon whether the temperature of the capillary tube and Bourdon tube is
above, or below, that of the bulb.

If the ambient temperature of the capillary and Bourdon tube is above
that of the bulb, then they will be full of vapour, which will transmit the
vapour pressure, as shown in *Figure 4.22(a)*. When the ambient temperature
increases, it will cause the vapour in the capillary and Bourdon tube to increase in pressure temporarily, but this will cause vapour in the bulb to
condense until the pressure is restored to the saturated-vapour pressure of
liquid at the temperature of the bulb.

If the temperature of the capillary and Bourdon tube is below that of the bulb, then the vapour will condense in the capillary and Bourdon tube, and the liquid so formed will transmit the pressure from the bulb to the Bourdon tube, as shown in *Figure 4.22(b)*. Again, increase in ambient temperature will cause the liquid in the tubes to expand and temporarily increase the pressure, but this again will cause vapour in the bulb to condense and restore the pressure to the characteristic saturated-vapour for the temperature of the bulb.

Measurement of temperatures near ambient temperatures

It can be seen that a difficulty will arise when the ambient temperature changes from below that of the bulb to a temperature above that of the bulb, or vice versa. A similar effect will be brought about when the temperature of the bulb crosses the ambient temperature.

Table 4.4

Liquid	Critical temp.	Boiling point	*Typical ranges available* °C
Argon	− 122° C	− 185·7° C	Used for measuring very low temperatures down to − 253° C in connection with the liquefaction of gases.
Methyl chloride	143° C	− 23·7° C	0 to 50
Sulphur dioxide	157° C	− 10° C	30 to 120
Butane (n)	154° C	− 0·6° C	20 to 80
Mehtyl bromide		4·6° C	30 to 85
Ethyl chloride	187° C	12·2° C	30 to 100
Di-ethyl ether	194° C	34·5° C	60 to 160
Ethyl-alcohol	243° C	78·5° C	30 to 180
Water	375° C	100° C	120 to 220
Toluene	321° C	110·5° C	150 to 250

Suppose the bulb temperature is just above the ambient temperature of capillary and Bourdon tube. The capillary and Bourdon tube will be full of liquid. Suppose the temperature of the bulb falls slightly, so that it is now just below the temperature of capillary and Bourdon tube. Vapour will now begin to condense in the bulb, but this vapour will be replaced by vapour produced by the evaporation of the liquid in the capillary tube. This will continue until all the liquid in the capillary and Bourdon tube has evaporated, and the vapour pressure in the system will then fall to the saturated-vapour pressure for the temperature of the bulb. This process will take a definite time to complete. The time required will depend upon the volume of liquid in the capillary and Bourdon tubes, and the area from which evaporation can take place. It will therefore increase with increase in length of the capillary tube.

If the temperature of the bulb rises, the vapour pressure in the system will increase to a value above the saturated vapour pressure for the temperature of the capillary and Bourdon tubes, and vapour will condense in these tubes. The pressure in the system will therefore not be the saturated-vapour pressure for the bulb temperature until the capillary and Bourdon tube are full of condensed vapour. There will be a very definite time lag in the response of a vapour-pressure thermometer every time the temperature of the bulb crosses the ambient temperature of the capillary and Bourdon tube.

In a similar way, there will be an inaccuracy in the reading of the thermometer when the ambient temperature crosses the temperature of the bulb; for this again will involve the distillation of liquid into, or out of, the capillary and Bourdon tube. If there is an appreciable difference in level between bulb and Bourdon tube, the reading will change even if the pressure at the bulb remains the same, for the pressure due to a capillary full of vapour will be less than the pressure due on a similar column of liquid. Thus, when the ambient temperature changes from below that of the bulb to above that of the bulb, the temperature reading will rise if the Bourdon tube is above the level of the bulb. For this reason a vapour-pressure actuated thermometer should not be used for measuring temperatures of about the same value as the ambient temperature of the capillary and Bourdon tube, particularly when the temperature fluctuates between values slightly above and slightly below the ambient temperature.

When the bulb of a vapour-pressure thermometer is suddenly cooled from a temperature above that of the capillary and Bourdon tube to a temperature below that of the Bourdon tube, the pointer may move downscale in rather an erratic manner. The movement downscale may be interrupted at intervals by sharp short 'kicks' upscale. This is due, not to any defect in the instrument, but to the fact that the evaporation of the liquid in the Bourdon tube and capillary under the reduced pressure brought about by the condensation in the bulb, does not take place at a uniform rate. The vapour is produced in a series of bubbles, just as water vapour is produced in boiling water. The production of one or more of these bubbles produces a temporary increase in pressure in the system which is shown as a kick of the pointer upscale.

In order to overcome the defects brought about by distillation of the liquid into, and out of, the capillary and Bourdon tubes, these tubes may be completely filled with a non-vaporising liquid which serves to transmit the pressure of the saturated vapour from the bulb to the measuring system. To prevent the non-vaporising liquid from draining out of the capillary tube, the capillary tube is continued well down into the bulb, as shown in *Figure 4.23*, and the bulb contains a small quantity of the non-vaporising fluid. The non-vaporising fluid will still tend to leave the capillary tube unless the bulb is kept upright.

Liquids used in vapour-pressure thermometers

A large variety of liquids is used in vapour-pressure thermometers; some of these, together with their boiling points, critical temperatures and typical ranges are given in Table 4.4.

Uses and limitations

The vapour-pressure thermometer is very widely used because it is cheaper than the mercury-in-steel thermometer, does not suffer from ambient

Table 4.5 COMPARISON OF THREE TYPES OF DIAL THERMOMETERS

	Liquid-in-metal	Gas expansion (constant volume)	Vapour pressure
Scale	Evenly divided.	Evenly divided.	Not evenly divided. Divisions increase in size as the temperature increases. Filling liquid chosen to give reasonably uniform scale in the neighbourhood of the operating temperatures.
Range (see tables)	Wide range is possible with a single filling liquid, particularly with mercury. By choice of suitable filling liquid, temperatures may be measured between −200°C and 570° C, but not with a single instrument.	Usually has a range of at least 50° C between −130° C and 540° C. Can be used for a lower temperature than mercury in steel.	Limited for a particular filling liquid, but by the choice of a suitable filling liquid almost any temperature between −50° C and 320° C may be measured. Instrument is not usually suitable for measuring temperatures near ambient temperatures owing to the lag introduced when bulb temperature crosses ambient temperature.
Power available to operate the indicator.	Ample power is available so that the Bourdon tube may be made robust and arranged to give good pointer control.	Power available is very much less than that from liquid expansion.	Power available is very much less than that from liquid expansion.
Effect of difference in level of bulb and Bourdon tube.	When the system is filled with a liquid at high pressure, errors due to difference of level between bulb and indicator will be small. If the difference in level is very large, a correction may be made.	No head error, as the pressure due to difference in level is negligible in comparison with the total pressure in the system.	Head error is not negligible, as the pressure in the system is not large. Error may be corrected over a limited range of temperature if the rates of pressure to deflection of the pointer can be considered constant over that range. In this case the error is corrected by resetting the pointer.
Effect of changes in barometric pressure.	Negligible.	May produce a large error. Error due to using the instrument at a different altitude from that at which it was calibrated may be corrected by adjusting the zero. Day to day variations in barometric pressure may be corrected for in the same way.	Error may be large, but may be corrected by resetting the pointer as for head error. Day to day errors due to variation in barometric pressure may be corrected by zero adjustment.
Capillary error.	Compensation for change in ambient temperature obtained as described in text (page 276).	Difficult to eliminate (see page 279).	No capillary error.
Changes in temperature at the indicator.	Compensation obtained by means of a bimetallic strip.	Compensation obtained by means of bimetallic strip.	Errors due to changes in the elasticity of the Bourdon tube are compensated for by means of a bimetallic strip.
Accuracy.	$\pm\frac{1}{2}\%$ of range to 320° C. $\pm 1\%$ of range above 320° C.	$\pm 1\%$ of differential range of the instrument if the temperature of the capillary and Bourdon tube does not vary too much.	$\pm 1\%$ of differential range even with wide temperature variation of the capillary and Bourdon tube.

temperature effects, and can have a smaller bulb than for the other types.

The range of an instrument using a particular liquid is limited by the fact that the maximum temperature for which it can be used must be well below the critical temperature for the liquid, and the range is further limited by the non-linear nature of the scale.

It cannot be used to measure a wide range of temperatures, or, except when specially designed, where the temperature to be measured fluctuates from slightly below to slightly above the ambient temperature of the capillary and Bourdon tube.

4.4.2 Pyrometric cones

At certain definite conditions of purity and pressure, substances change their state at fixed temperatures. This fact forms a useful basis for fixing temperatures, and is the basis of the scales of temperature.

For example, the melting points of metals give a useful method of determining the electromotive force of a thermocouple at certain fixed points on the International Practical Temperature Scale and this method will be described in the section on thermocouples.

In a similar way, the melting points of mixtures of certain minerals are used extensively in the ceramic industry to determine the temperature of kilns. These minerals, being similar in nature to the ceramic ware, behave in a manner which indicates what the behaviour of the pottery under similar conditions is likely to be. The mixtures, which consist of silicate minerals such as kaolin or china clay (aluminium silicate), talc (magnesium silicate), felspar (sodium aluminium silicate), quartz (silica) etc., together with other minerals such as calcium carbonate, are made up in the form of cones known as Seger cones. By varying the composition of the cones, a range of temperature between 600° C and 2000° C may be covered in convenient steps.

A series of cones is placed in the kiln. Those of lower melting point will melt, but eventually a cone is found which will just bend over. This cone indicates the temperature of the kiln. This can be confirmed by the fact that the cone of next higher melting point does not melt.

Since the material of the cone is not a very good conductor of heat, a definite time is required for the cone to become fluid, so that the actual temperature at which the cone will bend will depend to a certain extent upon the rate of heating. In order to obtain the maximum accuracy, which is of the order of $\pm 10°$ C, the cones must, therefore, be heated at a controlled rate.

4.5   ELECTRICAL METHODS OF MEASURING TEMPERATURE

4.5.1 Simple electronics

In order to understand electrical methods of determining temperature, and in particular, those involving the use of transistors and photo-electric cells, it is as well to have some idea of what constitutes an electric current.

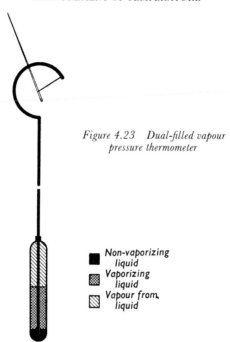

Figure 4.23   Dual-filled vapour
pressure thermometer

■ Non-vaporizing
    liquid
▨ Vaporizing
    liquid
▧ Vapour from,
    liquid

The structure of matter

All matter is made up of very small particles called atoms. In an element all these atoms are made up in the same way. A compound consists of groups of atoms called molecules. All the atoms in the group will not be identical; some will differ from others according to the nature of the compound. A water molecule is made up of two atoms of hydrogen and one of oxygen, while a molecule of cane sugar is made up of 12 atoms of carbon, 22 of hydrogen and 11 of oxygen.

Until the end of the nineteenth century it was believed that the atom was incapable of further subdivision, and that atoms of different elements were made up of different materials. It is now realized that the atom is not a fundamental particle but is made up of several component parts, and that atoms of different elements differ only in the number and arrangement of these component parts. The most important of these parts are the electron, the proton, and the neutron. The structure of all atoms is similar to that of a miniature solar system. They consist of a nucleus made up of protons and neutrons. Protons and neutrons have an approximately equal mass, and this is many times (1840 times) greater than that of the electron, so that the nucleus constitutes almost all the mass of the atom. Around the central nucleus, electrons move in orbits in the same way as the planets move around the sun, except that the orbits of the electrons are not all in the same plane. The electron has a negative charge of electricity, while the proton has an equal positive charge. The neutron has no electric charge.

In the normal state, an atom has as many electrons as it has protons, so that it is electrically neutral. If, however, it loses one or more of its electrons, it becomes positively charged and is called a 'positive ion', and the atom is said to be 'ionised'. If the atom acquires one or more electrons in addition to its normal complement, it is negatively charged, and is described as a 'negative ion'.

The simplest atom is the hydrogen atom. This consists of a nucleus of one proton around which a single electron moves. The electron is maintained in its orbit by the attractive force between the positively charged nucleus and the negatively charged electron. The helium nucleus consists of two protons and two neutrons, and around this moves two planetary electrons. The helium atom, therefore, has four units of mass. The oxygen atom consists of eight protons and eight neutrons in the nucleus, and eight planetary electrons. Heavier atoms have greater numbers of protons, neutrons and electrons, but are built up in the same way.

The number or orbital electrons associated with a given nucleus is equal to the atomic number of the element. The electrons revolve in a number of orbits or shells. The first orbit is complete when it contains two electrons, and the atom with a single complete shell is that of the inert gas helium. The second shell is complete when it has 8 electrons, and the atom with the first two shells complete is that of the inert gas neon. The third shell is complete when it has either 8 or 18 electrons. The inert gas argon (atomic number 18) has 3 shells complete. The elements having atoms with complete outer shells are very stable because it is difficult to remove an electron from the complete shell or to insert other electrons into it. Atoms combine by virtue of the electrons in the outermost shells. An element with one electron in the outermost shell will readily combine with an atom whose outermost ring is deficient in one electron. The number of electrons in the outermost shell, or the number of electrons required to complete the outermost shell, determines the chemical valency of the element, and for this reason the outermost shell is often called the valence shell.

Electrical conductors

The copper atom has 29 orbital electrons, 28 of which completely fill the three inner shells leaving one for the fourth shell. Similarly silver has 47 orbital electrons, 46 of which completely fill the four inner shells leaving one for the fifth shell. The electrons in the incomplete outer shell behave as if they were 'free' and pass readily from one atom to another. The movement of these 'free' electrons is entirely haphazard and in the normal way there is no movement in any particular direction. If, however, an electromotive force such as that produced by a battery is applied to a wire the electrons will be driven from the negative pole of the battery towards the positive pole, and there flows what is known as an electric current. Elements which possess 'free' electrons are termed electrical conductors.

Semiconductors

As the name suggests semiconductors are materials with a conductivity lying between that of a conductor and that of an insulator. These materials are of particular importance in the manufacture of the transistor, a small semiconducting device which can perform the function of a thermionic valve but with far greater efficiency. The most important elements in this field are germanium and silicon.

The germanium atom has 32 orbital electrons, 28 filling the three inner shells, with the remaining 4 occupying the outer shell. Likewise the silicon atom has 14 orbital electrons, 10 filling the two inner shells with the remaining 4 occupying the outer shell.

The significant feature of the two structures is that the outermost shells contain 4 electrons. If the atoms are considered individually it would be expected that the electrons in the valence shells of these elements could be readily displaced, and that these elements would be good electrical conductors. In fact, crystals of pure germanium and pure silicon are very poor conductors. The explanation of this poor conductivity lies in the relationship between the atoms and their neighbours in the regular geometric pattern of the crystal structure. Each atom is surrounded by similar atoms having four valence bonds and each completes its outer shell by sharing a valence electron with four of its neighbours. These bonds between atoms sharing valence electrons are called co-valent bonds. Thus, in pure germanium and pure silicon crystals, the atoms behave as if there were no free electrons and the materials are poor conductors.

In practice germanium and silicon are partial conductors owing to the presence of traces of impurity and due to the fact that the thermal energy is sufficient even at ambient temperature for a small number of electrons to break the co-valent bonds. If, however, the kinetic energy of the atoms is increased by subjecting them to light or by increasing the temperature, more valence electrons are freed and the conductivity increases. When the temperature of germanium is increased to 100° C the conductivity increases to such an extent that transistor action is impossible and for this reason germanium transistors should not be operated above say 50° C if a reasonable life is required. Silicon transistors will however give a satisfactory life even if operated at temperatures up to 150° C.

Effect of impurities

Suppose an atom of a pentavalent element such as arsenic is introduced into the crystal structure of the pure germainum taking the place of a germanium atom. The pentavalent atom has 5 electrons in the outer shell, and when 4 of these form co-valent bonds with the electrons of the neighbouring atoms the remaining electron is left unattached in the same way as the free electron in the copper and silver. By virtue of the added impurity the crystal has become conducting, and the added impurity is called a 'donor' impurity because it gives 'free' electrons to the crystal. Germanium treated with a pentavalent impurity is termed n-type because negatively charged particles are available to carry current through the crystal.

Suppose on the other hand a germanium atom in the crystal is replaced by an atom of a trivalent substance such as indium. The trivalent atom has 3 electrons in its outermost orbit so that when these electrons in the indium atom form co-valent bonds with outer electrons of neighbouring germanium atoms, the group is still deficient in one electron. The group will therefore behave in the same way as a positively charged particle with a charge equal in magnitude to an electron. In semiconductor theory this deficiency is called a 'hole', and the introduction of the trivalent impurity is said to give rise to holes in the crystal of pure germanium. These can carry current through the crystal and because the current carriers have a positive charge, germanium treated with a trivalent impurity is termed p-type. As the impurity takes electrons from the crystal structure it is called an 'acceptor' impurity. When a battery is connected across a crystal of p-type germanium a current can flow. The holes have an effective positive charge so that they tend to move under the influence of the electromotive force of the battery from the positive to the negative terminal of the battery. It is important to realise that the actual atoms do not move but the passage of an electron from one atom to another causes the deficiency or hole to pass through the material. Each time the deficiency reaches the negative terminal, an electron from the terminal neutralises it. At the same time an electron from a co-valent bond enters the positive terminal resulting in a hole in a crystal. This hole moves towards the negative terminal and thus a stream of holes flows from the positive to the negative terminal, and this flow of holes is equivalent to a flow of electrons in the opposite direction.

Transistors

It is important to realise that the concentration of impurity required in transistor manufacture is very small. The concentration of impurity is usually of the order of 1 part in 10^8 and this increases the conductivity by 16 times. Before the impurity is introduced, the germanium or silicon must be refined to a degree of purity beyond that attainable by normal chemical methods. The remaining impurity must not exceed about 1 part in 10^{10}. The process is called zone refining. An ingot of the metal is drawn slowly through a tube which is surrounded at intervals by radio-frequency heating coils. Only a few zones of the ingots are molten at any one time. The impurities become concentrated in the molten zones rather than in the solid portions. By drawing the ingot several times through the tube, the impurities are swept to one end which is discarded. The selected material is then added in the correct proportion and the metal grown as a single crystal in a furnace filled with an atmosphere of nitrogen and hydrogen to prevent oxidation.

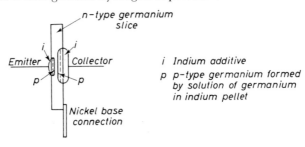

Figure 4.24   Cross-section of p-n-p transistor

Transistors may be made in a number of ways but most are made by the alloy-junction method. To manufacture a *p-n-p* transistor, a grown germanium crystal containing *n*-type additive is turned into slices about $4 \times 2 \times 0.12$ mm. A pellet of an acceptor impurity is placed on each side of the slice, the one which will form the collector being about three times the size of the one used for the emitter. The assembly is heated in a hydrogen atmosphere until the pellets melt and dissolve some of the germanium from the slice. On being cooled, the pellets with the dissolved germanium start to solidify and a crystal of *p*-type germanium grows at the solid–liquid interface. At a lower temperature, the rest of the pellet solidifies.

Figure 4.24 shows a cross-section of the resulting three-layer structure. Leads from the emitter and collector are soldered to the surplus material in the pellets, thus making non-rectifying contacts. A nickel tab is soldered to the slice to make a connection to the base. The assembly is etched to remove surface contamination, covered in moisture-proof grease and hermetically sealed into a glass envelope with the leads passing through the glass foot.

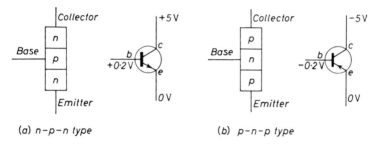

Figure 4.25   Representation, conventional symbols and typical voltages of transistors

The glass bulb is then coated with opaque paint to exclude light. *Figure 4.25* shows the representation of transistors, the conventional symbol and typical voltages. The arrow on the emitter points in the conventional direction $+$ to $-$.

Behaviour of a p-n or n-p junction

A crystal of *n*-type or *p*-type germanium is a linear conductor. It will carry a current equally well in either direction, i.e. reversing the applied voltage does not change the magnitude of the current. If, however, the crystal has *n*-type constitution at one end and *p*-type at the other, as indicated in *Figure 4.26(a)*, the crystal so produced has asymmetrical conducting properties, i.e. the magnitude of the current flowing through the crystal when an e.m.f. is applied between the ends depends on the polarity of the e.m.f., being small when the e.m.f. is in one direction, and large when it is reversed. Such as crystal has obvious applications as a detector or rectifier and is known as a junction diode.

Its behaviour is due to the potential barrier which exists at the junction of the two types of germanium. The pattern of electrical charges is as shown in *Figure 4.26(a)*. In the *n*-type germanium, if we assume the electrons are free, the atoms of the pentavalent impurity carry a positive charge and are fixed in space in the crystal lattice. Similarly, in the *p*-type germanium the atoms of the trivalent impurity carry a negative charge and these are also fixed. The ringed signs, therefore, represent the charges of the fixed impurity ions while the unringed signs represent the charges of the free electrons and holes. The free electrons being attracted by the fixed positive charges and repelled by the fixed negative charges move to the left, whilst the holes attracted by the fixed negative charges are repelled by the fixed positive charges tend to move to the right. The electrons and holes tend to move away from the junction and do not tend to neutralise each other. Rather, they behave as if a battery were connected with its positive terminal to the *n*-type region and its negative terminal to the *p*-type region. In other words the material behaves as if a potential barrier existed at the junction and this is represented by the battery shown dotted.

When an external battery is connected as shown in *Figure 4.26(a)*, the applied e.m.f. is in the same direction as the potential barrier tending to separate electrons and holes even more effectively, and no current is therefore

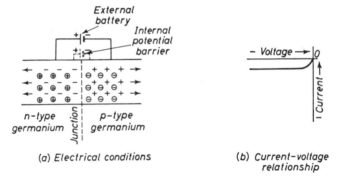

(a) Electrical conditions

(b) Current–voltage relationship

Figure 4.26   Reverse-bias condition of p-n junction

taken from the battery. This is an example of reverse bias. In practice there is usually a small current due to thermal dissociation of the co-valent bonds of the germanium atoms and the current–voltage relationship is as shown in *Figure 4.26(b)*. The significant feature is that the current is small and practically independent of voltage.

When the external battery is reversed as shown in *Figure 4.27(a)* the applied e.m.f. neutralises the effect of the internal potential barrier and a current can flow, the magnitude being shown in *Figure 4.27(b)*. The curve has a small slope for small applied voltages because the internal potential barrier is effective in preventing large migration of electrons and holes. Further increase in the applied e.m.f. completely offsets the internal e.m.f. and the current increases steeply with applied voltage. Junction diodes are very efficient, and generate very little heat in operation. Small diodes can rectify surprisingly large currents, e.g. a silicon junction diode less than 25 mm in diameter and 12 mm long can supply 5 amps at 100 volts.

Figure 4.28 illustrates the use of transistors as amplifiers. The current amplification factor which is the ratio of small change in collector current i_c

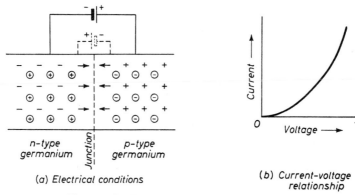

(a) Electrical conditions

(b) Current-voltage relationship

Figure 4.27 Forward-bias condition of p-n junction

to the change in emitter current i_e which gives rise to it is represented by α

i.e.
$$\alpha = \frac{i_c}{i_e}$$

For a junction transistor α is usually between 0·95 and 0·99.

The voltage gain can be calculated as follows. Suppose α is unity, then $i_c = i_e$. Suppose the input resistance is 50 ohm and the load resistance R_L is 4700 ohm. An input voltage of 1 mV will give rise to an input current i_e where

$$i_e = \frac{1 \times 10^{-3}}{50} \text{A} = 20 \times 10^{-6} \text{ A.}$$

Then, the output voltage $E_L = i_c R_L$
$$= 20 \times 10^{-6} \times 4700$$
$$= 94 \text{ mV.}$$

Thus, the voltage gain of a transistor circuit is approximately equal to the ratio of the load resistance to the input resistance.

Installation of Transistors

Particular attention should be given to the circuit diagram, as connecting a transistor with the incorrect polarity may change its characteristics permanently, or lead to its destruction.

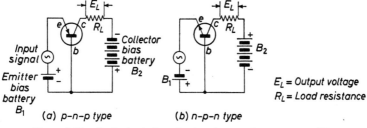

E_L = Output voltage
R_L = Load resistance

(a) p-n-p type   (b) n-p-n type

Figure 4.28   Basic circuit for using junction transistor as an amplifier

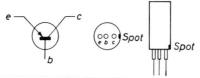

Figure 4.29   Examples of electrode connections of transistors

The electrode connections are given in the technical data in the form of diagrams. *Figure 4.29* shows those which apply to most small-signal low-power transistors. The central lead is for the base, and there is a greater space between the base and the collector than between base and emitter leads. The collector is also distinguished from the emitter by a spot on the adjacent part of the body.

Disconnect the supply while installing transistors, and if possible while making circuit adjustments. An accidental short circuit from base to the collector supply line may cause sufficient current to flow to damage the transistor. Also, in some circuits the supply voltage is higher than the rating of the transistor, the excess voltage being dropped in a series resistance in the collector and/or emitter circuits. Short-circuiting this resistance may

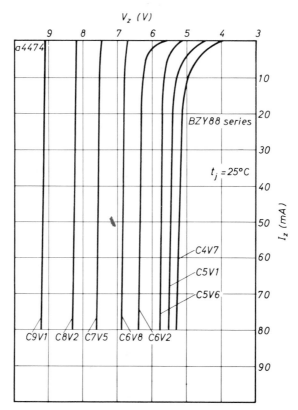

Figure 4.30   Typical dynamic characteristics of voltage regulator diode.
(By courtesy, Mullard Ltd.)

result in excessive voltage being applied to the transistor. Associated components and devices should not be inserted or replaced with the power supplies connected, because of the surges which can occur, for instance from the discharge of capacitors through the devices.

In order to facilitate soldering, leads are tinned. Soldering should be completed quickly and on no account should a transistor be allowed to heat up. The leads should be held in a cool pair of long jawed pliers which act as a thermal shunt and prevent heat from being conducted to the transistor. The electrical insulation between heating element and the bit of some electrical soldering irons is sufficiently poor to cause a dangerously high current to flow through the transistor. Such irons should be disconnected during soldering. Transistors may be dip soldered, usually at a solder temperature of 240° C for a maximum of 10 seconds to a point 2 mm from the seal.

Transistors may be mounted in any position. The only restriction on location may arise from the need to provide adequate cooling and ventilation.

Where the envelope of the transistor is provided with an opaque coating to exclude light, care must be taken to see that this is not damaged. High-energy radiation such as X-rays, γ-rays and neutrons affect the junction, usually adversely and permanently.

For most transistors the storage temperature must be within the range -55 to $+75°$ C, otherwise the transistor may be damaged.

Voltage regulator (Zener) diodes

Voltage regulator diodes are silicon semiconductor devices that have voltage-current characteristics similar to those in *Figure 4.30*. For a given type of diode the knee always occurs about the same voltage within close limits, the precise value being a constant for each individual diode.

Zener effect

When it is forward biased, the voltage regulator diode behaves as a normal diode. When the voltage is reversed, a leakage current of only a few micro-amps flows through the diode. This current is independent of the voltage over a wide range. If the voltage is increased it eventually reaches a value at which the current increases suddenly to a large value which may be as large as several amps. The current must however be limited to keep the heat dissipation within the power rating of the device. The voltage at which this occurs is known as the breakdown voltage V_{BR}. This effect was first discovered by Zener in his investigations of the breakdown phenomenon in dielectrics. When the reverse voltage across the diode is small, the current flowing is the surface leakage current plus the normal saturation current owing to the thermally generated holes and electrons. At first increasing the voltage does not increase the number of carriers and the current is therefore constant. However, as the voltage is increased, the electric field becomes sufficient to cause the disruption of the covalent bonds among the atoms close to the junction generating a large number of hole-electron pairs that produce the great increase in current. This action is known as the 'zener effect.'

The voltage at which the zener effect occurs depends upon how the silicon is doped and can be quite high. A sudden increase in current may be achieved below the voltage required to produce the zener effect by means of an 'avalanche effect'. If the covalent bonds are not broken the current remains at only a few microamps until the avalanche effect occurs. As the voltage across the diode is increased the velocity of the carriers is increased. Eventually the velocity is sufficient to cause ionisation when the electrons collide with semiconductor molecules. The carriers produced from the collisions take part in further collisions and produce even more carriers. Consequently the number of carriers and hence the current is rapidly increased by the avalanche effect.

Breakdown stability

Voltage breakdown in a regulator diode occurs at a definite stable level which is practically constant no matter whether the zener or avalanche effect is the cause. Provided the maximum permissible junction temperature is not exceeded the breakdown is reversible and non-destructive. The voltage across the diode after breakdown is known as the 'reference voltage' and is virtually constant. The voltage quoted in the published data indicate the voltage limits for a type of voltage regulator diode not an individual. Each individual diode will have its own constant reference voltage within the limits quoted.

When breakdown occurs, the large current change is accompanied by a small voltage change producing a drop in the dynamic resistance r_z. The published data usually quote the slope of the line relating r_z to current at three current levels.

Effect of temperature changes

By using pulsed currents when testing diodes, the temperature effects are minimised and the characteristics obtained relate to a fixed temperature so the published data applies when the junctions are initially at 25° C. The basic curve of the voltage regulator diode is its dynamic characteristic which is the variation of voltage with pulsed current. The slope of the characteristic is the

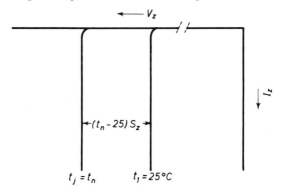

Figure 4.31   Typical dynamic characteristics of a voltage regulator diode showing the effect of temperature change

dynamic resistance r_z. *Figure 4.31* shows the characteristic at a temperature of 25° C and another ambient temperature t_n. The displacement between the curves is caused by the temperature sensitivity of the diode characteristics and equals a voltage of $S_z(t_n - 25)$ where S_z is the temperature coefficient of the diode.

The temperature coefficient is negative in diodes with a breakdown voltage below 5 V and positive in diodes with a higher breakdown voltage. Thus, by using two suitable diodes in series, it is possible to build stabilised supplies with a temperature coefficient approaching zero. In diodes with a breakdown voltage of 5 V, the temperature coefficient can be zero at one specific current over a limited range.

When a constant current is applied to a diode, thermal stability is not achieved immediately but may be delayed by as much as six minutes. During this period the voltage across the diode changes because of the temperature dependence of the reference voltage. This change in voltage is, however, a negligible fraction of the reference voltage.

In order to reduce the temperature change of a diode owing to heat generated within the diode it should be provided with an adequate heat sink.

Voltage reference circuits

Voltage reference circuits are used to provide highly stable voltage sources for potentiometeric and other similar types of instruments where a stable source is essential. The simplest form of stabilising circuit consists of a resistor and voltage regulator diode arranged as shown in *Figure 4.32*. The particular circuit shown has an output of 7·5 V at 100 mA when using a Mullard BZY96–C7V5 voltage reference diode, and has a fractional change coefficient S_F of 0·063 where S_F is the ratio of the percentage change in output to the percentage change in input causing the change in output assuming the load current and temperature are constant.

A higher voltage, or perhaps a voltage nearer the desired value, can be arranged by means of two voltage regulator diodes connected as shown in *Figure 4.32(b)*. By selecting diodes with opposite temperature coefficients this arrangement can be given a negligible temperature coefficient.

When a particularly stable output is needed it may be obtained from a two-stage regulator of the form shown in *Figure 4.32(c)*. The fractional change coefficient of this circuit is equal to the product of the fractional change coefficient for each stage. Thus, if the fractional change coefficient for each stage is 0·06 their combination will have a fractional change coefficient of 0·003 6. If the resistor between the two diodes is replaced by a potentiometer as shown in *Figure 4.32(d)* an adjustable stabilised output can be obtained.

By the use of the seven diodes and a suitable resistance network a secondary standard voltage source with a high temperature stability which has a performance greatly superior to that of the Weston Standard cell can be produced. It is undamaged by intermittent short circuits, is more resistant to mechanical shock and has an output which is practically constant from 30° C to 60° C. In a test of reliability such a source was switched on for one hour and off for twenty minutes repeatedly for one month. During this time no change of output was detected with a voltmeter having an accuracy of 0·01 per cent.

By using shunt or series transistors with the voltage regulator diodes the range of load currents of stabilised sources can be greatly increased.

Complete constant voltage unit

Figure 4.33(a) shows the schematic diagram of the Honeywell Constant Voltage Unit designed to replace manual or automatic standardisation in electronic potentiometers, eliminating the standard cell, standardising mechanism, dry cell battery and associated components. It is also available as a component unit for incorporation in any electric circuit.

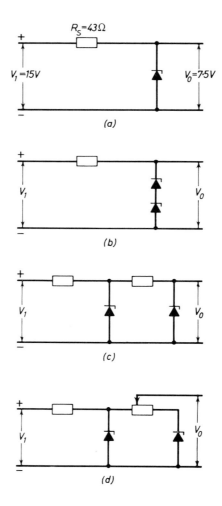

Figure 4.32
(a) Single-stage stabilising circuit with a single diode
(b) Single-stage stabilising circuit with more than one diode
(c) Two-stage stabilising circuit
(d) Stabilising circuit with adjustable output

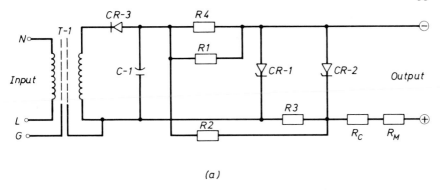

(a)

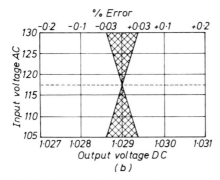

(b)

Figure 4.33   Constant voltage unit. (By courtesy, Honeywell Ltd.)

The Constant Voltage Unit consists of an a.c. d.c. power supply and an accurate voltage regulator. The a.c. supply is reduced by a step-down transformer T-1, converted to d.c. by the diode rectifier CR-3, and smoothed by capacitor C-1. The d.c. voltage is regulated by a two-stage Zener diode network, the first stage of which provides regulation in the order of $\frac{1}{2}$ per cent. This voltage is fed to the second stage which consists of a bridge network. In this, manganin wire-wound resistors and a Zener diode are employed for improved line voltage regulation and compensation for temperature effect of the first regulator stage. Cascading two Zener diode circuits provides a high degree of output stability. *Figure 4.33(b)* shows the percentage error of output voltage with changes of input voltage.

Small temperature changes have little effect on accuracy, and physically the unit will withstand temperatures from below zero up to 70° C without impairment of operation. The unit will not be damaged if submitted to temperatures from below freezing to 110° C.

4.5.1.1   THERMAL EMISSIONS OF ELECTRONS

The valve

When a metallic wire is heated sufficiently, electrons leave the wire through the surface. The rate of emission increases rapidly as the temperature rises but unless there is an electric field in the vicinity of the wire, a cloud of electrons will collect around the wire and some of the electrons will return to the wire. If a wire and a plate are placed in an evacuated bulb, as shown in *Figure 4.34*, and the plate is made positive relative to the wire, by connecting the plate and wire to a 'high tension' battery, then, when the wire is heated, the electrons leaving the wire will be attracted to the positive plate, and an electric current will flow from the plate to the wire across the space in the bulb. The current would be shown by a milliammeter in the plate circuit.

The value of the current which flows will depend upon the potential of the positive plate; the greater the potential difference between the wire and the plate, the greater will be the current. The current, however, will reach a maximum value when the potential difference between plate and wire is great enough to attract across the space all the electrons leaving the wire. If the wire is made of tungsten, it will have a high melting-point, and can be strongly heated and a measurable current obtained. The emission at any temperature can be greatly increased by coating the wire with a film of oxide of an alkaline earth. The effect of this coating is clearly seen from the comparative figures for the emission from certain surfaces under the same heating conditions as expressed in watts used: clean tungsten, 1 milliamp per watt; thoriated tungsten, 25 milliamp per watt; treated oxide coat 150–250 milliamp per watt.

The discovery of this effect was responsible for the disappearance of the old 'bright emitter' valve, and the development of the present-day efficient valves. The heater wire in the modern valve is usually used to heat a coated cylinder known as the cathode, which supplies the electrons which constitute the current.

Diode as a rectifier

The simplest type of valve is the diode. It consists of a plate or anode, and a coated cathode with its heater, all contained in an evacuated tube. When connected in a circuit as shown in *Figure 4.34*, a current will flow across the space in the valve only while the anode is at a positive potential relative to the cathode. When the anode is negative relative to the cathode, the electrons will be repelled towards the cathode. A positive current will, therefore, only flow from the anode to the cathode within the valve, never in the reverse

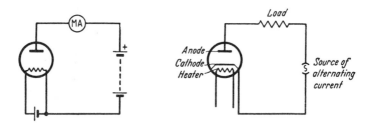

Figure 4.34   (left) The simple valve; (right) The diode as a rectifier

direction. This extremely valuable property operates when the valve is used to produce a unidirectional current, i.e. used as a rectifier.

The triode valve: control of the electron flow

When a third electrode consisting of a grid, or mesh, of wire is introduced into the valve and placed much nearer to the cathode than the anode, the voltage on this grid will have a much greater influence on the current through the valve than the voltage on the anode has. Thus, triode valves may be designed with a high 'amplification factor'; the amplification factor being defined as 'the change in anode voltage which will produce the same change in the anode current as a change of 1 volt in the grid voltage'.

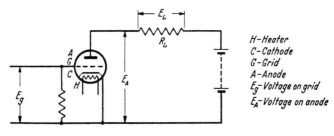

Figure 4.35   The triode

If a triode valve is connected to batteries and load resistance as shown in *Figure 4.35*, then the voltage drop, E_L, across the resistance, R_L, will be given by:

$$E_L = R_L i_a, \text{ where } i_a \text{ is the anode current.}$$

When the voltage on the grid varies, there will be a corresponding variation in the anode current and, consequently, a corresponding variation in the voltage drop, E_L. Owing to the amplification of the valve, however, the change in the voltage drop across R_L will be much greater than the change in the voltage on the grid. For example, suppose variation of 6 V in the voltage on the grid produces a change of 10 mA in the anode current; then, if $R_L = 10\,000$ ohm, the variation in the voltage drop across R_L will be $10\,000 \times {\cdot}01 = 100$ V.

A triode valve may, therefore, be used to amplify small variations in voltage. If further amplification is required, the output of the first valve may be fed on to the grid of a second valve through a transformer or condenser, which isolates the grid of the second valve from the high steady voltage of the anode, but transmits the varying voltage. The output of the second valve will then be the twice-amplified result of the original voltage variation.

Thyratrons or gas-filled triodes

When, instead of being completely evacuated, a triode valve contains an inert gas such as helium, argon, or neon, at an absolute pressure of about 10^{-3} mm of mercury, or when the valve contains mercury vapour, it behaves in a very different way from the evacuated valve. When the anode reaches a sufficiently high voltage relative to the cathode—known as the ionisation potential—the electrons traversing the valve will have sufficient energy to ionise the gas molecules in the valve when they collide with them. This produces further electrons together with positive ions. As a positive ion has the same charge as an electron, both will acquire the same kinetic energy in moving the same distance across the valve. The positive ion has, however, a much larger mass than the electron so it will move much more slowly than the electron, and will remain in the inter-electrode space for a much longer time. Thus, there is in the valve at any time, a considerable number of positive ions which will influence the behaviour of the valve.

If the ionisation produced is sufficient, the positive ions, which are attracted towards the cathode, will neutralise the negative space charge, or even produce a positive space charge. All electrons emitted by the cathode will, therefore, together with those produced by ionisation of the gas, move to the anode and the valve is capable of carrying a much larger current than a similar evacuated valve.

As in the evacuated valve, the value of the anode voltage at which the flow of current through the valve starts depends upon the voltage on the grid. By adjusting the voltage on the grid, the anode current may be arranged to start when the anode voltage reaches a definite value; or if the anode voltage is constant, the current will start when the grid voltage has some definite value. Once the anode current has started, however, there will be a large number of positive ions within the valve. If a negative potential is now applied to the grid, the grid attracts to itself a covering of positive ions which neutralise the effect of the negative voltage on the grid, and prevent the grid from exercising any controlling influence on the anode current. The current through the thyratron can, therefore, be controlled in the external circuit only, and can be stopped by reducing the value of anode voltage below the value of the ionisation potential, for a time which is long enough to allow the ions to recombine so that there are no longer any positive ions present.

If the anode voltage is applied to the valve when the cathode is cold, a discharge will occur in the gas in the valve. This is due to the fact that there will always be a few ions present in the gas owing to cosmic radiation. These ions will be accelerated within the valve and will produce further ionisation by collision. The ions produced will, in turn, be accelerated and produce further ionisation. The positive ions produced will move with high velocity towards the cathode which is without its protecting cloud of electrons. They will bombard the cathode and damage its emitting surface. It is, therefore, essential to delay the application of the anode voltage until the emission from the cathode has reached its full value. It is equally important that the cathode temperature should not be allowed to fall below the minimum value required to provide adequate emission during use, and when the valve is switched off, the anode voltage should be removed at the same time as, or slightly earlier than, the heating voltage.

4.5.1.2   PHOTO-ELECTRIC EFFECT

As was stated previously, the atom consists of a nucleus around which the orbital electrons move. These orbital electrons are considered to be arranged in orbits, or shells, according to the energy they possess. The electrons in the inner orbits are at a lower energy level than those in the outer orbits. Each shell has a definite quota of electrons, and in the normal state of the atom, the quota of the inner shells must be complete before electrons can occupy positions of a high energy level. In order to move an electron from one energy level to a higher one, or to remove it from an atom altogether, it is necessary to supply energy to the atom, usually in the form of electro-magnetic radiation such as heat or light. If an electron moves from an orbit of higher level to one at a lower level, the atom gives out energy in the form of radiation. This radiation will possess a definite quantity of energy, corresponding to the loss of energy of the electron when it moves from the higher to the lower energy level.

According to the quantum theory, suggested by Max Planck in 1900, radiation of a frequency v can only be emitted, or absorbed, in definite packets having an amount of energy which is a whole number multiple of a basic unit of radiant energy, hv, where h is known as Planck's constant ($h = 6.62 \times 10^{-34}$ Js).

In 1905, Einstein explained the photo-electric effect in terms of the quantum theory, and pictured light as a stream of 'light quanta' or 'photons' each having an energy of hv. When a photon of light falls on an atom it gives up its energy, hv, to the planetary electrons of the atom. The result of this increase in energy of a planetary electron will depend upon the initial energy level of the electron and the amount of additional energy given to it. The photo-electric effect therefore takes two forms:

1. In non-conductors which do not possess free electrons, the electrons may be moved from low energy levels to high energy levels where they may behave as 'free', or conduction, electrons. In semiconductors, free electrons and holes are produced. This results in an increase in the conductivity of the material and is known as the inner photo-electric effect. Photovoltaic and conductivity cells are based upon this effect.

2. In metals and other conductors, the electrons in the outer orbits at the higher energy levels are emitted by the metal provided the frequency of the incident light is high enough. (When a photon strikes the metal, part of its energy is used in liberating an electron and the rest in giving the electron kinetic energy.) Thus, when ultra-violet light falls on common metals like zinc and copper, electrons are emitted. Certain alkali metals like sodium, potassium and caesium will evince the same effect when illuminated with light of lower frequency, and will emit electrons when illuminated with visible or even infra-red radiation. The number of these electrons, known as photo-electrons, emitted per second is proportional to the intensity or brightness of the light. The velocity with which the electrons leave the surface of the metal does not depend upon the intensity of the light. The electrons leave the metal with a velocity which varies from zero to a sharply defined limit. The value of this upper limit depends upon the frequency of the light. If the frequency of the light falls below a certain minimum value, which is characteristic of the metal, then no photo-electrons are emitted however bright the light, as the photons have not sufficient energy to liberate the electrons. This form of the photo-electric effect is known as the outer photo-electric effect, and the emission type of photo-electric cell is based upon it.

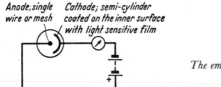

Anode; single   Cathode; semi-cylinder
wire or mesh   coated on the inner surface
with light sensitive film

Figure 4.36
The emission photo-electric cell

4.5.1.3   PHOTO-ELECTRIC CELLS

Several devices used as detectors of radiation are based on the photo-electric effect. These may be divided into three classes.

(1) Photo-conductive cells.
(2) Photo-emissive cells and photo-multipliers.
(3) Photo-voltaic cells and photo-transistors.

Photo-conductive cells

In this form of detector, the intensity of the radiation is measured as the increase in the conductivity produced in a layer of semi-conducting material owing to the inner-photo-electric effect. The direct effect of the illumination is to increase the number of mobile charge carriers in the material and hence increase the conductivity. Materials such as germanium, lead sulphide, lead selenide, lead telluride or indium antimonide are used. Since the cell resistance also changes with ambient temperature it is common practice to chop the radiation at a frequency of 800–1000 hertz, and to amplify the intermittent current by means of a valve amplifier. The sensitivity of the detector is dependent on the wavelength of the radiation, and the range of wavelengths for which it may be used is dependent on the material of the detector.

Photo-emissive cells and photo-multipliers

This type of cell, shown in *Figure 4.36*, is constructed in a similar way to the diode valve. The cathode is usually in the form of a semi-cylinder coated on the inner surface with a light sensitive film of materials such as sodium, potassium, caesium or caesium oxide on silver, potassium on silver, or antimony-caesium. The anode, which is maintained at a potential of about 100 volts above that of the cathode, is made in the form of a ring, or grid, so that light may pass through it on to the cathode. When the light-sensitive surface of the cathode is illuminated, electrons are emitted from the cathode and are attracted to the anode and a current flows in the external circuit. The sensitivity of the cell is increased by introducing some inert gas, such as neon or argon, at a pressure of a few millimetres of mercury into the envelope. The photo-electrons in these circumstances produce a further supply of electrons and ions by collision with the gas molecules. These electrons and ions in turn produce further ionisation so that a greatly increased current flows.

Emission types of cells have a large variety of uses. They are used to measure the blackness of the image on the sound-track of a film, and by controlling amplifying equipment will reproduce speech or music. They are used in industry for counting purposes, for opening doors at the approach of a truck, and for a variety of other purposes.

The sensitivity of a photo-emissive cell may be greatly increased by using the fact that if fast-moving electrons strike a target further electrons may be liberated. The number of electrons emitted depends upon the material of the target and upon the angle of incidence of the primary electrons. In photo-multipliers the electrons emitted are accelerated by an electrostatic field in such a way that they strike a number of targets in turn. A separate voltage of about 150 volt is applied to each stage. A final current as large as 1 mA may be obtained but for accurate quantitative work it is usually limited to about 10 per cent of this.

Photo-voltaic cells and photo-transistors

The operation of a photo-voltaic cell is dependent upon the existence of a potential barrier at the junction of a semiconductor and a metal, or at the junction between *n*-type and *p*-type semiconductor material (see *Figure 4.26*). The mobile carriers on the two sides of the junction will diffuse across the junction until a potential barrier is set up which limits the further exchange of carriers.

Some carriers receive sufficient random energy to cross the barrier and this constitutes the normal leakage, or dark, current. If light is allowed to fall on the junction, hole-electron pairs are created on both sides of the junction. The barrier potential sweeps the holes one way and the electrons the other. The current now flowing is the light current equal to the dark current plus the photo-electric current.

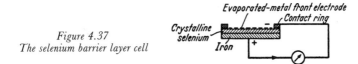

Figure 4.37
The selenium barrier layer cell

Two forms of photo-voltaic cells are used, the selenium barrier-layer cell shown in *Figure 4.37* and the silicon solar cell shown in *Figure 4.38*. The selenium cell consists of an iron base, roughened by sandblasting. On to this base, selenium, with the addition of about 2 per cent of thorium, zirconium, or cerium, is melted, and then annealed to produce the crystalline form. The top electrode is then applied by evaporation, sputtering, or spraying.

Formerly the selenium cell was the most common, but its industrial use is now limited to the measurement of the temperature of streams of molten iron from the cupola. Its place in radiation pyrometry has now been taken by the silicon solar cell which was developed to convert sunlight into electricity for use in space vehicles. It is basically a silicon diode, consisting of a thin layer of *n*-type silicon on a base of *p*-type silicon, or vice versa. The back and a strip on the front are metallised to form electrical contacts, as shown in *Figure 4.38*. Light falling on the front surface releases electrons and holes which can move through the crystal lattice. If the contacts are connected through a suitable resistive load a current flows without the use of any external source of e.m.f. Under suitable conditions, the current which flows is proportional to the intensity of the incident radiation.

Figure 4.38
Construction of silicon solar cell

When a photo-cell is required to supply a larger current, amplification is necessary, but this is present in a photo-transistor which may be regarded as a combination of a photo-diode and a transistor amplifier. In fact, the photo-transistor is a conventional transistor in which the emitter-base current is controlled by the incident radiation instead of a signal applied between the emitter and base. To facilitate this, light is permitted to fall on the emitter side of the crystal forming the transistor, instead of the envelope being completely opaque. *Figure 4.39* shows a typical photo-transistor circuit.

4.5.2　Instruments for temperature measurement—electrical method

In temperature measurement, as in the measurement of many other variables in industrial processes, the measured variable may, by means of a suitable piece of apparatus, be converted into a varying electrical quantity. The value of this quantity will bear a definite relationship to the measured quantity. This varying electrical quantity, such as resistance, current, electromotive force, may then be measured, either at its source, or at some remote point, and will give a measure of the magnitude of the process variable. For example, if it is required to measure temperature, a resistance bulb consisting of a coil of suitable wire, whose resistance varies with temperature, may be placed at the point at which it is required to measure the temperature, the value of the resistance of the bulb will then be a measure of the required temperature. This provides a convenient method of measuring temperatures up to about 600° C.

At higher temperatures, it is more convenient to use a thermocouple. This consists of a pair of dissimilar metals so arranged that the electromotive force produced by the couple depends upon the difference in temperature between the hot and cold junctions of the metals. The hot junction is placed at the point at which it is required to measure the temperature and compensation provided for variation in the cold junction temperature. By measuring the electromotive force produced by the thermocouple, the temperature of the hot junction can be measured.

At still higher temperatures, the radiation from the hot source whose temperature is required may be allowed to fall on a receiving element such as a thermocouple, a thermopile (consisting of a number of thermocouples in series), or a photo-electric cell. The radiation falling on the receiving element will cause its temperature to rise and there will be a corresponding rise in the electromotive force produced by the receiving element. By measuring this electromotive force, the temperature of the radiating source may be measured.

Electrical methods of measuring temperature will, therefore, be described under the following headings.

Elements for conversion of temperature variation into an electrical variable;
    Electrical resistance bulbs.
    Thermistors.
    Thermo-electric junctions and thermopiles.
Methods of detecting and measuring the electrical variable;
    Moving-coil indicators.
    Wheatstone's bridge methods.
      Unbalanced bridge.
      Balanced bridge; (*a*) Manual balancing, (*b*) Automatic balancing.
Potentiometric methods;
    Manual balancing.
    Automatic balancing.

4.5.2.1 ELECTRICAL RESISTANCE BULBS

Electrical units

The idea of linking mechanical and electrical units by adopting an electrical unit as a fourth defined unit was discussed for a long time. The m.k.s.a. system in which the ampere is the fourth unit was adopted in 1954 together with the kelvin and the candela. These six units were formerly given the title SI in 1960. Thus, the SI electrical units coincide with the 'absolute units' not the so-called 'international units' which existed prior to 1948 in which the ampere was defined in terms of its electrolytic effect and the ohm in terms of the resistance of a column of mercury.

The definitions of electrical units in the SI systems are:

Ampere (A). The ampere is that constant current which if maintained in two straight parallel conductors of infinite length, of negligible cross-section and placed 1 m apart in vacuum would produce between the conductors a force equal to 2×10^{-7} newton per metre of length.

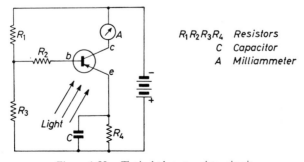

R₁R₂R₃R₄ Resistors
      C Capacitor
      A Milliammeter

Figure 4.39   Typical photo-transistor circuit

The ampere is a base-unit, the other units are derived units.

Volt (V). The difference of potential between two points of a conducting wire carrying a constant current of one ampere when the power dissipated between the points is equal to one watt.

Ohm (Ω). The resistance between two points of a conductor when a constant potential difference of one volt, applied between two points, produces in this conductor a current of one ampere, the conductor not being source of any electromotive force.

The definition of resistance is based on Ohm's Law. This law is extremely useful in calculations involving conductors, for if two out of the three quantities current, potential difference and resistance are known, the third can always be found for

$$\frac{V}{A} = \Omega, \quad V = A \times \Omega, \quad \text{and} \quad A = \frac{V}{\Omega}$$

By the application of Ohm's law, the measurement of one electrical variable enables the value of another to be obtained. Thus, with standard resistances and a standard cell, both ammeters and voltmeters may be calibrated by the use of a potentiometer.

A potentiometer, in its simplest form, consists of a length of uniform wire, AB, usually fixed over a scale of length. The resistance per unit length of the wire may be regarded as being constant, so that when a current flows, the voltage drop along the wire is uniform. The wire is connected with an accumulator, as shown in *Figure 4.40*. If a standard cell is connected in series

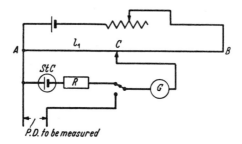

Figure 4.40   The simple potentiometer

with a large resistance and a galvanometer, a point C may be found in AB such that no current flows through the galvanometer when it is connected to C. The high resistance, R, is for the protection of the standard cell, which must never be permitted to give an appreciable current as this will alter its e.m.f. The cell gradually recovers its e.m.f. when it ceases to give a current. When the point C has been found it is known that the potential drop between A and C is equal to the e.m.f., or electromotive force, of the standard cell (E)

$$E = Kl_1, \text{ where } K \text{ is a constant and } l_1 = AC.$$

If now another source of potential difference is connected to A and the galvanometer, and a new point C_1 is found such that there again is no deflection of the galvanometer, then:

Applied potential difference $= Kl_2$ where $l_2 = AC_1$

$$\therefore \frac{\text{Applied potential difference}}{E} = \frac{l_2}{l_1} \qquad (4.9)$$

This enables any applied P.D. to be measured. To measure current, the P.D. across a standard resistance may be measured when the current is flowing.

The P.D., V volt, across the resistance, R ohm, when a current, i amp, is flowing is then given by:

$$V = iR \text{ volt}, \quad \text{or} \quad i = \frac{V}{R} \text{ amp}.$$

Resistances may be compared by placing them in series and passing a current through them. The ratio of the potential differences across them, as measured by means of the potentiometer, will be the same as the ratio of the resistances.

In industrial instruments, the potentiometer is always more complicated than this simple version but the principle is the same. These complications will be described later.

Resistance and temperature

It is found that the resistance of pure metallic conductors increases with temperature. In practical thermometry, the metals used are platinum, nickel and copper because they can be manufactured at a high degree of purity and have a high degree of reproducibility of resistance characteristics. Copper has the disadvantage that it has a low specific resistance (resistance between the faces of a cube of the metal having all edges 1 cm long) but it has a similar temperature coefficient to platinum and it is relatively inexpensive. Elements made of copper have a usable temperature range of $-50°$ to $+250°$ C, but copper is readily oxidised so that adequate steps must be taken to protect the element from oxidation. Nickel provides an inexpensive substitute for platinum at temperature between -200 and $+350°$ C, is not readily oxidised, and has a temperature coefficient which is about $1\frac{1}{2}$ times that of copper or platinum.

Platinum is the standard material used in the resistance thermometer that defines the International Practical Temperature Scale, not because it has a particularly high temperature coefficient of resistance, but because of its stability in use. In fact, a high temperature coefficient is not, in general, necessary for a resistance thermometer material as resistance values can be determined with a high degree of accuracy using suitable equipment and taking adequate precautions.

Platinum, having the highest possible coefficient of resistance, is considered the best material for the construction of thermometers as a high value of this function is an indication that the platinum is of high purity. The presence of impurities in resistance thermometer material is undesirable, as in service, diffusion, segregation and evaporation may occur resulting in a lack of stability of the thermometer. The temperature coefficient of resistance is also sensitive to internal strains so that it is essential that the platinum should be annealed at a temperature higher than the maximum temperature of service. The combination of purity and adequate annealing is shown by a high value of the ratio of the resistances at the steam and ice points. To comply with the requirements of the International Practical Temperature Scale of 1968 this ratio must exceed 1·392 50.

It is essential that the platinum element is mounted in such a way that it is not subject to stress in service.

Platinum is used for resistance thermometry in industry for temperatures up to 800° C. It does not oxidise, but must be protected from contamination. The commonest cause of contamination of platinum resistance thermometers is contact with silica, or silica bearing refractories, in a reducing atmosphere. In the presence of a reducing atmosphere, silica is reduced to silicon which alloys with platinum making it brittle. Platinum resistance thermometers may be used for temperatures down to about 20 K.

For measuring temperatures between 1 K and 40 K doped germanium sensors are usually used, while carbon resistors are used between 0·1 K and 20 K. Above 20 K platinum has a greater resistance to temperature coefficient and has greater stability. Between 0·35 K and 40 K a new resistance thermometer material 0·5 atomic % iron-rhodium is also used.

Calibration of resistance thermometers

To conform with IPTS 68 the resistance of the thermometer at temperatures below 0° C is measured at a number of defining points and the calibration is obtained by difference from a reference function W which is defined and tabulated in the scale. The differences from the function ΔW are expressed by polynomials the coefficients of which are obtained from calibration at fixed points for each of the ranges 13·81 K to 20·28 K, 20·28 K to 54·361 K, 54·361 K to 90·188 K and 90·188 K to 273·15 K. The last mentioned range was formerly defined by the Callendar-Van Dusen equation but the difference from the reference function given by the equation

$$\Delta W = A + B \left(\frac{t_{68}}{100° \text{ C}} - 1 \right) \frac{t_{68}}{100}$$

is now used where

$$t_{68} = T_{68} - 273 \cdot 15 \text{ K} \tag{4.10}$$

and the constants A and B are determined by measurements of W at 100° C and $-182 \cdot 962°$ C (90·188 K).

For the range 0° C to 630·74° C the Callendar equation is still used but a correction term is added so that the calibration procedure is to measure the resistance of the thermometer at 0° C (obtained by way of the triple point of water), the boiling point of water (100° C) and the freezing point of zinc (419·58° C on the 1968 scale, formerly 419·505° C′ (on the 1948 scale). The Callendar equation is then used to determine the intermediate value of t'

$$t' = \frac{1}{\alpha} \left[W(t') - 1 \right] + \delta \left(\frac{t'}{100° \text{ C}} \right) \left(\frac{t'}{100° \text{ C}} - 1 \right) \tag{4.11}$$

t' is then corrected by an amount which varies with temperature but is the same for all thermometers which meet the specification of the Scale

$$t_{68} = t' + 0.45 \left(\frac{t'}{100° \text{ C}} \right) \left(\frac{t'}{100° \text{ C}} - 1 \right) \left(\frac{t'}{419.58° \text{ C}} - 1 \right) \left(\frac{t'}{630.74° \text{ C}} - 1 \right) °\text{C}$$

$$(4.12)$$

The value of α for a given specimen of platinum is the same on the 1948 and 1968 scales but the value of δ changes because of the change in the assigned zinc point, e.g. a δ coefficient of 1·492 on the old scale becomes 1·497 on the new.

Calibration of industrial resistance thermometers

Industrial platinum resistance thermometers are in general based on B.S. 1904: 1964, but thermometers having a higher precision than is required to meet this standard are produced. The degree of precision attained is largely a function of the platinum purity and the care taken in the design and construction of the resistance element to ensure it is free from stress under all circumstances.

The calibration of a resistance thermometer is based on IPTS 68 and is usually carried out by comparison with a standard resistance thermometer.

A typical calibration table for a high precision element designed for a temperature range of $-200°$ C to $800°$ C is given in Table XIX in the appendix.

Resistance thermometer bulbs

Platinum resistance sensors may be designed for any range within the limits of 15 K and 800° C and may be capable of withstanding pressures up to 600 bar and vibration up to 60 g's, or more, at frequencies up to 2000 Hz. The size of the actual sensitive element may be as small as 2 mm diameter by 8 mm long, in the case of the miniature fast response elements, to 6 mm diameter by 50 mm long in the more rugged types. A wide range of sensor designs are available, the actual form used depending upon the duty, the speed of response required and some typical forms of construction are illustrated in *Figure 4.41*. *Figure 4.41(a)* shows a high temperature form in which the spiral platinum coil is bonded at one edge of each turn with high temperature glass inside cylindrical holes in a ceramic rod. In the high accuracy type used mainly for laboratory work the coil is not secured at each turn but is left free to ensue a complete strain-free mounting, *Figure 4.41(b)*. Where a robust form suitable for use in aircraft and missiles or any severe vibration condition is required the ceramic is in solid rod form and the bifilar wound platinum coil is sealed to the rod by a glass coating as shown in *Figure 4.41(c)*. Where the sensor is intended for use for measuring surface temperatures, the form shown in *Figure 4.41(d)* is used. In all forms, the ceramic formers are virtually silica-free and the resistance element is sealed in with high temperature glass to form an impervious sheath which is unaffected by most gases and hydrocarbons. The external leads which are silver or platinum of a diameter much larger than the wire of the resistance element are welded to the fine platinum wire wholly inside the glass seal.

The inductance and capacitance of elements are made as low as possible in order to allow their use with alternating current measuring instruments. Typically the elements shown will have self-inductance of 2μH per 100 Ω and element self-capacitance will not exceed 5 pF. It is recommended that the current permitted to pass through the elements should be limited by considerations of self-heating. The response time when plunged into water at 80° C moving at 1 m/s of the order of 0·1 s for the smallest type to approximately 1 s for the larger types. When enclosed in a metal sheath the response time will be correspondingly increased.

Figure 4.41(e) shows an open-wire element suitable for applications on clean electrically non-conducting liquids or gases where a very quick response is required.

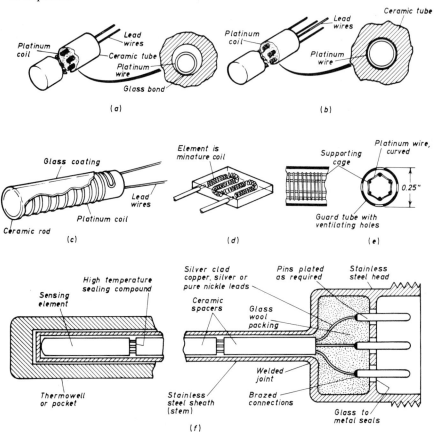

Figure 4.41   Typical construction of REC platinum resistance temperature sensors.
(By courtesy, Rosemount Engineering Co. Ltd.)

Figure 4.41(f) shows the construction of a complete thermometer bulb suitable for three-wire connection.

In the Foxboro Dynatherm Resistance-bulb, shown in *Figure 4.42*, the resistance wire, which is insulated nickel, is wound on a silver core. A spring presses the bulb down into the protective well so that this core is pressed on metal foil which is in contact with the bottom of the well. This metal to metal contact greatly increases the rate of heat transfer between the bottom of the well and the resistance bulb, and reduces the time lag in the response of the

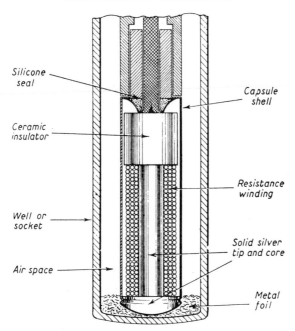

Silicone seal

Capsule shell

Ceramic insulator

Resistance winding

Well or socket

Solid silver tip and core

Air space

Metal foil

Figure 4.42   The dynatherm resistance bulb. (By courtesy, Foxboro-Yoxall Ltd.)

thermometer to changes of temperature of the medium whose temperature is being measured. It is used for temperatures up to 310° C.

Many other forms of resistance elements are available for special purposes. In fact the resistance thermometer has very many uses and is capable of measuring to a high degree of accuracy ($\pm 0 \cdot 75$ per cent of scale range) temperatures up to 600° C. If special precautions are taken to avoid strains due to vibration this range may be extended to 1100° C.

The methods adopted to measure the resistance of the bulb are described in section 4.7.

4.5.2.2 THERMISTORS

Whereas a resistance bulb consists of a coil of pure metal wire, a thermistor consists of an element of semiconductor which has a negative temperature coefficient about ten times greater than that of copper or platinum. Owing to the high temperature coefficient, a thermistor temperature element is considerably more sensitive than the conventional metal wire resistance element. As the resistivity of the thermistor material is also much higher than that of any metal, the size of the thermistor element can be very small so that the speed of response to change of measured temperature is very high. The upper temperature limit for a thermistor depends upon the make and is usually 300° C, but thermistors for use at temperatures up to 900° C are at present under development.

Owing to the negative temperature coefficient, increase in temperature may result in an increase in the current flowing in the thermistor and hence an increase in the power dissipated as heat. Precautions must therefore be taken to ensure that self-heating is kept negligibly small.

Thermistors are made from metal oxides or mixtures of metal oxides. The oxides used are the oxides of cobalt, copper, iron, magnesium, manganese, nickel, tin, titanium, uranium and zinc. The oxides in powder form are usually compressed into the desired shape and then heated to a temperature sufficiently high to recrystallise them, resulting in a dense ceramic body. Electrical contacts are made with the thermistor by means of wires embedded before the firing, by plating, or by metal ceramic coatings baked on.

Thermistors for resistance thermometry are usually in the form of a bead 0·25–0·5 mm in diameter. The beads may be covered with a thin coherent coat of glass to reduce composition changes at high temperature, and supported only by its leads, or sealed into the tip of a glass probe, or sealed to a copper disc.

Disc thermistors ranging from 5 to 25 mm in diameter, or rods from 0·75 to 6 mm in diameter and up to 50 mm in length, either self-supporting or mounted on a small plate, are used mainly for temperature control. The advantage of the larger thermistor is that it is capable of dissipating a larger quantity of heat to its surroundings than the smaller thermistor so that it can carry a current large enough to operate a relay without excessive self-heating.

The relationship between the absolute temperature $T(\mathrm{K})$ and the resistance R of a thermistor is

$$R = a\,\mathrm{e}^{\,b/T} \qquad (4.13)$$

in which a and b are nearly constant over small ranges of temperature.

It is not at present possible to make thermistor thermometers having characteristics as clearly reproducible as those of a platinum resistance thermometer. The measuring circuit design must therefore allow for tolerances of up to ± 20 per cent on resistance at a given temperature and temperature coefficient tolerances of up to ± 10 per cent. Although the resistance/temperature relationship is not linear it is possible to use a thermistor in combination with a resistor of negligible temperature coefficient in such a way that an approximately linear relationship is obtained.

Owing to the high temperature coefficient of the thermistor it is possible to measure temperature differences as small as $0·001°$ C but careful account must be taken of the tendency of the calibration of a thermistor to change with time.

In the early stages of their development thermistors were notorious for their instability of calibration but this has been considerably reduced, so that for certain types of thermistors a stability of $0·1°$ C per annum may be expected without special selection. To maintain this stability, however, it is essential to ensure that the permissible temperature limit is never exceeded and that the thermistor is not subject to severe mechanical stresses or to excessive vibration.

Specially selected silicon semiconductors having a high degree of linearity and stability are also available. These are normally 6 mm in diameter and they are mounted in the last 10 mm of a sealed sheath to give a temperature sensor which when incorporated in a suitable bridge circuit gives a sensitivity of $-2·00$ mV/° C from $-100°$ C to $+100°$ C. Sensors are available having up to a $200°$ C span anywhere between $-200°$ C and $+150°$ C.

4.5.2.3   THERMO-COUPLES

Thermo-electricity—Seebeck effect

In 1821, Seebeck discovered that if a closed circuit is formed of two metals, and the two junctions of the metals are at different temperatures, an electric current will flow round the circuit. Suppose a circuit is formed by twisting or soldering together at their ends (as shown in *Figure 4.43*) wires of two different metals, as, for example, iron and copper. If one junction remains at room temperature, while the other is heated to a higher temperature, a current is produced, which flows from copper to iron at the hot junction, and from iron to copper in the cold one.

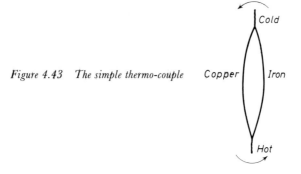

Figure 4.43   The simple thermo-couple

Seebeck arranged a series of 35 metals in order of their thermo-electric properties. In a circuit made up of any two of the metals, the current flows across the hot junction from the earlier to the later metal of the series. A portion of his list is as follows: Bi—Ni—Co—Pd—Pt—U—Cu—Mn—Ti—Hg—Pb—Sn—Cr—Mo—Rh—Ir—Au—Ag—Zn—W—Cd—Fe—As—Sb—Te.

Peltier effect

In 1834, Peltier discovered that when a current flows across the junction of two metals, heat is absorbed at the junction when the current flows in one direction, and liberated if the current is reversed. Heat is absorbed when a current flows across an iron–copper junction from copper to iron, and liberated when the current flows from iron to copper. This heating effect should not be confused with the Joule heating effect, which being proportional to i^2R, depends only upon the size of the current and the resistance of the conductor and does not change to a cooling effect when the current is reversed. The amount of heat liberated, or absorbed, is proportional to the quantity of electricity which crosses the junction, and the amount liberated, or absorbed, when unit current passes for a unit time is called the Peltier coefficient.

As heat is liberated when a current does work in overcoming the e.m.f. at a junction, and is absorbed when the e.m.f. itself does work, the existence of the Peltier effect would lead one to believe that the junction of the metals is the seat of the e.m.f. produced in the Seebeck effect. It would appear that an e.m.f. exists across the junction of dissimilar metals, its direction being copper-to-iron in the couple considered. The size of this e.m.f. depends upon the temperature of the junction. When both junctions are, at the same temperature, the e.m.f. at one junction is equal and opposite to that at the second junction, so that the resultant e.m.f. in the circuit is zero. If, however, one junction is heated, the e.m.f. across the hot junction is greater than that across the cold junction, and there will be a resultant e.m.f. in the circuit which is responsible for the current: i.e.

$$\text{e.m.f. in the circuit} = P_2 - P_1,$$

where P_1 is the Peltier e.m.f. at temperature T_1, and P_2 is the Peltier e.m.f. at temperature T_2, where $T_2 > T_1$.

Thomson effect

Reasoning on the basis of the reversible heat engine, Prof. W. Thomson (later Lord Kelvin) pointed out that if the reversible Peltier effect was the only source of e.m.f., it would follow that if one junction was maintained at a temperature T_1, and the temperature of the other raised to T_2, the available e.m.f. should be proportional to $(T_2 - T_1)$. It may be easily shown that this is not true. If the copper-iron thermo-couple, already described, is used, it will be found that on heating one junction while the other is maintained at room temperature, the e.m.f. in the circuit increases at first, then diminishes, and passing through zero, actually becomes reversed. Prof. Thomson, therefore, concluded that in addition to the Peltier effects at the junctions there were reversible thermal effects produced when a current flows along an unequally heated conductor. In 1856, by a laborious series of experiments, he found that when a current of electricity flows along a copper wire whose temperature varies from point to point, heat is liberated at any point, P, when the current at P flows in the direction of the flow of heat at P, i.e. when the current is flowing from a hot place to a cold place; while heat is absorbed at P when the current flows in the opposite direction. In iron, on the other hand, the heat is absorbed at P when the current flows in the direction of the flow of heat at P; while heat is liberated when the current flows in the opposite direction from the flow of heat.

Thermo-electric diagram

It will be seen therefore, that the Seebeck effect is a combination of both the Peltier and Thomson effects and will vary according to the difference of temperature between the two junctions, and with the metals chosen for the couple. The e.m.f. produced by any couple with the junctions at any two temperatures may be obtained from a thermo-electric diagram suggested by Professor Tait in 1871. On this diagram the thermo-electric line for any metal is a line such that the ordinate represents the thermo-electric power (defined as the rate of change of e.m.f. acting round a couple with the change of temperature of one junction) of that metal with a standard metal at a temperature represented by the abscissa. Lead is chosen as the standard metal as it does not show any measurable Thomson effect. The ordinate is taken as positive when, for a small difference of temperature, the current flows from lead to the metal at the hot junction. If lines *a* and *b* (*Figure 4.44*) represent the thermo-

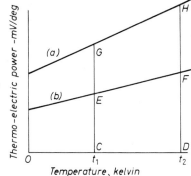

Figure 4.44   Thermo-electric diagrams

electric lines for two metals *A* and *B* then, the e.m.f. round a circuit formed of the two metals, when the temperature of the cold junction (t_1), is represented by *OC*, and that of the hot junction (t_2) by *OD*, will be represented by the area of the trapezium *GEFH*, i.e.

$$\frac{(EG + HF)}{2} (t_2 - t_1)$$

Thermo-electric inversion

Figure 4.45 shows the thermo-electric lines for several common materials. It will be seen that the lines for iron and copper cross at a temperature of 275° C. Thus, if the temperature of the cold junction of iron and copper is below 270° C, and the temperature of the other junction is raised, the thermo-electric e.m.f. of the circuit (represented by a trapezium) will increase until the temperature of the hot junction reaches 275° C (when the e.m.f. is represented by a triangle). Further increase in the temperature of the hot junction will result in a decrease in the thermo-electric e.m.f. (for the e.m.f. represented by the second triangle will be in the opposite sense). When the average temperature of the two junctions is 275° C, or what comes to the same thing, the sum of the two temperatures is 550° C, the areas of the two triangles will be equal and there will be no thermo-electric e.m.f.; 275° C is known as the 'neutral temperature' for the copper-iron couple. With circuits of other materials, the neutral point will occur at a different temperature. Further increase in the temperature of the hot junction will produce a thermo-electric e.m.f. in the opposite direction, i.e. from iron to copper at the hot junction, which will again increase with increasing temperature of the hot junction.

In choosing two materials to form a thermo-couple to measure a certain range of temperature, it is very important to choose two which have thermo-electric lines which do not cross within the temperature range, i.e. the neutral temperature must not fall within the range of temperature to be measured. If the neutral temperature is within the temperature range, there is some ambiguity about the temperature indicated by a certain value of the thermo-electric e.m.f., for there will be two values of the temperature of the hot junction for which the thermo-electric e.m.f. will be the same. For this reason tungsten v molybdenum thermo-couples must not be used at temperatures below 1250° C.

Tables giving the thermo-electric e.m.f. for the most frequently used thermo-couples are given in the Appendix, and will be found in detail in the following British Standards.

Platinum	v Rhodium-Platinum	B.S. 1826
Nickel-Chromium	v Nickel-Aluminium	B.S. 1827
Copper	v Constantan	B.S. 1828
Iron	v Constantan	B.S. 1829

The thermo-electric e.m.f. (E) developed by a standard platinum v platinum 10 per cent rhodium thermo-couple with junctions at 0° C and t° C, where t is between 630° C and the gold-point, is given by the equation

$$E = a + bt + ct^2 \tag{4.14}$$

The value of a, b and c are determined by calibrating the couple at the freezing point of antimony, silver and gold.

Laws of addition of thermo-electric e.m.fs.

In measuring the e.m.f. in any circuit due to thermo-electric effects, it is usually necessary to insert some piece of apparatus, such as a millivoltmeter, somewhere in the circuit, and since this generally involves the presence of junctions other than the two original junctions, it is important to formulate the laws according to which the e.m.f.s produced by additional junctions may be dealt with. These laws, discovered originally by experiment, have now been established theoretically.

1. *Law of intermediate metals.* In a thermo-electric circuit composed of two metals A and B with junctions at temperatures t_1 and t_2 respectively, the e.m.f. is not altered if one or both the junctions are opened and one or more other metals are interposed between the metals A and B, provided that all the junctions by which the single junction at temperature t_1 may be replaced are kept at t_1, and all those by which the junction at temperature t_2 may be replaced are kept at t_2.

   This law has a very important bearing on the application of thermo-couples to temperature measurement, for it means that, provided all the apparatus for measuring the thermo-electric e.m.f., connected in the circuit at the cold junction, is kept at the same temperature, the presence of any number of junctions of different metals will not affect the total e.m.f. in the circuit. It also means that if another metal is introduced into the hot junction for calibration purposes it does not affect the thermo-electric e.m.f., provided it is all at the temperature of the hot junction.

2. *Law of intermediate temperatures.* The e.m.f., $[E]_1^3$, of a thermo-couple with junctions at temperatures t_1 and t_3, is the sum of the e.m.f.s of two couples

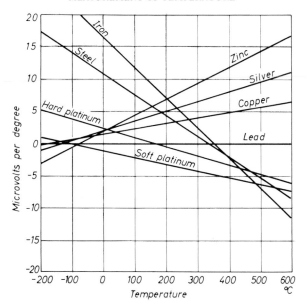

Figure 4.45   Thermo-electric diagrams

of the same metals, one with junctions at t_1 and t_2 (e.m.f. $= [E]_1^2$), and the other with junctions at t_2 and t_3 (e.m.f. $= [E]_2^3$).

$$
t_1 \underset{B}{\overset{A}{\longleftrightarrow}} t_2 \quad + \quad t_2 \underset{B}{\overset{A}{\longleftrightarrow}} t_3 \quad = \quad t_1 \underset{B \ t_2 \ B}{\overset{A \ t_2 \ A}{\longleftrightarrow}} t_3
$$

$$
\text{or} \quad [E]_1^2 \quad + \quad [E]_2^3 \quad = \quad [E]_1^3 \tag{4.15}
$$

This law is extremely useful for calculating the effect of change of temperature of the cold, or reference, junction. It makes it possible to correct the indication of a thermo-electric pyrometer for changes in the temperature of the cold junction. The e.m.f., $[E]_0^t$, for a thermo-couple with its cold junction at $0°$ C and hot junction at $t°$ C, will be equal to the e.m.f. of the same couple with cold junction at say $15°$ C, $[E]_{15}^t$, plus the e.m.f. of the same couple with junctions at $0°$ C and $15°$ C, $[E]_0^{15}$

$$
\text{i.e.} \ [E]_0^t = [E]_{15}^t + [E]_0^{15}
$$

$$
\text{or,} \ [E]_{15}^t = [E]_0^t - [E]_0^{15} \tag{4.16}
$$

A thermo-electric pyrometer, calibrated with its cold junction at $0°$ C, will read correctly provided it is arranged that an e.m.f. equal to that of a thermo-couple with junctions at $0°$ C and the actual cold junction temperature is added to the reduced thermo-electric e.m.f. Tables 5.5 to 5.9, give the required correction. This addition may be accomplished automatically by the method described on page 343.

Table 4.6   BASE METAL THERMO-COUPLES*

Name	Composition	Maximum temperature °C		Remarks
		Continuous	Spot reading	
Copper v Constantan or Eureka	Cu v Ni 40%, Cu 60% approx.	400	500	High resistance to corrosion by condensed moisture
Iron v Constantan	Fe v Ni 40%, Cu 60% approx.	850	1100	Low cost suitable for use in oxidis-ing or reducing atmosphere
Nickel chromium v Constantan	Ni 90%, Cr 10% v Ni 40%, cu 60% approx.	700	1000	Best service in oxidising atmos-phere
Nickel Chromium v Nickel Aluminium	Ni 90%, Cr 10% v Ni 94%, Al 2% + Si and Mn approx.	1100	1300	Best service in oxidising atmos-phere

*The wire mentioned first is the positive wire.

Industrial thermo-couples

Materials are suitable for use in industrial thermo-couples only if, in addition to developing a relatively large e.m.f., this e.m.f. does not depart in an appreciable degree from the calibrated value when the thermo-couple is in use. The material must not change in composition, or melt, at the highest temperature at which it is used, and the thermo-couple must be protected where necessary from contamination and mechanical strains which will influence its calibration. The material must be such that successive batches can be produced with the same thermo-electric characteristics so that thermo-couples may be replaced without the necessity of recalibrating the temperature scale of the indicating instrument. The wires should be 'aged' before use by heating to a temperature equal to the maximum temperature of use. This removes, to a very large extent, the tendency for the e.m.f. of the thermo-couple to change gradually with time.

Thermo-couples may be divided broadly into two classes; precious metal and base metal. Precious metal thermo-couples are used in a wide variety of industries the most common being the platinum *v* rhodium-platinum which is used for control of furnaces in the steel, glass, semiconductor and other industries. With suitable methods of measuring the thermo-couple output they can be used constantly from 200° C to 1500° C or intermittently up to 1650° C. Errors in thermocouple readings result from loss of homogeneity in one or both limbs, which may be caused by strain of the thermocouple wires, by transfer of rhodium from the alloy to the pure platinum limb ('rhodium drift'), or by pick-up of foreign elements by the thermo-couple from the surrounding atmosphere (contamination).

The effect of strain is to reduce the e.m.f. and hence provide a low reading. The strain may be removed by re-annealing the whole thermo-couple.

Thermo-couple installations should be designed so that there is no strain on the wires so that a horizontal sheath is preferable to a vertical one.

Rhodium drift occurs if a rhodium-platinum limb is maintained in air for long periods close to its upper temperature limit. Rhodium-oxide will form and volatilize, and some of this oxide can settle on, and react with, the platinum limb causing a fall in e.m.f. output.

This process is a slow one and, for most practical applications, its effect is insignificant. However, if high stability must be maintained for long periods at high temperatures rhodium drift must be avoided. This can be achieved either by using long separate single-bore insulators on each thermo-couple limb, or by substituting a 6 per cent rhodium-platinum v 30 per cent rhodium-platinum thermo-couple. The sensitivity of this thermo-couple is comparable with that of the platinum v 10 per cent rhodium-platinum and platinum v 13 per cent rhodium-platinum thermocouples, although it has a lower e.m.f. output and is slightly less accurate. Its e.m.f. output is much less affected however by rhodium transfer than those of the platinum v 10 per cent rhodium-platinum and platinum v 13 per cent rhodium-platinum thermocouples.

At temperatures below 1000° C the slow formation of rhodium oxide has a different effect. The oxide remains on the alloy limb as a black film and after long periods, some internal oxidation of the alloy takes place. This effectively reduces the rhodium content of the alloy and results in a low e.m.f. output. The rate of oxidation is extremely slow and is only significant in those installations that are expected to last for tens of thousands of hours (e.g. the thermo-couples used to measure creep-test temperatures). The effect is easily removed by heat-treating the thermo-couple at 1300° C; this causes dissociation of the rhodium oxide and restores to normal the composition of the alloy limb of the thermo-couple.

Contamination is by far the most common cause of thermo-couple error and often results in ultimate mechanical failure of the wires. Elements such as Si, Al, P, Pb, Zn, and Sn combine with platinum to form low melting-point eutectics and cause rapid embrittlement and mechanical failure of the thermo-couple wires. Elements such as Ni, Fe, Co, Cr, and Mn affect the e.m.f. output of the thermo-couple to a greater or lesser degree, but contamination by these elements does not result in wire breakage and can only be detected by regular checking of the accuracy of the thermo-couple. Contamination can be avoided by careful handling of the thermocouple materials before use and by the use of efficient refractory sheathing. Care should be taken to prevent dirt, grease, oil or soft solder coming into contact with the thermo-couple wires before use. If the atmosphere surrounding the thermo-couple sheath contains any metal vapour, the sheath must be impervious to such vapours.

Although the materials used for noble metal thermo-couples have a high intrinsic value, the cost of thermo-couple installations is reduced by the fact that wire from 0·050 to 0·50 mm (the latter provides the best balance between material economy and performance reliability) may be used depending upon the application, and the high scrap value and comparatively long life makes the running costs of the installation comparatively low. This makes the use of noble metal thermo-couples an economic alternative to base metal couples at the upper end of their operating ranges where the latter become inaccurate and have a short life in corrosive atmospheres.

To overcome these disadvantages Pallador I which has a thermal e.m.f. comparable with iron v constantan, and Pallador II which provides a noble metal alternative to nickel-chromium v nickel aluminum. The positive limb of Pallador I is 10 per cent iridium platinum and the negative limb is 40 per cent palladium gold, while the positive limb of Pallador II is 12½ per cent platinum palladium and the negative is 46 per cent palladium gold. The maximum operating range for Pallador I thermo-couples is 1000° C and the maximum operating temperature for Pallador II is 1200° C when protected by a 10 per cent rhodium-platinum sheath. (Pallador is the registered Trade Mark of Johnson Matthey Metals Ltd.)

The corresponding base metal thermo-couple wires may be used as the compensating lead and an accuracy of ±1 per cent will be attained on an instrument calibrated on the base metal characteristics. When the instrument is calibrated on the Pallador temperature e.m.f. relationship an accuracy of ±2° C over the whole operating range is attainable.

Noble metal thermo-couples may also be used for measuring cryogenic temperatures. Iron gold v nickel chromium or iron gold v silver normal (silver with 0·37 atomic per cent gold) may be used for temperatures from 1 K to above 300 K.

Noble metal thermo-couples are often used in the metal clad form with magnesia or aluminal powder as the insulant.

The following sheath materials are used, nickel, stainless steel, inconel in

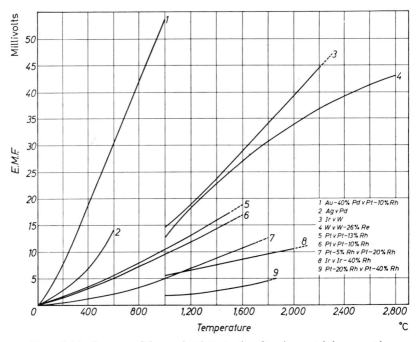

*Figure 4.46   Summary of thermo-electric properties of precious metal thermo-couples.
(By courtesy, Engelhard Industries Ltd.)*

The broken lines indicate the maximum temperatures at which the respective couples may be used for intermittent service.

1·6 and 3·2 mm sizes, and 5 per cent rhodium-platinum and 10 per cent rhodium-platinum both in 1·0 mm sizes. For high temperature work other special thermo-couples have been developed, Tungsten 5 per cent Rhenium *v* Tungsten 20 per cent Rhenium for use in hydrogen, vacuum and inert gas atmospheres up to 2320° C and Tungsten *v* Molybdenium and Tungsten *v* Iridium for temperatures up to 2100° C.

The 5 per cent *v* 20 per cent rhodium-platinum has been replaced by the 6 per cent *v* 30 per cent couple which has the great advantage that its thermo-electric e.m.f. approaches zero at room temperature so that it may be used without compensating leads in many applications. The thermo-electric properties of the precious metal thermo-couples are summarised in *Figure 4.46.*

The e.m.f.-temperature relationship for Platinum *v* 13 per cent Rhodium-Platinum thermo-couples is given in Table 5.6. This table is based on the calibration of a representative number of thermo-couples, and thermo-couples from different sources may vary slightly from these figures. However, manufacturers in this country normally guarantee that Platinum *v* Platinum-Rhodium alloy couples supplied to British Standard Specifications agree with the Standard Reference Tables to the following tolerances.

Up to 1100° C $\pm 1°$ C
1100 to 1400° C $\pm 2°$ C
1400 to 1600° C $\pm 3°$ C.

Table 4.6 lists the common base metal thermo-couples and indicates the maximum temperatures for which they are suitable when protected by a suitable sheath. The tolerances on base metal thermo-couples varies more widely than on precious metal ones, and depends upon the materials used and the matching of the wires. A general idea of the tolerances to be expected is given in Table 4.7 which applies to regular grade wires. Premium grade thermo-couples with closer tolerances are sometimes available at an increased cost.

Table 4.7   TOLERANCES ON BASE METAL THERMO-COUPLES

Type	Temperature range	Tolerance
Copper v Constantan	$-180°$ C to $-60°$ C	$\pm 2\%$
	$-60°$ C to $-100°$ C	$\pm 1°$ C
	$+100°$ C to $+400°$ C	$\pm \frac{3}{4}\%$
Iron v Constantan	Up to 400° C	$\pm 2°$ C
	400° C to 800° C	$\pm \frac{1}{2}\%$
Nickel Chromium v Nickel Aluminium	Up to 300° C	$\pm 2°$ C
	300° C to 1100° C	$\pm \frac{3}{4}\%$

Types of thermocouple

The same precautions must be observed with thermo-couples as with other forms of temperature-measuring instruments to ensure that the hot junction of the thermo-couple is at the temperature of the medium whose temperature is being measured. To facilitate this, thermo-couples are made in three forms: the 'Surface Contact' type; the 'Insertion or Immersion' type; and the 'Suction' type.

The 'surface contact' type

This type is used to measure the temperature of solids, and thermal contact of the hot junction with the solid is usually made by means of some form of contact plate. When using this form of thermo-couple, care should be taken to see that there is no exchange of heat between the contact plate and adjacent surfaces other than those of the hot body whose temperature is being measured.

The 'insertion or immersion' type

This form is used mainly to measure the temperature of liquids and gases, and is immersed in the medium whose temperature is being measured. It is constructed by twisting together the ends of a suitable pair of thermo-couple wires. The end of the twisted wires is then welded, and forms the hot junction. Insulators suitable for the temperature range of the instrument are then threaded on to the wires, if bare wire has been used. Rubber may be used as an insulator up to 50° C, silk, cotton or impregnated cambric sleeving, or tape, up to 120° C, some enamels up to 200° C, suitably impregnated asbestos (except in damp conditions) up to 600° C. Above 600° C, refractory insulators are used. These should have been kilned, at a temperature higher than the maximum temperature of service of the couple, in order to remove all volatile impurities. The wires are brought out to a pair of terminals on an insulating head as shown in *Figure 4.47*.

Protection of the thermo-couple

In order to preserve the thermo-electric characteristics of the thermo-couple wires and to protect the wires from the effects of rapid oxidation, or other forms of deterioration due to the action of the gas or liquids in which they are immersed, the thermo-couple is protected by a sheath whose form and com-position depend upon the use to which the thermo-couple is put.

For temperatures below 500° C protection may be omitted, particularly when a rapid response is required, but the mechanical protection of a sheath is desirable, and in many cases a sheath of mild steel is suitable. For tem-peratures below 900° C a sheath of steel suitably treated, e.g. by 'Calorising' can be used, or, for prolonged exposure temperatures near 900° C a sheath of nickel-chromium used. Special heat-resisting steels may be used for tem-peratures up to 1100° C, but above this temperature a refractory sheath is essential. Precious metal thermo-couples are liable to become contaminated by.metal vapours and gases in the hot zone and must, therefore, be protected at all temperatures above 500° C, and even below, by a gas-tight refractory sheath.

Refractory sheaths are more impervious to gases at higher temperatures than metal sheaths, but they are more liable to fracture if heated or cooled suddenly or unequally. Where the conditions demand, the refractory sheath may be protected by an outer sheath of a material such as fireclay, more resistant to thermal shock, but this of course reduces the speed of response of the thermo-couple.

Owing to its low coefficient of expansion fused silica withstands rapid and unequal heating but tends to devitrify at temperatures above 1000° C. It cannot be used under reducing conditions at temperatures in excess of 900° C as the silica may be reduced to silicon which combines with the platinum seriously affecting its thermo-electric properties and rendering it brittle and very liable to fracture. Where sheaths may be used, fine bore silica tubes are suitable for insulating the thermo-element.

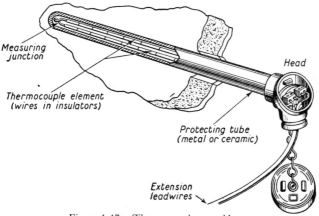

Figure 4.47 Thermo-couple assembly

Pure alumina is one of the most useful refractory materials for use in conjunction with platinum-metal thermo-couples. Impervious recrystallised alumina is suitable under both oxidising and reducing conditions for service up to 1850° C as sheath and insulating material. Its thermal shock resistance is, however, insufficient for the sheath to withstand direct insertion into a very hot furnace or melt. The sintered alumina has a superior resistance to thermal shock but it is not impervious to volatile contaminants.

Alumina-Silica refractories based on mullite $(3Al_2O_3\ 2SiO_2)$ and sillimanite $(Al_2O_3\ SiO_2)$ have a superior resistance to thermal shock than alumina. An impervious form of mullite produced by firing at very high temperature is suitable for use in oxidising atmospheres at temperatures up to 1700° C. The silica-bearing refractories like pure silica cannot be used with platinum in reducing atmospheres at temperatures in excess of 900–1000° C.

For use with the higher temperature thermo-couples such as iridium *v* iridium rhodium much more refractory materials are required. Beryllia is satisfactory under oxidising conditions up to 2300° C, and *in vacuo* or reducing conditions at temperatures up to 2000° C. It has good resistance to thermal shock and retains its electrical insulating properties at high temperatures. In the presence of water vapour at temperatures above 1650° C it becomes volatile and presents a health hazard.

Magnesia may be used under oxidising conditions at temperatures up to 2400° C but is readily reduced by carbon at temperatures above 1800° C. Under reducing conditions the maximum temperature of use is 1600° C and its thermal shock resistance is only fair.

Zirconia stabilised by the addition of 3 per cent calcium oxide is suitable for use in oxidising atmospheres up to 2300–2400° C and in reducing atmospheres up to 2200° C but its thermal shock resistance is only fair. At elevated temperatures its electrical insulating properties are impaired.

Thoria is suitable for use up to 2600° C under most conditions but it is expensive and its resistance to thermal shock is not good.

Mineral Insulated Thermo-couples

The mineral insulated base metal thermo-couple, which is robust and has good heat resistant properties, is being used increasingly in industry. The conductors are insulated from each other and from the sheath by a layer of compressed magnesium oxide. Thermo-couples with an external diameter

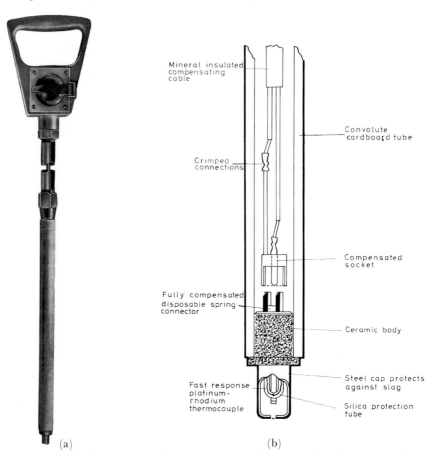

Mineral insulated compensating cable

Convolute cardboard tube

Crimped connections

Compensated socket

Fully compensated disposable spring connector

Ceramic body

Fast response platinum-rhodium thermocouple

Steel cap protects against slag

Silica protection tube

(a)   (b)

Figure 4.48 'Dipstick' type of thermo-couple probe. (By courtesy, Land Pyrometers Ltd.)

as small as 0·1 mm suitable for use without further protection are available ensuring a rapid response rate to temperature changes.

Copper v constantan thermo-couples protected by either a mild steel or cupro-nickel sheath are suitable for use up to 400° C.

Iron v constantan thermo-couples protected by a mild steel sheath may be used up to 600° C.

Nickel chromium v nickel aluminium protected by a 25/20 chromium steel sheath are suitable for temperatures up to 1100° C.

Mineral insulated compensating leads are also available.

Liquid Metal Temperature Measurement

The temperature of liquid steel is usually measured by means of the 'Dipstick' type of thermo-couple probe illustrated in *Figure 4.48 (a)* and having the disposable tip illustrated in *Figure 4.48 (b)*. The platinum-rhodium thermo-couple is protected from the liquid steel by threading it through a silica tube which acts both as an insulator and protective sheath. The wire is only 0·08 mm diameter, so the tip can be made so inexpensive that it can be thrown away after it has been used once.

The thermo-couple unit is cemented into a ceramic housing containing a plastic plug and mounted on the end of a cardboard tube. The silica tube is covered by a thin steel cap which protects it from the slag and melts off in the liquid steel. The cardboard tube slides over the long tubular arm that carries the extension lead and has a concentric connecting socket at its end, which engages with the plug in the disposable unit. The cardboard is dipped in the liquid steel and survives for the 5 s needed to obtain a measurement, acting as an effective thermal insulator protecting the arm from the heat of the liquid steel. As the tip is thrown away after each measurement, maintenance is greatly reduced and an uncontaminated hot junction is used for each measurement. Each batch of thermo-couple wire used for the tips is tested at the melting point of paladium 1554° C ensuring a high degree of accuracy.

The 'suction' type*

This form of thermo-couple is useful for measuring the temperature of a gas in a furnace, or other enclosure, whose walls are cooler than the gas. If the normal thermo-couple is placed in the hot gases, it will receive heat from the gases but will also radiate heat to the cooler walls. It will, therefore, indicate a temperature which is too low. In this type of measurement, however, the gas under measurement is extracted from the furnace and flows continuously past the hot junction during the measurement. The couple is shielded by a series of concentric shields so that it does not radiate more heat than it receives by radiation, and owing to the fact that the gas is moving rapidly past the thermo-couple the reading is not influenced by the low thermal capacity of the gas.

The furnace gases are drawn through the sensing head at the appropriate

*Land, T. and Barber, R. Design of Suction Pyrometers. *Trans. Soc. Inst. Tech.* 6, No. 3 Sept. 54 Suction Pyrometers in Theory and Practice. *J. Iron St. Inst.* 184, Nov. 56.

rate (95–110 kg/hr) for the head by means of an eductor operated by air or steam at a pressure of 2 bar. *Figure 4.49* illustrates a water-cooled unit which may be used at temperatures up to 1600° C. Three interchangeable shield systems are available, the one shown in the figure being an all-metal unit fitted outside the probe and suitable for temperatures up to 1150° C. Between 1150° C and 1400° C, a refractory radiation shield is used, and above 1400° C a special external shield of alumina is used.

Calibration of thermo-couples

The accuracy of a thermo-couple depends to a very high degree upon the standard of maintenance and upon the tolerance permitted in the calibration. Well-maintained base metal thermo-couples may be expected to give an accuracy of $\pm 3°$ C for temperatures up to 540° C. For temperatures above 540° C, the accuracy for iron-constantan is $\pm\frac{1}{2}$ per cent of the scale range, while that of nickel chromium v nickel aluminium is $\pm\frac{3}{4}$ per cent of the scale range. Precious metal thermo-couples will give an accuracy of $\pm 2°$ C for temperatures below 600° C, and $\pm 3°$ C for temperatures above 600° C. If great care is taken with the complete installation, the accuracy for precious metal thermo-couples may be so increased that temperatures may be measured with an accuracy of $\pm 1°$ C up to 1063° C.

In making tests of thermo-couples for purposes of pyrometer maintenance, or of works control, it is usually sufficient to test at one standard temperature, particularly if that temperature is near the average measured in service, for in general any alteration of the calibration curve will affect the values of the thermo-electric e.m.f. at all temperatures.

The thermo-couple may be checked by immersing the couple together with a standard thermo-couple in a zone which is at a uniform temperature and comparing the readings. To ensure that both couples are at the same temperature they may be inserted into holes drilled close together in a pure metal block.

If the range of temperature is suitable, the thermo-couple may be immersed in a sufficiently large mass of some standard substance, while the substance is cooling through its 'freezing-point'*. When a substance is changing from a liquid to the solid state, 'latent heat' is released which maintains the substance at a constant temperature for some time even though the substance is giving up heat to its surroundings. Care must be taken to see that the thermo-couple is sufficiently immersed in the test substance, and that it is well away from the walls of the containing vessel. A list of suitable substances, for this type of test together with their freezing-point on the International Practical Temperature Scale is given at the beginning of the temperature section.

Another method which is suitable for calibrating precious-metal thermo-couples at the gold-point (1064° C), the palladium-point (1554° C), and, at the platinum-point (1772° C) is to support the thermo-couple wires about 5 mm apart in the centre of a rhodium-platinum-wound tube furnace. The free ends of the wires are bridged by a short length of pure gold, pure palladium or pure platinum (except in the case of a thermo-couple which has

*'The Calibration of Thermometers.' C. R. Barber, *HMSO 1971*.

pure platinum as one of the limbs) wire which may be twisted around the
ends of the thermo-couple wires or fused on. The cold junction of the couple
is placed in melting ice in a vacuum flask, and the temperature of the hot
junction raised to within 20° C of the melting point to be determined, and
then raised at about $\frac{1}{2}$ to 1° C per minute. Usually the e.m.f. rises steadily
to a maximum and then drops very slightly to a steady value which persists
until the wire melts and breaks the circuit.

It should be realised that deterioration sufficient to affect the calibration
of a thermo-couple may occur, particularly in a precious-metal thermo-
couple, without the development of a mechanical weakness sufficient to
attract the user's attention to the deterioration. It is, therefore, important to
calibrate a thermo-couple at regular intervals, these intervals being shorter
for couples used near the maximum permissible temperature than for couples
used at lower temperatures. When a thermo-couple has been in use for a
time which has been found in practice to be its useful life, many users discard
the whole thermo-couple or remove a portion at the hot end and remake the
hot junction.

Compensating leads

In the simplest arrangement, the thermo-couple is connected directly to the
indicating instrument which measures the thermo-electric e.m.f. of the
couple, and the terminals of this instrument then form the cold junction of
the couple. This arrangement is used for instruments measuring the temperature of small muffle furnaces, and in certain forms of portable pyrometers.

In very many installations, however, it is impracticable to connect the
terminals on the head of the pyrometer stem directly to the indicating instrument, so that leads must be inserted between the terminals of the stem and the
terminals of the measuring instrument. If both terminals on the head are at
the same temperature, they may be connected to the measuring instrument
by copper leads, but the temperature of the thermo-couple head will then be
the temperature of the cold junction. This is very unsatisfactory for the
temperature at the head may fluctuate widely, and the measured e.m.f. will
be considerably less than that which the thermo-couple would develop if its

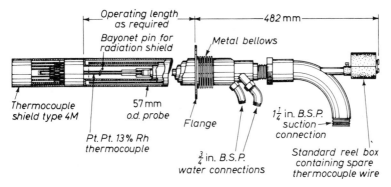

*Figure 4.49   Suction pyrometer for temperatures up to 1600° C.
(By courtesy, Land Pyrometers Ltd.)*

cold junction were at a lower temperature. If the leads are made of the same materials as the couple, then the cold junction is transferred to the terminals of the measuring instrument where the temperature will be more consistent, and arrangements can be made to compensate for temperature changes. Often, however, the thermo-couple wire is expensive so that the same effect is achieved by using leads made of materials which, although different from those of the thermo-couple, have the same thermo-electric characteristics. Such leads are called 'compensating leads'. Sometimes compensating leads are made of the same materials as the thermo-couple but the specification is considerably relaxed so that the cost per foot of the leads is less than that of the thermo-couple material.

When compensating leads are used, care must be taken to see that the temperature at the head of the thermo-couple does not become too high. Compensating leads have similar characteristics to the thermo-couple wires for a limited range of temperature, usually between 0° C and 100° C, and the compensation will not be accurate if the temperature of the leads exceeds the figure specified by the manufacturer.

With copper v constantan, iron v constantan and nickel-chromium v nickel-aluminium couples, the lead wires are usually made of the same material as the couples, although with the latter couple, 55 per cent copper-45 per cent nickel alloy with the other wire of standard high conductivity copper is also often used. With platinum v platinum (10 or 13 per cent) rhodium, compensating leads of copper and 0·6 per cent nickel copper are used, the copper being connected to platinum-rhodium wire.

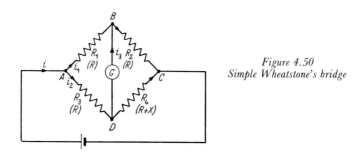

Figure 4.50
Simple Wheatstone's bridge

Where an installation involves very long runs of compensating cable, which would prove expensive, the compensating lead may be used up to a point at which the temperature can be maintained at a steady known value, and the remainder of the connection made by means of copper leads. Several methods are adopted for achieving a steady cold junction temperature. One method is to house the junction between the compensating leads and copper cables in a small chamber which is heated to a temperature slightly above the highest value of the ambient temperature, and maintained at this temperature by a thermostatically controlled heater. Alternatively, the temperature of the junction may be allowed to vary, and some form of electrical cold-junction temperature compensator, such as that shown in *Figure 4.58*, inserted in the circuit at the junction. Methods of dealing with cold-junction temperature variation at the instrument will be dealt with in the section on measuring instruments for thermo-electric e.m.f.

4.5.2.4 THERMOPILES

The e.m.f.s produced by a number of thermo-couples are additive. In order to detect very small changes of temperature several thermo-couples may be arranged in series to form a thermopile. The thermo-couples are made of fine silver and bismuth wires, or chromel and constantan wires. Used with an instrument which is capable of measuring small e.m.f.s, a thermopile provides a sensitive method of measuring radiation and is used as described later to measure the roof temperature of furnaces, and the temperature of moving masses of metal such as slabs of steel.

4.6   MEASUREMENT OF THE ELECTRICAL VARIABLE

The resistance of a thermometer bulb may be measured in several ways. The most commonly used method is to employ a Wheatstone's bridge circuit. Another method is to compare the potential drop across the thermometer coil with that across a standard resistance of manganin wire when the same current flows through both. A potentiometer is used to compare these potential differences and is employed in the same way as it is to measure the e.m.f. of a thermo-couple.

A third method is to measure the current which flows through the resistance coil when a known potential difference is applied to its ends.

4.6.1   Wheatstone's bridge

Four resistances, R_1, R_2, R_3, R_4, are connected to form a bridge network as shown in *Figure 4.50*. The points A and C are connected to a battery, while B and D are connected to a galvanometer. The values of the arms of the bridge are adjusted until no current flows through the galvanometer.

When this is so, the values of the currents and resistances are as shown, then:

Potential at B = potential at D

i.e.   potential drop in AB = potential drop in AD

$$i_1 R_1 = i_2 R_3 \tag{4.17}$$

Also,   potential drop in BC = potential drop in DC

$$i_1 R_2 = i_2 R_4 \tag{4.18}$$

Dividing equation (4.17) by equation (4.18),

$$\frac{i_1 R_1}{i_1 R_2} = \frac{i_2 R_3}{i_2 R_4} \quad \text{or} \quad \frac{R_1}{R_2} = \frac{R_3}{R_4} \tag{4.19}$$

Many methods of measuring resistance are based on the use of the balanced bridge, but in some methods, the deflection of the galvanometer or a milli-voltmeter is used to measure the resistance. Suppose the resistances of the arms of the bridge shown in *Figure 4.50* are all equal to R ohm. Suppose the resistances in the arms AB, BC, and AD, are made of manganin wire and remain constant, and the value of the resistance in the arm DC increases by a small amount x ohm, the change being so small that the change in the current through AC can be neglected. If the total potential difference between A and C is V volt, then:

$$\text{Potential drop in } AB = \frac{V}{2} \text{ volt}$$

$$\text{Potential drop in } AD = \frac{VR}{2R+x} \text{ volt}$$

$$\text{Difference in potential between } B \text{ and } D = \frac{V}{2} - \frac{VR}{2R+x}$$

$$= V\frac{(2R+x-2R)}{2(2R+x)}$$

$$= \frac{Vx}{2(2R+x)} \tag{4.20}$$

As R is much larger than x, this potential difference is approximately equal to

$$\frac{Vx}{4R}$$

If a millivoltmeter, having a resistance so large that it does not influence the currents ABC and ADC, is connected between B and D, its deflection will be proportional to the change in resistance x. This principle is used in the millivolt type of resistance thermometer. The actual millivoltmeter used will be similar to that used in measuring the e.m.f. produced by a thermo-couple and will be described later.

4.6.2   Kirchoff's laws

In order to find the value of the deflection of the millivoltmeter without making any approximations, and when it is required to find the value of the currents and potential differences in more complicated networks, it is necessary to use Kirchoff's laws. These laws state:

In a network of electrical conductors;
(1) There is no accumulation of current at any junction of the conductors, i.e. total current flowing into any point = total current flowing out of the point, e.g. in *Figure 4.51(a)* $i_1 + i_2 = i_3 + i_4 + i_5$.
(2) In any closed circuit, the algebraic sum over the whole circuit of the products of the resistance and current for each portion of the circuit is equal to the total e.m.f. in the circuit; e.g., in *Figure 4.51(b)* $i_1 R_1 + i_1 R_2 + i_2 R_4 + i_2 R_5 + i_2 R_3 = V$.

Straightforward application of these laws to the problem of finding the actual current through the galvanometer in an unbalanced bridge circuit leads to a large number of equations which must be solved. The work may be simplified by adopting Maxwell's 'cyclic current' method of solving the problem. In this method, currents x, y and z are assumed to flow in the 'cyclic circuits' ABD, BCD and ADC, as shown in *Figure 4.52*.

Application of Kirchoff's first law gives the values of the currents in each branch as indicated in the figure.

From Kirchoff's second law

For circuit ABDA $\qquad R_1 y + Rg(y-z) + R_3(y-x) = 0$ $\qquad\qquad$ (4.21)

For circuit BCDB $\qquad R_2 z + R_4(z-x) - Rg(y-z) = 0$ $\qquad\qquad$ (4.22)

For circuit ADCA $\qquad R_3(x-y) + R_4(x-z) + xB = E$ $\qquad\qquad$ (4.23)

The values of the known resistances may be substituted in these equations and the values of x, y and z found. The value of the current through the galvanometer is then given by $(y-z)$.

4.6.3   Moving-coil instruments

The construction of a moving coil instrument is the same whether it be a galvanometer, milliammeter or millivoltmeter. In general the galvanometer is more sensitive than the other instruments. The resistance of the millivolt-meter is usually large, while that of a milliammeter must be very small. A milliammeter is placed in a circuit, and its resistance should be low so that placing it in the circuit does not materially change the total resistance of this circuit. A millivoltmeter, on the other hand, is connected to two points in a circuit in order to measure the difference of potential between the two points. Its resistance should be high so that it will take very little current and not, by its presence, change the current in the original circuit.

Moving coil instruments used as indicators in temperature measurement have to be specially designed for this purpose, for they must be robust enough for industrial use; the effects of ambient temperature changes must be negli-gible; and the instrument must be sensitive enough to indicate small changes of current. The sensitivity and resistance of the instrument should be such that it is suitable for use with the rest of the measuring circuit.

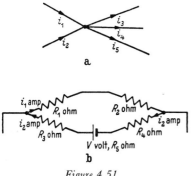

Figure 4.51
Simple circuits Kirchoff's laws

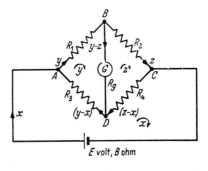

Figure 4.52
Maxwell's cyclic currents

The moving coil instrument consists of a coil of insulated wire which is suspended or pivoted so that it may turn freely in the field due to a permanent magnet. In moving coil galvanometers, which measure small currents of a few microamperes, the suspension is a fine phosphor-bronze wire, and the torsion in the suspension provides the controlling force. In the other moving coil instruments a pivoted coil is used, unless the position of the pointer is detected mechanically, when the coil is usually suspended. When a current flows through the coil there will be forces acting on the sides of the coil which are at right-angles to the lines of force of the magnetic field. The direction of these forces will be given by Fleming's left-hand rule, and are as shown in *Figure 4.53*. These forces form a couple whose magnitude will be proportional to the strength of the magnetic field, the number of turns in the coil, and the current flowing through the coil, provided the magnetic field is always parallel to the plane of the coil. In order to make the magnetic field parallel

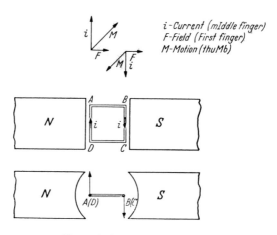

Figure 4.53   Forces on moving coil

to the plane of the coil over a large range of movement, the magnetic field is made radial by placing a cylindrical soft-iron core (6) between curved pole pieces (7). These are shown in *Figure 4.54* which shows the movement of an industrial millivoltmeter.

The restoring force on the coil is provided by two control hairsprings (10 and 15), which provide a restoring force which is proportional to the deflection of the coil. The coil, therefore, rotates through an angle which is proportional to the current flowing. The springs also conduct the current to the coil, and care must be taken to see that excessive currents are not passed through the instrument or the hairsprings will be heated, their elastic properties changed and the calibration of the instrument destroyed, even if the coil is not burnt out. The zero of the instrument can be adjusted by moving the zero arm (12) to which the upper control spring is attached.

As the current which flows through an industrial measuring galvanometer is very small, it is essential that the friction at the pivots should be at a minimum, yet the instrument must withstand the vibration which is met in industrial use. This is achieved by carrying the pivots (2) in conical jewels

(3) which are mounted on springs (5). The springs keep the jewels and pivots in continuous contact so that they cannot strike each other owing to vibration. The coil (1) is mounted slightly eccentric to the axis (4) so that its weight balances the weight of the pointer (8), a final adjustment of balance being made by the position of the sliding weight (9). The balancing should be such that the reading is reliable whether the instrument is level or not.

Other features of the instrument may be gathered from *Figure 4.54*.

4.6.4   Wheatstone's bridge circuits in practice

There are a large number of different electrical circuits used to measure the resistance of a thermometer coil but the most common are forms of the Wheatstone's bridge circuit.

4.6.4.1   UNBALANCED WHEATSTONE'S BRIDGE

In this method, the deflection of the instrument connected to B and D gives an indication of the change of resistance of the thermometer coil. The instrument, unlike the millivoltmeter for use with thermo-couples, need not necessarily have a high internal resistance, provided the current flowing through it is taken into account in calibrating the bridge circuit. In fact, the best operating accuracy is determined by the resistances in the other parts of the circuit.

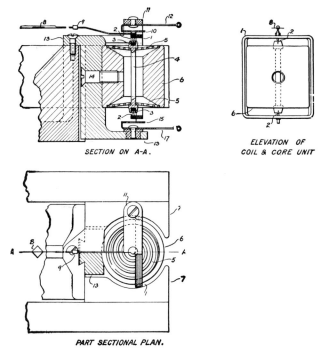

SECTION ON A-A.

ELEVATION OF
COIL & CORE UNIT

PART SECTIONAL PLAN.

Figure 4.54   Moving coil galvanometer movement. (By courtesy, George Kent Group)

The arms R_1, R_2, R_3 of the bridge are made of wire having a negligible temperature coefficient of resistance, such as manganin, while T, the thermo-meter bulb, has a large temperature coefficient. The values of R_1, R_2, R_3 and T are made so that the bridge is balanced for one value of the temperature of T, and the pointer of the indicating instrument will be undeflected. As the temperature of T increases, the bridge will become unbalanced. This produces a movement of the pointer of the instrument over the scale which is calibrated in temperature degrees.

Effect of supply voltage

As the deflection of the indicating instrument will depend upon the voltage of the battery as well as upon the variation of resistance of the bulb, it is necessary to arrange some method of compensating for variation of battery voltage.

One method of doing this is to place a variable resistance and a milli-ammeter in the battery circuit, and to adjust the resistance so that the current through the bridge is maintained at a constant value.

If it is required to use the indicating instrument of the bridge as a check-meter of the voltage applied to the bridge, a resistance R_4, also made of wire of negligible temperature coefficient, is switched into the bridge circuit in place of the thermometer bulb T. The variable resistance, F (*Figure 4.55(a)*), is then adjusted so that the indicator gives a full scale deflection. This adjust-ment should be carried out at regular intervals.

This system is now only used on portable instruments. On permanent plant instruments the bridge is supplied with the rectified output of a constant voltage step-down transformer, or some other voltage stabilised source of the form already described.

Another method of overcoming the difficulty of variation of supply voltage is to employ the 'cross-coil' type of indicator. In the conventional form of moving coil instrument the controlling force is provided by a hairspring. In the 'cross-coil' type, the controlling force is provided by a second coil rigidly attached to the main moving coil of the indicator movement. A fixed portion of the main current passes through this second, or crossed, coil which is set at a definite angle to the main coil. The force on the crossed coil is such that it would restore the pointer to zero if there were no deflecting force on the main coil. Any variation in the supply voltage will result in equal changes in both the deflecting and restoring forces on the indicator movement so that its deflection will remain unchanged. In these circumstances, it is not necessary to provide the standard resistance R_4, and the rheostat F may be replaced by a fixed resistance G as shown in *Figure 4.55(b)*.

Multipoint temperature indicators

In measuring the temperature at several points which may be distributed geographically such as in several holds of a ship, at several points at different levels in grain silos, or on assembled chemical plants etc., it is often convenient to have several resistance temperature coils which can be connected in turn to a single indicator *(Figure 4.55(c))*. Single or double pole switching may be used, but double pole switches are less likely to suffer from troubles due to 'earthing'.

In order to give the same indication on the instrument for the same temperature at each testing point, the value of the thermometer bulb resistance and of the resistance for each thermometer circuit must be the same. A balancing resistance is, therefore, provided in the leads of each thermometer bulb. This resistance is made of wire having a negligible temperature coefficient, and is adjusted so that the resistance of the bulb at a standard temperature, leads and balancing resistance add up to the same total for each circuit. In order to make this adjustment, the leads are placed in position between bulb location and the indicator and the bulb is replaced by a standard resistance of wire having a negligible temperature coefficient. Wire is then removed from the balancing resistance until a standard deflection is obtained on the indicator. This is done with each portion of the installation in turn so that the total resistance for any particular bulb temperature will be the same for each circuit.

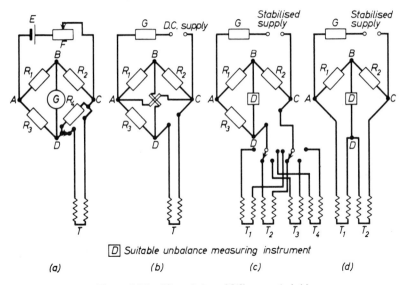

Figure 4.55   The unbalanced Wheatstone's bridge

Where it is required to measure the mean temperature of the contents of a tank in which temperature stratification may occur, a number of bulbs are inserted at different levels. The bulbs are then connected in series in the arm *CD* of the bridge. The resulting total resistance will be a measure of the mean temperature of the tank contents.

Measurement of temperature difference

In certain types of chemical plant it is important to measure the change which takes place in the temperature of the reacting substances at each stage of the process. In order to do this, two identical resistance-thermometer bulbs are used. These bulbs are connected in two arms of the Wheatstone's bridge, as shown in *Figure 4.55(d)*, and the bridge is balanced when the bulbs are at the same temperature. The indicator is then calibrated in terms of the difference of temperature between the two bulbs. Temperature difference may be measured equally well by means of the balanced Wheatstone's bridge type of instrument.

Thermometer leads

When the thermometer bulb is situated some distance away from the bridge, the connecting cable, which is usually made of copper, must be of generous section so that the resistance of the leads is only a small fraction (less than 2 per cent) of the resistance of the bulb. The actual diameter of the wire used will depend upon the resistance of the bulb and the length of the leads; the longer the leads the greater must be their conductivity.

If the change of resistance of the leads with temperature is very small in comparison with the change in the resistance of the bulb with temperature, then ambient temperature effects will be negligible. When, however, the leads from the bulb are very long, it might become expensive to put in leads whose resistance is small enough to fulfil this requirement. The difficulty can be overcome by using multiple-wire connections to the bulb.

The effect of change of lead resistance upon the temperature reading of the indicator can be reduced or even eliminated by using more than two leads to the bulb. There are two systems in common use; the three-lead system, and and the four-lead system.

Three-lead connection

This system of connection was originated by Sir W. Siemens in 1871, and is shown in *Figure 4.56(a)*. Three leads, l_1, l_2, l_3, are connected to the resistance coil and bridge circuit. The lead l_1 is in the same arm of the bridge as the resistance R_3 while l_3 is in the thermometer arm. The lead l_2 connects D to one terminal of the bulb. It will be seen that if l_1 and l_3 are made of identical wire and have the same length, they will have equal resistances, and changes of ambient temperature will affect both leads equally. If $R_1 = R_2$, and $R_3 = T$ when the bridge is balanced, then changes in the resistance of the leads will affect both arms AD and CD equally, and the bridge will remain balanced.

Four-lead connection

This method of connection was introduced by Callendar in 1886, and is shown in *Figure 4.56(b)*. Two identical pairs of leads are connected in the arms AD and CD of the bridge. The pair in the arm AD are connected together at a point near the thermometer bulb; while the second pair, which is in the arm CD, is connected to the thermometer resistance bulb. Both pairs of leads are enclosed in the same outer cover so that the leads in AD compensate for any changes in the resistance of the leads in *CD* due to ambient temperature variations.

4.6.4.2 THE BALANCED WHEATSTONE'S BRIDGE

When the highest accuracy and sensitivity is required the balanced form of the Wheatstone's bridge is used to measure the resistance of the resistance-thermometer bulb. This method is described as a null method, because when the bridge is balanced the voltage across *BD* is reduced to zero. The sensitive galvanometer was generally used as a detector of unbalance, but in most modern instruments automatically balancing bridges involving electronic methods of detection are used.

No special precautions need be taken to keep the voltage applied to the bridge constant as this does not alter the values of the resistances at which the bridge will balance. The method is also independent of the calibration of the unbalance detector. Any detecting circuit is suitable if it is sufficiently sensitive to detect small voltages, has a stable zero, and is robust enough for industrial use.

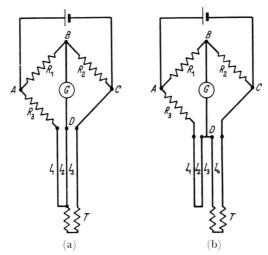

(a)   (b)

Figure 4.56   Resistance thermometer lead systems

In this method, the resistance of one arm, or in some cases more than one arm, of the bridge is adjusted until the bridge is balanced. This adjustment is made by hand in the manually operated instruments, but in others it is done automatically. The adjustable resistance is calibrated so that the position of the moving contact on it is a measure of its resistance. By arranging for a pointer to move over a scale as the adjustment of the resistance is made, it is possible to produce an instrument which indicates temperature directly.

Figure 4.57 shows five possible methods of setting up a bridge circuit.

In the simplest form shown in *Figure 4.57* (*a*) the resistance R_3 is adjusted until the bridge is balanced.

Figure 4.57 (*b*) shows the Callendar-Griffiths bridge in which a slide wire is placed between the resistance R_1 and R_2, and the bridge is balanced by moving the point of contact B. By this method, the contact resistance of the adjustable resistance is removed from the arms of the bridge and does not affect the point of balance. High contact resistance will, however, affect the sensitivity of the bridge as the increased resistance in the detector circuit will reduce its sensitivity. Using this bridge with the three-lead connection will result in completely accurate compensation for changes in the lead resistance due to ambient temperature changes, only when $R_1 = R_2$ and the bulb is at one temperature (usually arranged to be at the middle of the temperature range). At other temperatures there will be a slight error due to ambient temperature changes. If, however, the four-lead connecting system is used, and the slide wire is connected between one thermometer lead and a compensating lead as shown in *Figure 4.57 (c)*, then if $R_1 = R_2$, complete ambient temperature compensation will be obtained at all bulb temperatures.

Another method of improving the ambient temperature compensation and keeping contact resistance out of the bridge arms is shown in *Figure 4.57 (d)*. Two slide wires are used, one connected between R_1 and R_2 and the other between R_1 and R_3. The contacts of the slide wires are linked so that they move together. The bridge is balanced by moving the linked contacts, and their position when the bridge is balanced indicates the temperature.

The bridge circuits already described may be used for a.c. or d.c., provided the resistances are wound non-inductively, and a suitable method is adopted to detect when the bridge is balanced.

Figure 4.57 (e) shows a bridge circuit suitable for a.c. only. In this form of bridge the resistances R_1 and R_2 are replaced by two condensers C_1 and C_2; C_1 has a fixed capacity, and C_2 is variable and is adjusted to balance the bridge.

The bridge is balanced automatically, an electronic method being used to detect when balance is attained.

Automatic balancing of the Wheatstone's bridge may be reduced to the same problem as balancing a potentiometer. The current, or e.m.f., produced owing to a bridge, or potentiometer, being out of balance, can be arranged to set in motion a mechanism which moves the contact on the slide wire of bridge or potentiometer towards the balance point, and to continue to move it, either continuously or in steps, until a balance is obtained. These mechanisms will be described later.

Sources of error

The sensitiveness of a bridge is increased by increasing the current through the bridge. The heating effect of the current in any portion of the bridge is proportional to the square of the current. In order to have a large resistance in a small space the wire of the resistance-bulb must be very fine. In such fine wire, the heating effect of the current will be appreciable unless the current through it is kept small. This heating effect must be taken into account when very accurate measurements are required.

One method of overcoming the self heating effect of the current when electronic detecting methods are used is to energise the bridge with pulses of current of short duration.

Owing to the difference of temperature between various parts of the bridge circuit and the variety of metals used, thermo-electric effects can become appreciable unless precautions are taken to reduce these effects. These effects may be reduced to a minimum if it is arranged that corresponding pairs of junctions of dissimilar metals are mounted together, or are at the same temperature wherever possible.

Care must be taken to see that all connections in the circuit are clean and tight so that contact resistances do not develop. It is essential also, that the insulation between the leads should be sufficient to prevent any partial short-circuiting owing to condensation of moisture, or other causes. These errors may be prevented by eliminating all unnecessary joints and connections,

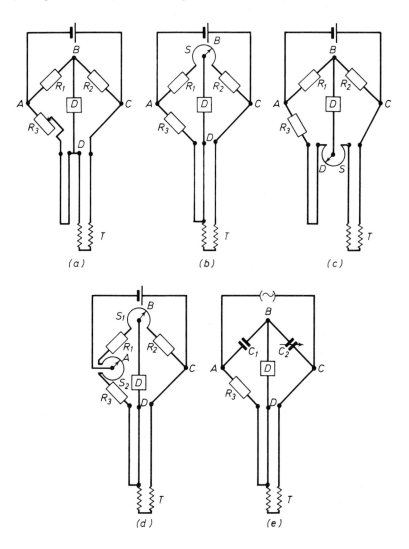

Figure 4.57   The balanced Wheatstone's bridge

making all leads continuous throughout their length as far as possible, and soldering or welding all joints. Cables should be renewed when they show signs of damage or weathering.

Use, range and accuracy

Resistance thermometers provide an accurate and sensitive means of measuring temperature or temperature difference. They are extremely useful where the temperatures at a number of measuring points are required as these can be obtained at one central position. They are used for measuring a number of temperatures at points which are remote from the indicator and where the cost of a capillary tube type of instrument would be excessive. They are useful where the mean temperature over a large zone is required, but are not suitable for measuring the temperature in a position where the space is too small for the whole of the resistance bulb to be immersed.

Resistance thermometers may be used for measuring temperatures for any range within the limits $-240°$ C to $1000°$ C, but for industrial use thermo-couples are preferable for temperatures above $540°$ C, or even $300°$ C in certain circumstances. This is because at temperatures approaching $600°$ C and above, elaborate precautions are necessary to prevent any distortion or strain of the resistance wire on its supports.

Balanced bridge types of measuring circuits are more suitable than the direct deflection type instrument when the range of temperature to be measured is small. At atmospheric temperature, a temperature range of $20°$ C may be indicated on a scale of 300 mm span with a balanced bridge instrument.

The limits of error of the balanced bridge instrument are usually of the order of ± 0.35 per cent of the scale range, and for the direct deflection type perhaps a little more. In the automatically balanced bridge a certain minimum out-of-balance is necessary to make the balancing mechanism operate. This is usually between 0.05 and 0.15 per cent of the scale range, thus a dead band of this magnitude is possible. The reproducibility is usually of the order of ± 0.15 per cent of the scale range.

4.6.4.3 DIRECT DEFLECTION METHOD OF MEASURED E.M.F. OF A THERMO-COUPLE; MILLIVOLTMETERS

The e.m.f. produced by a thermo-couple may be measured by using a moving coil galvanometer arranged as a millivoltmeter. The deflection of the galvanometer will depend upon the current flowing through it, or upon the potential difference across its terminals. This will depend, not only upon the e.m.f. produced by the thermo-couple, but also upon the total resistance in the circuit. The electrical resistance of the thermo-couple wires will increase when placed in the furnace, owing to their rise in temperature. This increase in resistance will depend upon the depth of insertion and upon the actual temperature of the furnace. The resistance of the connecting cable and the coil of the measuring instrument, which are usually made of copper, will also change with temperature. To ensure that these changes of resistance shall have a negligible effect upon the accuracy of the reading, it is necessary to insert in the instrument circuit a 'ballast' resistance of a material having a negligible temperature coefficient, such as manganin, and whose value is large enough to reduce the changes in the circuit resistance to a very small fraction of the total resistance.

Making the total resistance of the circuit large, however, reduces the size of the current producing the deflection of the galvanometer, and makes it essential to have a very sensitive instrument. If the resistance external to the instrument can be reduced to small proportions, the instrument resistance may be reduced. The instrument may, in these circumstances, be made more robust as a larger deflecting torque will be available. Thermo-couples of base metal are, therefore, made of heavy gauge wire and used with a low resistance type of millivoltmeter. The use of heavy gauge wire for precious metal thermo-couples would prove too expensive so that the higher resistance type of milli-voltmeter must be used.

The millivoltmeters are, in general, not calibrated with the thermo-couple with which they are going to be used, for in industrial work the thermo-couple wires have to be replaced from time to time. Instead, they are calibrated with a fixed external resistance by applying a range of e.m.f.s. The values of the e.m.f.s of various thermo-couples at various temperatures are given in the tables in the Appendix.

The value of the external resistance with which the instrument was cali-brated should be marked on the scale, and the instrument should be used only with this resistance.

As the temperature of the instrument is the cold-junction temperature of the thermo-couple in most cases, this temperature should be marked on the scale. The internal resistance of the instrument is also marked on the scale.

4.6.4.4   AMBIENT TEMPERATURE COMPENSATION

Changes of ambient temperature at the instrument have two effects. Owing to increased temperature, the resistance of the instrument coil will increase. The effect of this is made negligible by the presence of the 'ballast' resistance whose value is not affected by temperature changes. A second and more serious effect exists in instruments where the temperature of the instrument is the temperature of the cold-junction, or the reference temperature. Increase in the temperature of the cold-junction will cause a decrease in the e.m.f. of the thermo-couple. This decrease is equal to the e.m.f. of the thermo-couple with one junction at the temperature at which the instrument was calibrated, and the other at the new temperature of the instrument. The instrument will, therefore, read low by an amount which depends upon the increase of ambient temperature.

The simplest method of allowing for the increase in cold-junction tempera-ture is to add the required correction to the temperature reading. Where the temperature-e.m.f. relationship is linear, as it is with base metal thermo-couples, this correction is equal to the difference between the actual cold-junction temperature and the temperature of the cold-junction for which the instrument was calibrated. For platinum v platinum 10 per cent rhodium thermo-couples with the hot-junction at about $1600°$ C, the correction to be added is approximately equal to half the increase in temperature of the cold-junction. For platinum *v* platinum 13 per cent rhodium thermo-couples in similar circumstances, the correction is about 0·4 of the increase in the cold-junction temperature.

There are also methods of automatically compensating for the increase in cold-junction temperature. In the method devised by Darling, what is normally the fixed end of the control spring of the moving coil of the measuring instrument is connected to a bimetal strip, and this is arranged to move the pointer over the scale a number of divisions equal to the increase in cold-junction temperature. In this way a total deflection is always obtained which is equivalent to the sum of the ambient temperature and the temperature indicated by the thermo-couple, so that the reading is independent of ambient temperature. This form of compensator is shown in *Figure 4.58 (a)*. In instruments where automatic cold-junction compensation is not provided, and

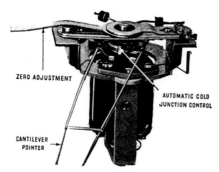

Figure 4.58(a)
Mechanical cold-junction compensation.
(By courtesy, Negretti & Zambra Ltd.)

where the cold-junction temperature is fairly constant, an adjustment equivalent to that produced automatically by the bimetal strip, may be made manually. This is done by disconnecting the thermo-couple, and setting the pointer of the indicating instrument to the value of the ambient temperature, by adjusting the zero setting. The actual temperature to which the instrument is set is measured by a reliable mercury-in-glass thermometer placed in the instrument case.

An electrical method of achieving the same result is illustrated in *Figure 4.58 (b)*. A Wheatstone's bridge circuit is introduced into one of the thermo-couple leads. This bridge consists of four equal resistances R_1, R_2, R_3, and R_4. Three of these resistances are made of manganin wire, which has a low temperature coefficient; while the fourth is made of wire, such as nickel, which has a large temperature coefficient. When the ambient temperature is the

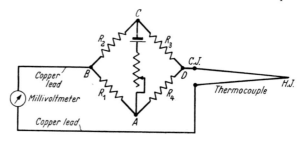

Figure 4.58(b)   Electrical cold-junction compensation

same as that at which the instrument was calibrated, the bridge is balanced, and no potential difference exists between B and D. The values of the resistances are so arranged that when the ambient temperature rises, the bridge becomes unbalanced, and produces between B and D a potential difference equal to the reduction in the e.m.f. of the thermo-couple. In this way, the indication of the instrument is made independent of the ambient temperature.

4.6.5 Potentiometric methods of measuring the e.m.f. produced by a thermo-couple

Development of the potentiometer.

As has already been explained an e.m.f. may be measured by means of a simple potentiometer connected up as shown in *Figure 4.59 (a)*.

The e.m.f. developed by a thermo-couple is very small however, being about 10 millivolts for precious metal thermo-couples, and about 40 millivolts for base metal thermo-couples, with the hot-junctions at 1000° C and the cold-junctions at 0° C. If a potential difference of $1\frac{1}{2}$ volts is applied to a uniform wire 100 cm long then the fall of potential per cm is 15 millivolts.

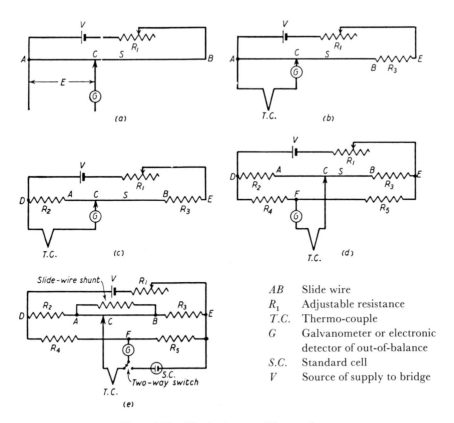

AB	Slide wire
R_1	Adjustable resistance
T.C.	Thermo-couple
G	Galvanometer or electronic detector of out-of-balance
S.C.	Standard cell
V	Source of supply to bridge

Figure 4.59   The development of the potentiometer

If such a potentiometer wire was used to measure the e.m.f. of a thermo-couple, then the point of balance would be very near the end A of the wire.

If a resistance R_3 is connected to the end B of the wire as shown in *Figure 4.59(b)*, its value can be chosen so that the fall of potential per centimetre of the slide wire is much reduced, and the potential difference across practically the whole of the slide wire is required to balance the e.m.f. of the thermo-couple. The e.m.f. of the thermo-couple can then be measured much more accurately.

In many uses of the thermo-couple pyrometer, it is not necessary to measure all temperatures from the temperature of the cold-junction upwards, but only temperatures in a limited range. If, therefore, a resistance R_2 is connected to the end A of the slide wire as shown in *Figure 4.59(c)*, and there is across R_2 a fall of potential equal to that produced by the thermo-couple when its hot-junction is at a temperature T_1; and across R_2 and S a fall of potential equal to the e.m.f. of the thermo-couple with its hot-junction at T_2; the instrument produced will be such that positions of the contact on the slide wire between A and B represent temperatures between T_1 and T_2.

As has already been explained, increase in ambient temperature, and consequent increase in the temperature of the cold, or reference, junction, will lead to a reduction in the e.m.f. of the thermo-couple. In order to compensate for this, two further resistances R_4 and R_5 are connected to D and E, as shown in *Figure 4.59(d)*. If R_2, R_3, and R_5 are made of manganin wire whose temperature coefficient is extremely small, and R_4 is made of a wire such as nickel wire whose resistance increases with temperature, then, when the temperature at the instrument increases, the fall of potential in R_4 will increase. The point F will, therefore, become less positive relative to the point D, moving the balance point C upscale, and having the same effect as adding a small potential difference to the e.m.f. of the thermo-couple, thus compensating for the increase in ambient temperature. If the resistance R_4 is housed at the same point as the effective cold-junction of the thermo-couple, then this compensation can be made very accurate, provided the e.m.f. temperature relationship of the thermo-couple is linear.

Figure 4.59(e) shows a further development now only used in portable instruments. A standard cell is connected to E and to one pole of a two-way switch. The galvanometer can, therefore, be connected to either the standard cell or the thermo-couple. The value of R_5 is made such that when the potential difference between D and E has the value required for the correct functioning of the potentiometer, the fall of potential across R_5 is equal to the e.m.f. of the standard cell. The galvanometer is connected to the standard cell, and the resistance R_1 adjusted until a balance is obtained. It is then known that the voltage applied to the network is correct. When adjustment of the resistance R_1 will no longer balance the bridge, this is an indication that the dry battery supplying the voltage to the potentiometer needs replacing.

In addition, the slide wire is often provided with a shunt which is adjusted by the manufacturer so that the resistance of slidewire and shunt together have the correct value for the network.

The circuit of the potentiometer is often complicated still further by other refinements. Further resistances may be added so that the indicator moves up-scale when the thermo-couple burns out. In other cases, a double slidewire is used so that the total resistance in the galvanometer circuit remains constant and the damping of the galvanometer movement remains at a critical value. Circuits of this nature are often used on the automatic-balancing potentiometers which are used in conjunction with temperature controllers.

4.6.5.1 MANUAL BALANCING

The simplest form of balanced potentiometer is operated manually. The circuit will have a similar form to that shown in *Figure 4.59(e)*. The instrument

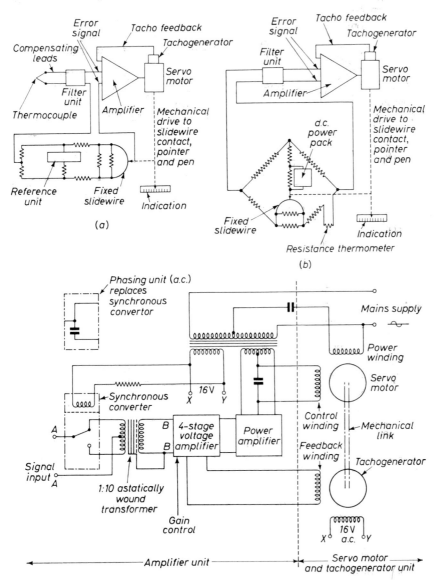

Figure 4.60   Electronic continuous measuring system. (By courtesy, Kent Instruments Ltd.)

may be used to measure the temperature at one point, when a single thermo-couple is connected to the instrument; or it may be used to measure the temperature at several points, when a number of thermo-couples, all identical, are arranged so that they may be switched into the circuit in turn, by means of a single or double pole multiway switch.

With the appropriate thermo-couple in the circuit, the contact on the slide-wire of circular form, is adjusted until no current flows in the galvanometer. The temperature is then read off on the appropriate scale from the position of the slide-wire contact, where the instrument has been calibrated in degrees of temperature. Where the indicating scale of the instrument is calibrated in millivolts, the reading is taken and the temperature found from the appropriate millivolt-temperature table for the type of thermo-couple used.

There are several forms of manually balanced potentiometers for laboratory and industrial use. For industrial use the instrument may be fixed or portable.

4.6.5.2   SELF-BALANCING POTENTIOMETERS

Mechanisms for balancing potentiometers may have a variety of forms. Whatever form the instrument takes the principle is the same, an electronic circuit, is arranged to detect the fact that the potentiometer is out of balance. The output of the detecting circuit controls the direction of rotation of a reversible motor which moves the slidewire contact towards the point of balance and positions the pen arm or pointer.

Several manufacturers produce equipment which is very similar in principle. The continuous system manufactured by George Kent, illustrated in *Figure 4.60*, is a good example of a modern system and will be described.

In potentiometric instruments a voltage is produced across the slidewire and associated circuit by an all-mains-operated d.c. reference unit, or a.c. power pack. The voltage between a point in the circuit and the slidewire contact is compared with the voltage from the detecting element and any out-of-balance voltage is fed through the high-gain electronic amplifier to the control winding of a two-phase induction motor. This servo motor then drives the slidewire contacts by means of a steel cord to a new position in which the out-of-balance voltage is zero. The servo motor also repositions the pen and pointer to give an indication of the measurement.

A highly stable reference potential is required across the slidewire. This is provided in d.c. instruments by a small mains-operated device which main-tains a precise 5 mA d.c. current through the stable 1000 ohm load of the instrument circuit. The output is stabilised by a Zener diode circuit*.

Two Zener diodes are selected or matched for low temperature coefficient and incorporated in a two-stage regulating system which can provide an output of about 8 volts at 5 mA stable within $\pm 0 \cdot 1$ per cent for temperature changes of $\pm 30°$ C, supply changes of ± 10 per cent and load variations of 100 μA.

The unbalanced direct voltage is converted into an alternating voltage by means of a synchronous converter (*Figure 4.61*) which is basically a single-pole double throw switch operating continuously at mains frequency. A soft iron armature, carried on a beryllium copper reed, is magnetised by the

*G. B. Marson, 'Reference Sources for Industrial Potentiometer Instruments'. *Brit. Comm. & Electronics*, Oct. 1959.

alternating current flowing in the energising coil. It is therefore attracted alternately to one and then the other pole of the permanent magnet thus causing the reed to vibrate at mains frequency. Contacts attached to the reed engage with two others mounted on light flexible blades and preloaded inwards against adjustable stops. Such an arrangement eliminates contact bounce and enables the 'make' period to be independently pre-adjusted to any setting. As the reed moves from one contact to the other, any unbalanced voltage will cause direct current to flow, first in one direction through one half of the primary winding of the input transformer, then in the opposite direction through the other half of the primary winding. This will result in an alternating voltage being induced in the secondary winding of the transformer which alternates in step with the a.c. supply.

The entire rebalancing action of the system is based upon the timing relationship between the alternating voltage formed by the converter and the a.c. supply voltage wave; this timing depends upon the direction of the unbalance in the potentiometer circuit. For example, when the measured temperature rises above the temperature at which the potentiometer circuit is in balance, the e.m.f. of the thermo-couple will cause a difference of potential across AA of a particular polarity. For this condition, the converter will induce in the input transformer secondary, an alternating voltage wave whose positive areas correspond with the positive areas of the a.c. supply voltage and the two voltages are said to be 'in phase'.

If on the other hand the temperature falls below the temperature at which the potentiometer circuit is in balance, the polarity across AA will be reversed, and so will the positive and negative areas of the output from the input transformer. When this timing relationship exists, the output of the transformer is said to be 180° out of phase with the supply voltage.

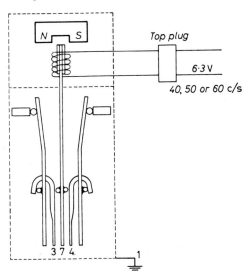

*Figure 4.61   Diagrammatic arrangement of synchronous converter.
(By courtesy, Kent Instruments Ltd.)*

Numbers refer to pins on base: contacts 3, 7 and 4 are the single pole, double throw switch

Thus, the voltage input into the amplifier for the two conditions of un-balance will be either in phase or 180 degrees out of phase with the a.c. supply voltage. This relationship provides the means, following the amplification of the transformer output, of driving the balancing motor in the proper direction to rebalance the potentiometer circuit. One motor winding is supplied with 100 V a.c. leading the mains frequency by 90°; the other winding is energised by the amplifier output signal in phase (or anti-phase) with the mains. The two windings thus set up a rotating magnetic field which generates eddy currents in the squirrel-cage rotor, causing the motor to rotate in one direction or the other (depending on the phase) and hence adjusts the slidewire contact position, and also actuates the indicating or recording mechanism.

In more recent instruments the mechanical chopper is being replaced by an electronic chopping system using field effect transistors.

In order to prevent the motor overshooting and hunting, an a.c. tacho-generator is also driven by the servo motor. This tacho-generator produces an output whose magnitude is proportional to the speed of rotation of the motor and which reverses in phase when the motor reverses. This output is fed into the grid of the third stage of the amplifier so as to oppose the error signal.

When the instrument is some way off balance the amplified error signal is much greater than the feedback signal and the latter has no effect. As balance is approached the error signal diminishes and a point is reached where feed-back becomes effective. The motor torque is then reduced and the overshoot and hunting prevented.

An upscale/downscale device is built into the amplifier and can be used for most applications. This device ensures that in the event of certain types of failure in the amplifier or external circuit, the servo motor drives the pen upscale or downscale, as desired. Selection of upscale, downscale or no-drive is made by means of a three-position plug.

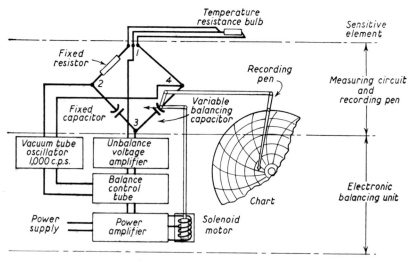

Figure 4.62   The Dynalog principle. (By courtesy, Foxboro-Yoxall Ltd.)

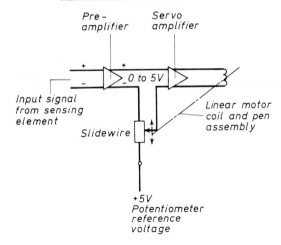

Pre-amplifier   Servo amplifier

Input signal
from sensing
element

Slidewire

0 to 5V

Linear motor
coil and pen
assembly

+5V
Potentiometer
reference
voltage

(a) Principle of the measuring system.

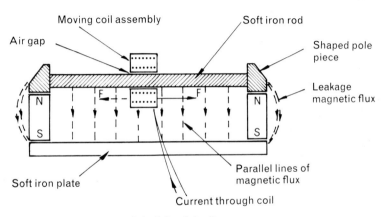

Moving coil assembly   Soft iron rod

Air gap

Shaped pole
piece

Leakage
magnetic flux

N   F   F   N

S   S

Soft iron plate

Parallel lines of
magnetic flux

Current through coil

(b) Principle of the linear motor.

Figure 4.63   The Clearspan recorder. (By courtesy, Foster Cambridge Ltd.)

Two-second balancing speed is standard; for d.c. potentiometric applications a one-second balancing speed can be provided.

The measuring-circuit current for the a.c. potentiometer is provided by a
transformer mounted in place of the d.c. reference unit. In order to eliminate
any difference in phase between the measured potential and the slidewire
potential owing to phase shift in transmission lines, transformers, etc., a
transistor-thermistor feedback quadrature suppressor is used*. If the phase
of the input were to differ from the phase of the balancing by as little as one
or two degrees quite substantial errors in the size of balancing could arise.

For a.c. measurements the synchronous converter is replaced by a phasing
unit.

In bridge-circuit instruments the steady smoothed current for the measuring

*I. C. Hutcheon and D. N. Harrison, 'A transistor-thermistor feed-back quadrature suppressor',
Electronic Engineering, 1960.

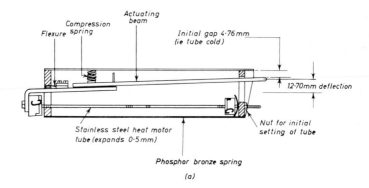

(a)

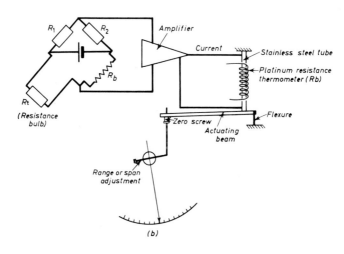

(b)

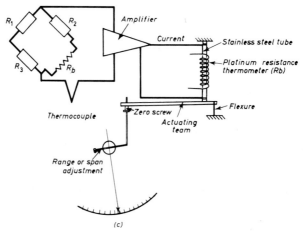

(c)

Figure 4.64   The heat motor self-balancing system.
(a) The heat motors
(b) Use with resistance bulbs
(c) Use with thermo-couples

(Courtesy Negretti and Zambra Ltd.)

circuit is provided by a mains power pack using a selenium rectifier and a four stage resistance-capacity filter.

The principle of operation of the Foxboro Dynalog instrument is completely different from the other self-balancing instruments. The principle of the method of operation is illustrated in *Figure 4.62*.

When the measuring circuit becomes unbalanced, it sets up a voltage. This voltage is amplified by the unbalance voltage amplifier; is passed through the unbalance detector circuit to determine the direction of change of the measured variable; and upsets the balance of equal and opposite voltages on two coils of the Dynapoise drive. This unbalance voltage on the drive coils causes the cores to move, thus rotating the moving vanes of the variable capacitor and the pen arm, or pointer, which is linked to it. The vanes of the variable capacitor are moved until the measuring circuit is rebalanced, when the unbalanced voltage becomes zero, and the pen, or pointer, stops at the new reading.

Figure 4.63(a) shows the principle of operation of an advanced form of electronic recorder.

The signal from the primary sensing element is amplified by an advanced transistorised preamplifier providing a very high input impedance, and the current passing through the signal source is 0·5 mA maximum.

The preamplifier output signal of 5 V nominal, is compared with a reference voltage picked off the slidewire, energised by a stabilised supply, and the difference is amplified by the servo amplifier, the output of which drives a linear motor.

The principle of the linear motor is shown in *Figure 4.63(b)*. When a current is passed through the coil of the linear motor, the coil will become a magnet with a N pole on one end and a S pole on the other. The S pole of the coil will be attracted by the pole piece at one end of the N pole element while the N pole of the coil will be repelled by the pole piece at the other end, resulting in the coil moving with the S pole leading. Reversing the current will reverse the direction of magnetism and hence the direction of motion. Thus, the linear motor produces a highly responsive system of low inertion without any gearing or linkages which are subject to wear.

Attached to the motor carriage is the indicating pointer, pen, and the slidewire contact which will be moved along the slidewire until the voltage it picks up from the slidewire is exactly equal and opposite to the output from the preamplifier so that there is no voltage applied to the servo amplifier, and the pointer indicates the measured value.

If a resistance thermometer (or some other sensing element giving a resistance change) is used, it is connected to form one arm of a resistance bridge from which the millivolt output is fed into the preamplifier.

This instrument has an accuracy of ± 0.25 per cent of the span, a response time of 1 s for full scale travel, and will respond to 0·15 per cent of the scale span.

A method of balancing a Wheatstone bridge without a slidewire or conventional balancing motor is provided in the Tensor self-balancing system illustrated in *Figure 4.64*.

The motive power of the system is provided by a stainless steel tube in the heat motor as shown in *Figure 4.64(a)*. The change in the length of the tube, which is limited to 0·5 mm, rotates an actuating beam about a flexure pivot. This beam, through a 25:1 magnifying linkage, provides the pointer movement. Increase in tube length is resisted by a compression spring which ensures the tube is in tension and provides overload protection. The tube is at 100° C when the pointer is at the zero of the scale, and at 300° C at full scale deflection. The initial mechanical setting is vital and must not be disturbed as it will affect the linearity of the system. The tube temperature range is also important as too high a temperature would reduce the life of the tube, while too low a temperature would limit the rate of heat loss from the tube thus slowing the rate of downscale movement. The time for full scale deflection is 3 to 5 s depending upon the stainless steel tube diameter. A platinum resistance element is wound on, but electrically insulated from, the steel tube.

In the resistance bulb application shown in *Figure 4.64(b)* the change in resistance with temperature of the platinum resistance element on the tube is used to balance the bridge. If the resistances R_1 and R_2 are equal, the bridge will be balanced when R_b is equal to R_t. If R_t increases the bridge will be unbalanced and the out-of-balance voltage will be applied to the high gain amplifier which will increase the current flowing through the steel tube until the resistance R_b increases sufficiently to be equal to R_t, and the pointer moves upscale. Likewise a fall in the resistance R_t will reduce the current through the tube reducing R_b until balance is restored, and moving the pointer downscale.

It is essential that the gain of the amplifier should be high so that even the smallest deviation from balance will produce a sufficiently large change in current so that the departures from true null balance will be less than $\frac{1}{2}$ per cent of the full scale range. The system accuracy claimed is 1 per cent of full scale range.

In the thermo-couple application shown in *Figure 4.64(c)*, the thermo-couple e.m.f. is opposed by the out-of-balance e.m.f. produced by the Wheatstone bridge circuit. When the e.m.f.s are equal and opposite, no e.m.f. will be applied to the amplifier. Thus as with the resistance bulb application, when the thermo-couple temperature changes the system will be balanced by changes in the temperature of the stainless steel tube which results in pointer movement indicating the temperature.

4.7 RADIATION OR INFRA-RED THERMOMETERS

Although temperatures above 1064° C on the International Practical temperature scale are defined in terms of the radiance of a hot body, radiation or infra-red thermometers are used mainly where the temperature exceeds the limit of a sheathed thermo-couple or because the corrosive conditions at the temperature are too severe for a thermo-couple to be satisfactorily used. Radiation thermometers are also used when it is not feasible to use a thermo-couple either because the hot object is moving, or is otherwise inaccessible, or if the body is in cold surroundings and a thermo-couple cannot be brought to the temperature of the hot body. Because of the extremely high speed of response of a suitable radiation thermometer (a few microseconds) they are very useful for measuring temperatures under certain transient conditions.

Radiation pyrometers do not require to be in contact with the hot object in order that a temperature measurement may be made. They are based upon the measurement of radiant energy emitted by the hot body. (B.S. 1041 Code for temperature measurement. Part 5, Radiation Pyrometers.)

There are three methods by which the transference of heat takes place.

They are conduction, convection and radiation. The first two methods require the presence of matter, but radiation takes place independently of it. It is method by which heat reaches us from the sun, and the fact that at an eclipse the heat and the light of the sun are cut off simultaneously indicates that radiant heat and light travel through space with the same velocity.

It is found that there are many other forms of radiant energy which have the same velocity as light and radiant heat. All these forms of radiation have the same nature, and are associated with electro-magnetic waves which are propagated through space with the common velocity of 299 774 km per second. Radiation has a variety of forms which range from long wireless waves, having a wavelength of about 2000 m, to visible light, having a wavelength of about 0.5×10^{-6} m; and very penetrating X-rays having a wavelength which is less than 10^{-8} m. The wavelength, when very small, is usually expressed in terms of the 'micron' (μm) which is 10^{-6} m, or the Ångstrom unit (Å) which is 10^{-10} m. The different forms of radiation are distinguished by their method of generation and detection. Visible light forms only a very small fraction of the whole gamut, having a wavelength of from 0.4 μm to 0.8 μm. 'Temperature Measurement Radiation Pyrometers' applies to this type of equipment.

For the purposes of temperature measurement the types of radiation measured are limited to the ultra-violet, visible and infra-red radiations, i.e. to wavelengths of from 0.1 μm, in the ultra-violet band; to about 10 μm, in the infra-red band. All forms of radiation within these wavelengths behave in the same way as light, i.e. the radiation travels in straight lines, may be reflected or refracted, and the amount of radiant energy falling upon a unit area of a receiver is inversely proportional to the square of the distance of the receiver from the radiating source.

Absorptive and emissive powers of surfaces

Radiation pyrometers are usually calibrated to measure correctly when sighted upon a full-radiator or 'black body'. A black body is a body which at all temperatures absorbs all radiation falling upon it, without transmitting or reflecting any, whatever the wavelength, the direction of incidence or the polorisation. It would appear absolutely black when cold, as it would absorb all light falling upon it. The absorptive power of its surface, defined as the fraction of the incident radiation which it absorbs, will, therefore, be unity. Other surfaces will not absorb all the incident radiation, but will reflect or transmit some wavelengths. The colour of a body when cold depends upon wavelengths which it reflects. These surfaces will have an absorptive power of less than one.

It is also found that a black body is a perfect radiator. When hot, it emits more radiation than any other body at the same temperature. The ability of a surface to radiate is, therefore, defined in terms of the ability to radiate of a black body. The 'emissive power', or 'emissivity' of a surface, is the ratio of the radiation emitted, to that which the surface would emit if it were a perfectly black body. When practically all wavelengths emitted from a hot surface are considered, the emissive power is called the 'total emissivity' and represented by e_t. When only a small band of wavelengths is considered the term 'spectral emissivity' is used and a subscript representing the effective wavelengths is added to the symbol, e.g. $e_{0\cdot65}$ represents the emissivity at a wavelength of $0\cdot65$ μm.

The emissivity of the surfaces of several common substances are given in Tables 5.10 to 5.13 (Appendix).

An important fact from the point of view of practical pyrometers which is not shown by the tables is that the spectral emissivity of metals is usually greater at shorter wavelengths than at longer ones. The opposite may be true for refractory surfaces. The spectral emissivity of glass at $0\cdot65$ μm is almost zero. Roughening the surface of a metal, or oxidising it, usually increases the emissivity.

A black body is realised in practice by having a large enclosure, A, maintained at a constant temperature, and having a small opening, B. Provided the inside walls of A are not perfectly reflecting, the opening B, if sufficiently small, is perfectly black. A ray of radiation falling on B passes inside the chamber and is reflected so many times on the inside surface before it reaches B that it is completely absorbed. Since the emissive power and the absorptive power of a surface are equal, the enclosure will also be a perfectly black body when acting as a radiator.

It is easily seen that the absorptive power of a body for a particular radiation must equal its emissive power for that radiation, since, if a body were able to absorb radiant energy more freely than it could emit, or re-radiate, that energy, the body would soon become hotter than its surroundings. If a body radiated energy more rapidly than it received it, it would cool below the temperature of its surroundings. It is well known that a body takes up the temperature of its surroundings, therefore, emissive power 'e' = absorptive power 'a'.

When the metal in a furnace has reached the furnace temperature, a radiation pyrometer sighted through a small hole in the back of the furnace on to the inside of the roof will measure the roof temperature under approximately black body conditions provided the effects of the flame may be ignored. The emissivity of the brickwork is less than unity, but the radiation will be supplemented by radiation radiated from other parts of the furnace and reflected from the roof, so that the total radiation will be approximately equivalent to that radiated by a black body.

When the cold metal has just been put into the furnace, the metal will be absorbing more radiation than it emits. Therefore, the radiation reflected from the roof will be less than that required to bring the total radiation from the roof up to that which would be radiated by a black body. A correction will, therefore, be required for the emissive power of the roof. The conditions can only be regarded as being black body conditions when the whole of the

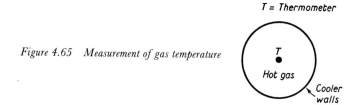

Figure 4.65   Measurement of gas temperature

enclosure is at the same temperature. This condition will exist only when all lines of demarcation between various objects contained in the furnace, completely disappear.

The temperature equilibrium within an enclosure is dynamic and not static. Even when all within an enclosure is at the same temperature, each portion remains at the same temperature because it receives exactly as much radiation as it loses.

Prevost's theory of exchanges

Consider a perfectly heat-insulated space containing two bodies A and B, both bodies will be radiating heat, and both will be absorbing heat. If A is hotter than B it will radiate more heat than B, so that B will receive more heat than it radiates, and its temperature will rise. A, on the other hand, will radiate more heat than it receives so its temperature will fall. This process will go on until A and B are at the same temperature, when the amounts of heat exchanged will be equal, and the bodies will remain at the same temperature.

The fact that bodies radiate heat, explains why it is difficult to measure the temperature of a hot gas contained in an enclosure, the walls of which are at a lower temperature than the gas (*Figure 4.65*). When the thermometer is placed in the gas it receives heat from the gas. The rate at which it receives heat from the gas is not very great because of the low specific heat of the gas. The thermometer will also receive heat by radiation from the walls of the enclosure, but owing to its temperature being greater than that of the walls it will radiate more heat to the walls than it receives from them. It will, therefore, indicate a temperature which is less than the true temperature of the gas.

Black body radiation: Stefan-Boltzmann law

The total power or radiant flux of all wavelengths (R) emitted into the frontal hemisphere by a unit area of a perfectly black body is proportional to the fourth power of the temperature kelvin.

i.e.
$$R = \sigma T^4 \tag{4.24}$$

where σ = Stefan-Boltzmann constant, having an accepted value
of $5 \cdot 669\ 7 \times 10^{-8}\ \mathrm{Wm}^{-2}\ \mathrm{K}^{-4}$

T = Temperature kelvin
(i.e. degrees Celsius $+ 273 \cdot 15$)

This law is very important, for most total radiation pyrometers are based upon it. If a receiving element at a temperature T_1, is arranged so that radiation from a source at a temperature T_2 falls upon it, then, it will receive heat at the rate of σT_2^4, and emit it at the rate of σT_1^4. It will, therefore, gain heat at the rate of $\sigma(T_2^4 - T_1^4)$. If the temperature of the receiver is small in comparison with that of the source, then T_1^4 may be neglected in comparison with T_2^4, and the radiant energy gained will be proportional to the fourth power of the temperature kelvin of the radiator.

The distribution of energy in the spectrum: Wien's laws

Optical pyrometers are based upon the measurement of the spectral concentration of radiance at one wavelength emitted by a source. They depend upon Wien's laws.

When a body is heated it appears to change colour. This is because the total energy, and distribution of radiant energy between the different wavelengths, is changing as the temperature rises. When the temperature is about 500° C, the body is just visibly red. As the temperature rises, the body becomes dull red at 700° C, cherry red at 900° C, orange at 1100° C, and finally white

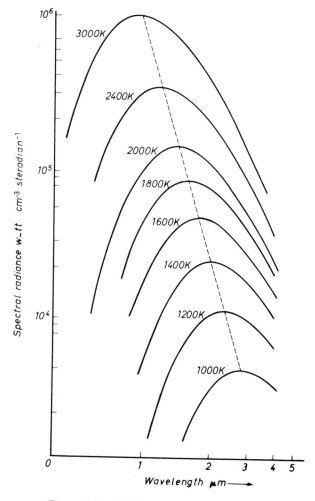

Figure 4.66   Distribution of energy in the spectrum

hot at temperatures above $1400°$ C. The body appears white hot because it radiates all colours in the visible spectrum.

It is found that the wavelength of the radiation of the maximum intensity gets shorter as the temperature rises. This is expressed in Wien's displacement law

$$\lambda_m T = \text{Constant} \tag{4.25}$$
$$= 2898 \ \mu\text{m K}$$

where λ_m is the wavelength corresponding to the radiation of maximum intensity, and T is the temperature kelvin. The actual value of the spectral radiance at the wavelength λ_m is given by Wien's second law

$$L\lambda_m = \text{Constant} \times T^5 \tag{4.26}$$

where, $L\lambda_m$ is the maximum value of the spectral radiance at any wavelength, i.e. the value of the radiance at λ_m. T is the temperature kelvin. The constant has not the same value as the constant in Equation (4.25). It is important to realise that it is only the maximum radiance at one particular wavelength which is proportional to T^5, the total radiance for all wavelengths is given by Stefan-Boltzmann's law, i.e. it is proportional to T^4.

Wein deduced that the spectral concentration of radiance, i.e. the radiation emitted per unit solid angle per unit area of a small aperture in a uniform temperature enclosure in a direction normal to the area in the range of wavelengths between λ and $\lambda + \delta\lambda$ is $L\lambda \ \delta\lambda$ where

$$L\lambda = \frac{c_1}{\lambda^5 e^{\frac{c_2}{\lambda T}}} \tag{4.27}$$

where T is the temperature kelvin, e the base of Naperian logarithms and c_1 and c_2 are constants. This formula is more convenient to use and applies with less than 1 per cent deviation from the more refined Plank's radiation law used to define IPTS(68) provided $\lambda T < 3 \times 10^3$ metre kelvin.

In 1900 Plank obtained from theoretical considerations based on his quantum theory, the expression

$$L\lambda = \frac{c_1}{\lambda^5 \left(e^{\frac{c_2}{\lambda T}} - 1 \right)} \tag{4.28}$$

where the symbols have the same meaning and $c_2 = 0.014\ 388$ metre kelvin.

These laws also enable the correction to be calculated for the presence of an absorbing medium such as glass in the optical pyrometer, and also the correction required for changes in the spectral emissive power of the radiating surface.

The variation of spectral radiance with wavelength and temperature of a black body source is given by *Figure 4.66*.

The temperature of a hot body may be measured in one of three ways.

Firstly; the total radiance of the body may be measured by means of a total radiation pyrometer. This form of pyrometer is intended to receive the maximum amount of radiant energy at all possible wavelengths. In general, the detector is capable of measuring radiation of a wide range of wavelengths, but the actual wavelengths received is usually restricted by the materials used in the optical system. The energy received is represented by an area under the curve given in *Figure 4.66* and is given by the Stefan-Boltzmann law. The temperature of the source can, therefore, be calculated.

Secondly; the spectral radiance of the hot body at a given wavelength may be measured by means of the optical pyrometer. This form of pyrometer is intended to receive the radiance at one wavelength only. The practical pyrometer usually responds to a band of about 0·01 μm width, at about 0·65 μm, i.e. in the red portion of the visible spectrum. From the curves given in *Figure 4.66*, the variation of energy at this wavelength with the temperature of the source is known so that the temperature of the source is known.

Thirdly; the radiance of a hot body over a certain band of wavelengths may be measured by means of a receiving element such as a photo-electric cell, the band depending upon the actual detector used. This type of pyrometer is usually calibrated by sighting it on the filament of a tungsten-strip lamp whose temperature is known in terms of the lamp current.

Radiation pyrometers will, therefore, be described under three headings:

Total radiation pyrometers.
Optical pyrometers.
Photo-electric pyrometers.

4.7.1 Total radiation pyrometers

In this type of instrument, the radiation emitted by the body whose temperature is required, is focused on to a suitable thermal-type receiving element. This receiving element may have a variety of forms. It may be a resistance element, which is usually in the form of a very thin strip of blackened platinum; or a thermo-couple, or thermopile. The change in temperature of the receiving element is then measured as has already been described.

In a typical radiation thermo-pile a number of thermo-couples made of very fine strips are connected in series and arranged side by side, or radially as in the spokes of a wheel, so that all the hot junctions, which are blackened to increase the energy absorbing ability, fall within a very small target area. The thermo-electric characteristics of the thermo-piles are very stable because the hot junctions are rarely above a few hundred degrees Celsius, and the thermo-couples are not exposed to the contaminating atmospheres of the furnace. Stability and the fact that it produces a measurable e.m.f. are the main advantages of the thermo-pile as a detector. In addition, they have the same response to incoming radiant energy regardless of wavelength within the range 0·3 to 20 μm. The main disadvantage of the thermo-pile is its comparatively slow speed of response which depends upon the mass of the thermo-couple elements, and the rate at which heat is transferred from the hot to the cold junctions. Increase in this rate of transfer can only be attained by sacrificing temperature difference with a resultant loss of output. A typical industrial thermo-pile of the form shown in *Figure 4.67* responds to 98 per cent of a step change in incoming radiation in 2 s. Special thermo-piles which respond within one half s are obtainable but they have a reduced e.m.f. output.

In order to compensate for the change in the thermo-pile output owing to

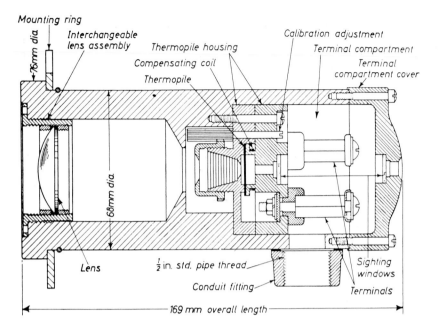

Mounting ring

76mm dia

Interchangeable
lens assembly

Thermopile housing

Compensating coil

Thermopile

Calibration adjustment

Terminal compartment

Terminal
compartment cover

Lens

68mm dia

$\frac{1}{2}$ in. std. pipe thread

Conduit fitting

Sighting
windows

Terminals

169 mm overall length

(a) Cross-section showing construction

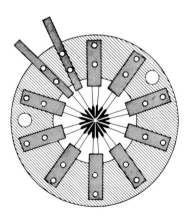

(b) Enlarged view of thermopile

Figure 4.67   Lens type of radiation pyrometer with thermal type of radiation detector.
(By courtesy, Honeywell Ltd.)

changes in the temperature of the cold junctions an ambient temperature compensating coil is located in close proximity to these junctions. A calibrating pinion enables the calibrating diaphragm, which limits the area of the beam falling on the detector, to be adjusted. As thermal type detectors are capable of detecting the total radiation emitted by a radiant source, instruments using such receiving elements are described as total radiation pyrometers although, in fact, for the reasons given later, they should to be accurate be described as 'wide-band' radiation pyrometers.

It is desirable that the indication of the pyrometer should be independent, over a considerable range, of the distance of the source, and should also be independent, within reasonable limits, of the size of the source. To secure this, a lens system, or mirror, is used. The lens, or mirror, will also have the effect of concentrating the radiation on to the receiving element, thereby increasing the sensitivity of the instrument. Instruments may be of the fixed focus type when a mirror is used to form an image of the open end of the pyrometer tube in the plane of the receiving element (*Figure 4.68*), and the instrument is independent of the distance from the source provided the source is large enough to fill the cone formed by the diaphragms of the instrument.

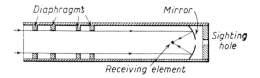

Figure 4.68   Mirror type of radiation pyrometer

Other forms employ either a movable mirror or a movable lens to focus an image of the source on to the receiving element. The pyrometer requires a minimum size of hot object in order that an image large enough to cover the whole of the thermal detector may be produced. The size required varies with the distance of the object from the pyrometer and the value is given in a table provided by the manufacturer. A typical pyrometer would require a hot object 63 mm diameter at a distance of 1·2 m from the pyrometer.

When a lens is interposed between the source and the receiver, as shown in *Figure 4.67(a)*, this lens will limit the bandwidth of the radiation received. In general, the lens transmits the shorter wavelengths only, and absorbs the infra-red region. This will reduce the quantity of energy received, and hence reduce the sensitivity of the pyrometer. This is not serious when measuring high temperatures but becomes a difficult problem at lower temperatures for two reasons. A body at 100° C emits less than one hundredth of the radiant heat emitted at 1000° C, and the radiation at 100° C is mainly of such a long wavelength that it is absorbed by a glass lens. As a result, the radiation which gets through the glass is two thousand times less at 100° C than at 1000° C. Lenses of other material must therefore be used. Table 4.8 gives details of the lens materials used in pyrometry, their transmission bands and the temperature ranges in which they are essential.

In addition to its wide transmission band, arsenic trisulphide has a very high refractive index and very little dispersion so that the lens can be used at a large numerical aperture, and is comparatively thin. In addition, the

Table 4.8

Lens material	Transmission band (μm)	Temperature range °C
Pyrex	0·3–2·7	800–1800
Fused silica	0·3–3·8	500–800
Calcium fluoride	0·1–10·0	300–500
Arsenic trisulphide	0·7–12·0	0–300

use of a thermistor bridge makes it possible to provide accurate compensation for the effect of temperature changes at the detecting head for any temperature between 0 and 80° C, making the provision of a radiation pyrometer for measuring temperatures below 100° C a practical proposition. A sighting hole is provided so that the instrument may be sighted on the hot source, and the image brought into focus at the plane of the receiving element. For use with sources at very high temperatures, the sighting hole is often fitted with a screen which protects the eye of the person focusing the instrument.

The instrument is sighted directly into a furnace through a hole in the wall, or, where this is not desirable, a refractory tube with a closed end is fixed into the furnace wall. The bottom of the tube projects into the furnace, and the pyrometer is sighted upon it. The sighting tube is usually made of silicon carbide, or sometimes of sillimanite. Silicon carbide has a high emissivity and is the best material for most applications. It is not satisfactory for highly oxidising atmospheres nor should it be used in glass tanks because of the danger of contaminating the glass. The tube should protrude into the furnace a distance equal to at least two or three tube diameters if the tube is of silicon carbide, of five tube diameters if the tube is sillimanite. If it is not possible to provide sufficient depth of immersion a shallow recess should be provided in the hot surface of the wall around the tube. The pyrometer must be carefully aligned to sight directly on to the bottom of the tube.

When the temperature of the housing might become too great (above 80° C say), air or water cooling is usually provided. *Figure 4.69* shows the complete

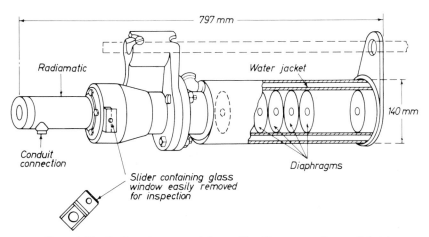

Figure 4.69   Radiamatic water-cooled assembly. (By courtesy, Honeywell Ltd.)

assembly of the Brown Radiamatic receiver with its water-cooled sighting tube as used for high temperature measurement, e.g. roof temperature measurement on furnaces.

Furnace temperature

Conditions in a furnace which might otherwise be considered as perfectly black body conditions may be upset by the presence of flame, smoke or furnace gases. In these conditions, a total radiation pyrometer generally indicates a temperature between that of the furnace atmosphere and the temperature which would be indicated if such an atmosphere were not present.

A thick luminous flame may shield the object almost completely. Non-luminous flames radiate and absorb energy only in certain wavelength bands, principally because of the presence of carbon-dioxide and water vapour. The error due to the presence of these gases can be reduced by using a lens of Pyrex which does not transmit some of these wavelengths, so that the instrument is less affected by variations in quantity of these gases. Where appreciable flame, smoke and gas are present it is advisable to use a closed ended sighting tube, or provide a purged sighting path by means of a blast of clean, dry air.

Errors in temperature measurement can also occur owing to absorption of radiation in the cold atmosphere between a furnace and the pyrometer. To ensure that the error owing to this source does not exceed 1 per cent of the measured temperature, even on hot damp days, the distance between pyrometer lens and the furnace should not exceed 1·5 m if a glass lens is used, 1 m if the lens is silica, and 0·6 m if it is of fluorite.

Calibration of total radiation pyrometers

A total radiation pyrometer may be calibrated by sighting it through a hole into a black body enclosure of known temperature. A special spherical furnace was developed by the British Iron and Steel Research Association for this purpose. The furnace consisted of a sphere 0·3 m diameter consisting of a diffusely reflecting material. For temperatures up to 1300° C stainless steel, 80 Ni 20 Cr alloy, or nickel may be used. For temperatures up to 1600° C silicon carbide is necessary, and for temperatures up to 3000° C graphite may be used provided it is filled with argon to prevent oxidation. The spherical core is uniformly wound with a suitable electrical heating element, such as Kanthal, embedded in Sillimanite and supported on diatomite brick and completely enclosed in a box containing Sil-o-cel insulating powder. For calibration of pyrometers up to 1150° C a hole of 65 mm diameter is required in the cavity, but above this temperature a 45 mm hole is sufficient.

Where the larger hole is used a correction for the emissivity of the cavity may be required for very accurate work. Two sheathed thermo-couples are usually placed in the furnace one near the back and the other just above the sighting hole. Comparison of the two measured temperatures indicates when the cavity is at a uniform temperature.

Calibration may be carried out by comparing the pyrometer and thermo-couple temperature, or the test pyrometer may be compared with a standard pyrometer when both are sighted on to the radiating source which may or may not be a true black body.

Cylindrical furnaces may also be used with a thermo-couple fitted in the sealed end of the cylinder, which is cut on the inside to form a series of 45° pyramids.

A choice of three aperture sizes are available at the open end. For temperatures up to 1100° C the furnace is made of stainless steel but for higher temperatures refractory materials are used. For further details see 'The Calibration of Thermometers' (H.M.S.O. 1971).

Objects in the open

When the temperature of a hot object in the open is being measured, due regard must be given to the correction required for the difference between the emissivity of the surface of the object and that of a perfectly black body.

The total radiant flux emitted by the source will be given by

$$R = e\sigma A T^4 \tag{4.29}$$

where e is the total emissivity of the body; A, the area from which radiation is received; σ, Stefan-Boltzmann constant; and T the actual temperature of the body.

This flux will be equal to that emitted by a perfectly black body at a temperature T_a, the apparent temperature of the body.

$$\therefore \qquad R = \sigma A T_a^4 \tag{4.30}$$

Equating the value of R in equations (4.29) and (4.30)

$$e\sigma A T^4 = \sigma A T_a^4$$

$$\therefore \qquad T^4 = \frac{T_a^4}{e}$$

or

$$T = \frac{T_a}{\sqrt[4]{e}} \tag{4.31}$$

The actual correction to be applied to the apparent temperature is given in Table 5.15 in the Appendix.

The radiation from a hot object can be made to approximate much more closely to black body radiation by placing a concave reflector on the surface. If the reflectivity of the reflecting surface is r, then it can readily be shown that the intensity of the radiation which would pass out of a small hole in the reflector is given by the formula

$$\frac{e}{1-r(1-e)}\,\sigma T^4$$

where e is the emissivity of the surface; σ, Stefan's constant; and T the temperature in kelvin. With a gold-plated hemisphere, the effective emissivity of a surface of emissivity 0·6 is increased by this method to a value of 0·97.

This is the basic idea of the Land Surface Pyrometer manufactured by Land Pyrometers Ltd., which employs a thermopile sighted on a small hole in a gold-plated hemisphere mounted on the end of a telescopic arm (*Figure 4.70*).

Gold is chosen for the reflecting surface because it is the best reflector of infra-red radiation known, and is not easily tarnished. The hole in the reflector is closed by a fluorite window which admits a wide range of radiation to the thermopile but it excludes dirt and draughts. This pyrometer will give accurate surface temperature readings for most surfaces, other than bright or very lightly oxidised metals, without any significant error due to surface emissivity changes. The standard instrument covers a temperature range of from 100 to 1300° C on three scales. A special low temperature version is available for the range 0 to 200° C. The indicator gives a reading in 5 to 6 seconds, and the pyrometer should not be left on the hot surface for more than this length of time, particularly at high temperatures. The thermistor bridge provides compensation for changes in the sensitivity of the thermo-pile at high temperatures, but if the head is too hot to touch it is in danger of damage to soldered joints, insulation etc.

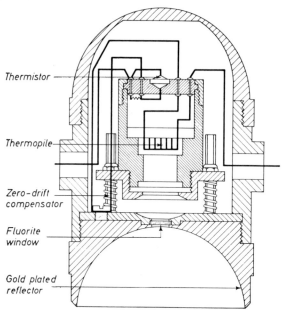

Figure 4.70   Land surface pyrometer. (By courtesy, Land Pyrometers Ltd.)

The instrument may be used to measure the mean emissivity of a surface for all wavelengths up to about 10μm. This value can be used for the correction of total radiation pyrometer readings. A black hemispherical insert is provided with the instrument which can be clipped into the hemispherical reflector to cover the gold. If two measurements are made, one with the gold covered and the other with the gold exposed, the emissivity can readily be deduced from the two measurements. A graph provided with the instrument enables the emissivity to be derived easily from the two readings, while a second graph gives an indication of the error involved in the temperature measurement of the hot body.

For continuous measurement, instruments with water cooling and air purging, which sweeps the hemisphere and window free of dirt, are available.

4.7.2 Optical pyrometers

Optical pyrometers are not suitable for recording or controlling temperatures, but they provide an accurate method of measuring temperatures between 600° C and 3000° C, and are very useful for checking and calibrating total radiation pyrometers.

They may be divided into two groups. In the first group, the light of a given wavelength from the hot body is optically matched with the light from a constant comparison lamp in the instrument by means of an optical wedge or polarising system.

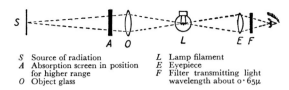

S	Source of radiation	L	Lamp filament
A	Absorption screen in position	E	Eyepiece
	for higher range	F	Filter transmitting light
O	Object glass		wavelength about 0·65μ

Figure 4.71   Optical system of disappearing filament pyrometer

In the second group, which has now become by far the most popular, the brightness of the light from the calibrated comparison lamp is varied to match the brightness of the light from the hot body. The brightness of the lamp is judged to be the same as that of the source when it merges into the image of the source. The instrument is, therefore, known as the Disappearing Filament Pyrometer.

In construction the Disappearing Filament Pyrometer is similar to an ordinary telescope (*Figure 4.71*). At the point where the object glass (*O*) forms an image of the source (*S*), the ribbon filament (*L*) of a standard tungsten lamp is placed. Both the filament and the image of the source are viewed through the eyepiece (*E*), which is fitted with a light filter (*F*) which transmits light of a wavelength of about 0·65μm only. By this means, the intensity of the light of wavelength 0·65μm from the source, may be compared with the intensity of light of the same wavelength from the lamp.

The appearance of the image as seen through the telescope is shown in *Figure 4.72*. If the current through the lamp is insufficient, the lamp filament

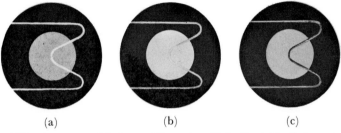

(a)   (b)   (c)

Figure 4.72   Appearance of the image. (By courtesy, Foster Cambridge Instruments Ltd.)

appears as a dark line on a lighter background as shown in *Figure 4.72(c)*. If the current is more than is necessary, the filament will appear as a bright line on a darker background as in *Figure 4.72(a)*. When the current has the correct value, the tip of the filament merges into the background and cannot be distinguished from it (*Figure 4.72(b)*).

The temperature of the filament of the lamp may be obtained in one of two ways. The temperature of the filament will depend upon the current flowing through it. If an ammeter is put in the circuit with the filament, variable resistance and battery, then the size of the current, when adjusted to the correct value, will be an indication of the temperature. The scale of the ammeter may, therefore, be calibrated directly in degrees Celsius. In the other method, the comparison filament is made one arm of a Wheatstone's bridge circuit and its resistance measured. The change of resistance is indicated by a galvanometer which is used as in the unbalanced type of Wheatstone's bridge (page 335). This method is illustrated in *Figure 4.73*.

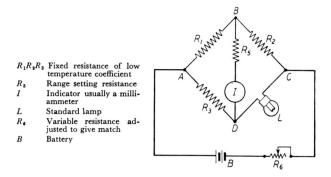

$R_1 R_2 R_3$	Fixed resistance of low temperature coefficient
R_5	Range setting resistance
I	Indicator usually a milli-ammeter
L	Standard lamp
R_4	Variable resistance adjusted to give match
B	Battery

Figure 4.73   Lamp bridge circuit

The advantage of measuring the lamp resistance instead of the current passing is that, by arranging for the bridge to be balanced at the lowest temperature to be measured, the whole of the indicator scale may be utilised for the useful range of the instrument.

A still further method is to measure the voltage across the lamp directly by means of a voltmeter with a scale calibrated as a temperature scale.

Range of the instrument

Lamps for optical pyrometers are not usually run at temperatures above 1400° C. In order to increase the usefulness of the instrument an absorption screen can be placed in front of the objective of the telescope. The transmission factor for the screen is known so that the fraction of the incident light which passes through it is known. The instrument can, therefore, be calibrated for a second range.

Pyrometers with a range of 800° C to 1250° C have an accuracy better than ±5° C while those having a range of 1100° C to 1900° C have an accuracy better than ±10° C.

Correction for non-black body conditions

Like the total radiation pyrometer, the optical pyrometer will be affected by the emissivity of the radiating source, and by any absorption which may take place between the radiator and the instrument.

The spectral emissivity of bright metal surfaces at 0.65μm is greater than the total emissivity e, representing the average emissivity over all wavelengths. The correction required for departure from black-body conditions is, there-fore, less in the optical pyrometer than in the total radiation pyrometer.

Owing to the fact that a given change in temperature produces a much larger percentage change in the radiant energy at 0.65μm than is produced in the average of the radiant energy at all wavelengths, the readings on an optical pyrometer require a smaller correction even with sources of the same emissivity.

The relationship between the apparent temperature T_a and the true temperature T is given by the equation based on Wien's law.

$$\frac{1}{T} - \frac{1}{T_a} = \frac{\lambda \log_{10} e_\lambda}{6245} \tag{4.32}$$

where λ is the wavelength in microns usually 0.65μm, and e_λ the spectral emissivity at a wavelength λ. Corrections based upon this equation are given in the table on page 392.

International Practical Temperature Scale (1968) above the gold-point 1337·58 K (1064·43° C)

Temperatures on the 1968 International Practical Scale above 1064·43° C are based on the measurement of monochromatic visible radiation. The tempera-ture is defined by the ratio of the spectral concentration $L_\lambda T_{68}$ of mono-chromatic visible radiation of wavelength λ emitted by a black body at temperature T_{68} to the spectral concentration $L\lambda\,T_{68}(Au)$ emitted by a black body at the gold-point $T_{68}(Au)$ by means of the formula

$$\frac{L_\lambda(T_{68})}{L\lambda[T_{68}(Au)]} = \frac{e^{\frac{c_2}{\lambda T_{68}(Au)}} - 1}{e^{\frac{c_2}{\lambda T_{68}}} - 1} \tag{4.33}$$

where $c_2 = 0.014\,388$ metre kelvin.

Since $T_{68}(Au)$ is close to the thermodynamic temperature of the freezing point of gold and C_2 is close to the second radiation constant of the Planck equation, it is not necessary to specify the value of the wavelength to be employed in the measurements (see 'Metrologia', Vol. 3, p. 28, 1967).

Calibration of optical and photoelectric pyrometers

The most satisfactory method of calibrating an optical or photoelectric pyrometer is by sighting it on the filament of a tungsten-strip lamp or black body lamp whose temperature is known in terms of the lamp current. Such lamps are calibrated at the National Physical Laboratory in terms of current and luminance temperature, the temperature scale being derived from the standard photoelectric pyrometer using a wavelength of about 0.66μ m.

The calibration of a strip lamp is only valid when the instrument under test uses approximately the same wavelength, but this condition is fulfilled by the red glass filters used in most commercial instruments. Tungsten filament lamps are of two kinds. The vacuum lamp having a filament 1.3 mm wide and 70 mm long suitable for temperatures from $700°$ C to $1550°$ C, and the gas-filled lamp having a filament 50 mm long with a temperature range of $1300°$ C to $2200°$ C. Calibrations are given by the N.P.L. in terms of the current at each $100°$ C, the accuracy being $\pm 5°$ C from $700°$ C to $900°$ C, $\pm 3°$ C from $900°$ C to $1300°$ C, $\pm 5°$ C from $1300°$ C to $1550°$ C and $\pm 7°$ C above this temperature.

If the area of the lamp filament does not provide a sufficiently large source, a lens may be used to provide a magnified image of the filament, but a correction must be applied for the absorption of the lens. A good quality glass lens will absorb about 8 per cent of the radiation in the visible range resulting in a correction of about $6.5°$ C at $1000°$ C and $14°$ C at $1600°$ C for a wavelength of 0.65μ m.

Tungsten filament lamps may be run for short periods at temperatures up to $2500°$ C but beyond this the black body lamp may be used up to $2700°$ C and the carbon arc point at $2520°$ C.

The black body lamp was developed by the N.P.L. in collaboration with the Hirst Research Centre. It consists of a tungsten tube 2 mm diameter and provides a black body source 1 mm diameter at temperatures up to $2700°$ C.

4.7.3 Photo-electric pyrometers

It can be shown that over a large part of the spectrum, the emissivity error for a given temperature and a given emissivity is proportional to the wavelength of the radiation used to make the measurement. A reading on an optical pyrometer sighted on a hot surface is nearer to the true temperature than a reading taken on a total radiation pyrometer. On an oxidised steel surface, emissivity 0.8, a radiation pyrometer with silica lens gives a reading of $1000°$ C when the true temperature is $1060°$ C; but an optical pyrometer reads about $1000°$ C when the true temperature is $1017°$ C. In addition the variation of emissivity of metals due to oxidation is also much less, and the emissivity for most metals slightly higher at shorter wavelengths.

Unfortunately, the optical pyrometer requires the presence of a skilled operator and cannot be used in conjunction with a recording or controlling instrument. Measurement of radiation at shorter wavelengths can, however, be achieved by the use of 'photon detectors' in the form of photo-electric cells. Owing to the fact that the output from photo-electric cells is brought about by changes which take place on the molecular scale rather than on a change in the temperature of the detector, these devices have a very high speed of response, in some cases taking only microseconds to respond to a step change in input. In addition, these devices are capable of giving a high output in terms of volts output per watt of incoming radiation.

Some photo-cells may change characteristics as a result of changes in composition, contamination, or changes in crystal structure with time, but silicon solar cells are very stable. Where the cell stability is in doubt it must be used in a design where it functions as a null balance detector.

The output of a photo-cell may have a temperature coefficient which is dependent not only upon the type of detector but upon the circuit in which the cell is used. This difficulty is overcome by means of an ambient-temperature compensating bridge using suitable thermistors.

The effectiveness of radiant energy in freeing electrons in a photo-electric device increases as the wavelength decreases, and a discrete minimum energy is required to release an electron. Thus, each type of detector material will only respond to wavelengths which are shorter than a certain maximum, and the response to incoming radiation of shorter wavelengths is dependent upon the wavelength. By careful choice of a suitable detector and lens system for a given temperature range, used in conjunction with a suitable electrical measuring system, photo-electric pyrometers having a very high accuracy and stability can be produced.

Such a pyrometer, known as the Land Continuous Optical Pyrometer is shown in *Figure 4.74*. This pyrometer uses a silicon solar cell (see page 305) as the detector. This cell is sensitive to radiation of wavelengths between 0·5 and 1·2 μm but cuts off fairly sharply at 1·0 μm. The instrument is fully compensated for ambient temperatures between 0 and 80° C and where this upper limit is likely to be exceeded the measuring head may be protected by a water-cooled housing. The time of response of the detector is about one thousandth of a second, so that the speed of measurement is limited only by the speed of the recorder. Its output is roughly equal to that of a comparable total radiation pyrometer at 700° C, ten times as great at 1100° C, and one hundred times as great at 2000° C. Its high sensitivity makes it very suitable for measuring the temperature of small hot objects. The temperature of a target 25 mm in diameter can be measured at a distance of 3 m with a suitable optical system.

The output from the detector may be measured by means of either a deflectional galvanometer or a self-balancing potentiometer. In either case, the measuring circuit must have a resistance of 500 ±5 ohms, in order to obtain the correct calibration and compensation for variation in ambient temperature. When a potentiometer is used it is essential to connect a resistance of 500 ±5 ohms across the input terminals. By using an additional slide wire

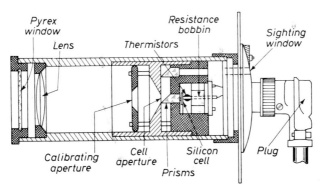

Figure 4.74   Land continuous optical pyrometer. (By courtesy, Land Pyrometers Ltd.)

on the potentiometer, which acts as a potential divider, it is possible to extend the scale so that a range of, say, 700 to 1200° C can be displayed on a single scale.

It is possible with a large enough target to design a pyrometer to read temperatures as low as 450° C but below 700° C reflected sunlight may affect the reading, and below 600° C, the surface must be shaded from direct shop lighting. There is no theoretical upper limit to the temperature which can be measured, but it is not possible at the present to produce good black-body sources at temperatures above 2000° C. Consequently, the calibration above 2000° C must be based entirely on theoretical considerations, and such calibrations may be in error by as much as ± 1 per cent of the measured temperature. For temperatures up to 1500° C the accuracy of calibration achieved is approximately ± 0.2 per cent of the temperature.

Owing to the fact that the effective wavelength of the pyrometer at 1000° C is $0.91 \mu m$ compared with the value for the optical pyrometer of $0.65 \mu m$, the emissivity error with a silicon cell will be about one-third greater than with an optical pyrometer. However, the human error of an operator may be of the same order of magnitude as this small difference.

Another photo-electric cell which is used for measuring the temperature of streams of molten iron from the cupola is the selenium barrier-layer cell described on page 305.

Other cells which are sometimes used are the cadmium sulphide cell which detects radiation between 0.5 and $1.0 \mu m$, and lead sulphide which is sensitive to radiation between 1.0 and $3.8 \mu m$. Cadmium selenide and selenium sulphide cells have also been used.

Glass surface temperature measurement

In the glass-making industry it is important to be able to measure the surface temperature of glass sheets, and it is often an advantage to make these measurements at high speed. This is a difficult problem as glass transmits wavelengths below $4.5 \mu m$, so that a pyrometer sensitive to these wavelengths measures not the temperature of the glass but the temperature of the surface behind the glass. Also, glass reflects radiation strongly in the region between 8 and $12 \mu m$. The ideal pyrometer should therefore be sensitive to wavelengths between 4.5 and $8 \mu m$.

This wavelength limitation can be achieved by using a filter which consists of a thin layer of lead selenide deposited on mica, which transmits wavelengths between 4.0 and $8.5 \mu m$. To make use of this waveband it is necessary to use an arsenic trisulphide lens. The radiation detector is usually a very fast thermopile.

Pyrometers using arsenic trisulphide lenses and filters are now used to control the annealing of sheet and plate glass, and in various glass forming and toughening processes. Care must be taken in the calibration of the pyrometers to make allowance for the humidity of the atmosphere, because the chosen band of wavelengths includes one of the most important water-vapour absorption bands in the infra-red region. This can cause variations of up to 10° C in the calibration of the pyrometer over a path length of 0.6 m. Fortunately, it is usually important to measure temperature variations between different places on the glass with high precision, rather than the absolute temperature level. It has proved possible to maintain a uniformity of calibration from batch to batch of glass pyrometers of about $\pm 2°$ C at $700°$ C.

Ratio pyrometers

This type of radiation pyrometer has been developed principally to reduce the emissivity error in surface temperature measurement. The principle of the method is that the ratio of the intensities of radiation in two narrow wavelength bands is measured. If the emissivity of the target changes then both intensities will be affected, but if the emissivity is the same at both wavelengths, the pyrometer will indicate the true temperature. Unfortunately the emissivity varies with wavelength to a greater or lesser degree for most surfaces. Ratio pyrometers will, therefore, still have residual errors owing to emissivity changes.

In addition, the ratio pyrometer is not affected if part of the radiation is lost by partial blockage of the sighting tube, or absorption in a dirty lens, or by smoke or fume in the line of sight, provided that the absorption of radiation is the same at all wavelengths, and is used mainly for measuring the temperature of the burning zone in a cement kiln. The optical path is obscured by falling particles and lumps of cement inside the rotating kiln which produce an equal loss of radiation at both wavelengths so that the ratio is unaffected.

In other applications the single wavelength instrument is nearly always as good or better than the ratio instrument and is cheaper and being simpler is less liable to go wrong.

5

APPENDIX

THICK WALL BELL-TYPE OF METER

The formula for the rise of the bell in terms of the applied pressures and the dimensions of the meter may be obtained as follows.

Suppose the dimensions of the meter are:

Radius of the chamber	$= R$
Inner radius of the bell	$= r_1$
Outer radius of the bell	$= r_2$
Density of the liquid in the chamber	$= \rho$
Applied differential pressure	$= p_1 - p_2$

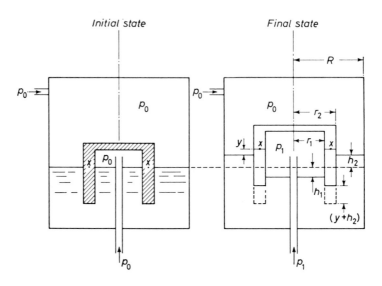

Figure 5.1 *Principle of thick wall bell-type meter. Radius of the central pipe is very small compared with r_1. X is the datum mark on the bell*

Referring to *Figure 5.1*, consider the final position of the bell.

Since the pressure at the level of the surface of the liquid in the bell is the same both inside and outside the bell:

$$p_1 = p_0 + (h_1 + h_2) \, \rho g \qquad (5.1)$$

Since the total quantity of liquid remains constant:

Vol. of liquid above the original level		Liquid filling the space left by bell		Liquid displaced from inside the bell	
$h_2 \pi (R^2 - r_2^2)$	$+$	$+ \pi(y + h_2)(r_2^2 - r_1^2)$	$=$	$= \pi h_1 r_1^2$	(5.2)

Since the float is still in equilibrium, the force owing to the increased pressure within the bell must be equal to the increase in effective weight of the bell, i.e. equal to the reduction in weight of liquid displaced by the bell.

Force due to increased pressure $= (p_1 - p_0) \pi r_1^2$.

Reduction in volume of float immersed $= y\pi(r_2^2 - r_1^2)$.

Reduction in weight of liquid displaced $= \rho g y \pi (r_2^2 - r_1^2)$.

$$(p_1 - p_0) \pi r_1^2 = \rho g y \pi (r_2^2 - r_1^2) \qquad (5.3)$$

Substituting from (5.1)

$$(h_1 + h_2) \rho g \pi r_1^2 = \rho g y \pi (r_2^2 - r_1^2)$$
$$(h_1 + h_2) r_1^2 = y(r_2^2 - r_1^2) \qquad (5.4)$$

Substituting in (5.2)

$$h_2(R^2 - r_2^2) + (h_1 + h_2) r_1^2 + h_2(r_2^2 - r_1^2) = h_1 r_1^2$$
$$h_2 R^2 - h_2 r_2^2 + h_1 r_1^2 + h_2 r_1^2 + h_2 r_2^2 - h_2 r_1^2 = h_1 r_1^2$$
$$h_2 R^2 = 0$$
$$h_2 = 0$$

$\therefore$ vertical movement of the bell $= y$

From (5.4)

$$y = \frac{(p_1 - p_0)}{\rho g} \frac{r_1^2}{r_2^2 - r_1^2}$$

i.e. the movement of the bell for a given pressure change depends only upon the sizes of the inner and outer radii of the bell and not upon the dimensions of the chamber.

Measurement of pulsating flow

When pulsating flow is being measured, damping will cut out the error between the pen-trace and the average differential pressure, but the main source of error is the fact that the square root of the mean of the differential pressure does not represent the mean flow. The reason for this error is illustrated in *Figure 5.2*. Suppose, for simplicity, the flow consists of a series of pulses in which the flow increases uniformly from zero to 100 per cent and then falls to zero. This type of flow is represented by the top graph. The differential pressures corresponding to each value of the flow will have the values shown in the lower graph. The mean value of the differential pressure is $33\frac{1}{3}$ per cent of the maximum differential pressure. This represents a flow of 57 per cent of the maximum flow. The damped differential pressure measuring instrument will, therefore, indicate a value of the flow which is 7 per cent too high.

If the magnitude of the flow fluctuates between 33 per cent and 66 per cent of the full flow, the error will be reduced to less than 1 per cent too high.

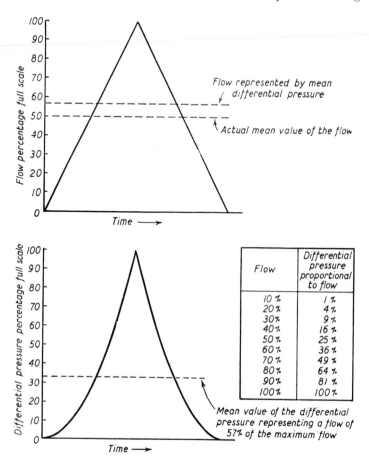

Figure 5.2 *Flow and differential pressure*

Table 5.1 APPROXIMATE DIFFERENCES $(t_{68} - t_{48})$, IN KELVINS, BETWEEN THE VALUES OF TEMPERATURE GIVEN BY THE IPTS OF 1968 AND THE IPTS OF 1948

t_{68} °C	0	−10	−20	−30	−40	−50	−60	−70	−80	−90	−100
−100	0·022	0·013	0·003	−0·006	−0·013	−0·013	−0·005	0·007	0·012		
− 0	0·000	0·006	0·012	0·018	0·024	0·029	0·032	0·034	0·033	0·029	0·022

t_{68} °C	0	10	20	30	40	50	60	70	80	90	100
0	0·000	−0·004	−0·007	−0·009	−0·010	−0·010	−0·010	−0·008	−0·006	−0·003	0·000
100	0·000	0·004	0·007	0·012	0·016	0·020	0·025	0·029	0·034	0·038	0·043
200	0·043	0·047	0·051	0·054	0·058	0·061	0·064	0·067	0·069	0·061	0·073
300	0·073	0·074	0·075	0·076	0·077	0·077	0·077	0·077	0·077	0·076	0·076
400	0·076	0·075	0·075	0·075	0·074	0·074	0·074	0·075	0·076	0·077	0·079
500	0·079	0·082	0·085	0·089	0·094	0·100	0·108	0·116	0·126	0·137	0·150
600	0·150	0·165	0·182	0·200	0·23	0·25	0·28	0·31	0·34	0·36	0·39
700	0·39	0·42	0·45	0·47	0·50	0·53	0·56	0·58	0·61	0·64	0·67
800	0·67	0·70	0·72	0·75	0·78	0·81	0·84	0·87	0·89	0·92	0·95
900	0·95	0·98	0·01	1·04	1·07	1·10	1·12	1·15	1·18	1·21	1·24
1000	1·24	1·27	1·30	1·33	1·36	1·39	1·42	1·44			

t_{68} °C	0	100	200	300	400	500	600	700	800	900	1000
1000		1·5	1·7	1·8	2·0	2·2	2·4	2·6	2·8	3·0	3·2
2000	3·2	3·5	3·7	4·0	4·2	4·5	4·8	5·0	5·3	5·6	5·9
3000	5·9	6·2	6·5	6·9	7·2	7·5	7·9	8·2	8·6	9·0	9·3

Table 5.2 RESISTANCE IN OHMS/TEMPERATURE (I.P.T.S. 68) FOR PLATINUM RESISTANCE TEMPERATURE SENSORS BASED ON BS 1904.
(By courtesy, Rosemount Engineering Co. Ltd.)

°C	0·0	−1·0	−2·0	−3·0	−4·0	−5·0	−6·0	−7·0	−8·0	−9·0
−200·00	18·5996									
−190·00	22·3415	22·4162	21·9906	21·5657	21·1404	20·7161	20·2914	19·8681	19·4445	19·0222
−180·00	27·0989	26·6735	26·2479	25·8223	25·3964	24·9708	24·5448	24·1192	23·6932	23·2675
−170·00	31·3410	30·9161	30·4930	30·0696	29·6458	29·2218	28·7974	28·3727	27·9477	27·5224
−160·00	35·5553	35·1301	34·7098	34·2891	33·8681	33·4468	33·0252	32·6034	32·1812	31·7587
−150·00	39·7390	39·3186	38·9012	38·4835	38·0655	37·6473	37·2288	36·8100	36·3909	35·9716
−140·00	43·8925	43·4783	43·0639	42·6492	42·2342	41·8190	41·4035	40·9878	40·5718	40·1556
−130·00	48·0179	47·6093	47·1976	46·7857	46·3735	45·9611	45·5485	45·1356	44·7224	44·3090
−120·00	52·1190	51·7135	51·3042	50·8947	50·4849	50·0750	49·6647	49·2543	48·8436	48·4327
−110·00	56·1977	55·7948	55·3877	54·9803	54·5728	54·1650	53·7571	53·3489	52·9405	52·5319
−100·00	60·2562	59·8549	59·4497	59·0444	58·6388	58·2331	57·8271	57·4210	57·0146	56·6081
−90·00	64·2964	63·8961	63·4927	63·0891	62·6854	62·2815	61·8774	61·4731	61·0686	60·6639
−80·00	68·3199	67·9199	67·5182	67·1163	66·7142	66·3119	65·9095	65·5069	65·1041	64·7012
−70·00	72·3277	71·9280	71·5278	71·1275	70·7269	70·3262	69·9254	69·5244	69·1232	68·7218
−60·00	76·3209	75·9214	75·5227	75·1237	74·7246	74·3254	73·9260	73·5265	73·1268	72·7269
−50·00	80·2995	79·8011	79·5038	79·1063	78·7086	78·3108	77·9129	77·5148	77·1166	76·7182
−40·00	84·2649	83·8670	83·4710	83·0748	82·6784	82·2820	81·8854	81·4886	81·0917	80·6947
−30·00	88·2173	87·8203	87·4255	87·0306	86·6356	86·2404	85·8451	85·4496	85·0541	84·6584
−20·00	92·1570	91·7612	91·3677	90·9740	90·5803	90·1864	89·7923	89·3982	89·0039	88·6095
−10·00	96·0844	95·6898	95·2975	94·9051	94·5125	94·1198	93·7270	93·3341	92·9411	92·5479
0·00	100·0000	99·6866	99·2155	98·8242	98·4328	98·0413	97·6497	97·2580	96·8661	96·4741

°C	+0·0	+1·0	+2·0	+3·0	+4·0	+5·0	+6·0	+7·0	+8·0	+9·0
0·00	100·0000	100·3908	100·7816	101·1722	101·5627	101·9531	102·3433	102·7335	103·1235	103·5135
10·00	103·9033	104·2929	104·6825	105·0719	105·4612	105·8504	106·2395	106·6284	107·0173	107·4060
20·00	107·7946	108·1831	108·5715	108·9597	109·3478	109·7358	110·1237	110·5115	110·8992	111·2867

°C	+0·0	+1·0	+2·0	+3·0	+4·0	+5·0	+6·0	+7·0	+8·0	+9·0
30·00	111·6741	112·0814	112·4486	112·8356	113·2226	113·6094	113·9961	114·3827	114·7691	115·1555
40·00	115·5418	115·9278	116·3138	116·6997	117·0854	117·4711	117·8566	118·2420	118·6273	119·0124
50·00	119·3975	119·7824	120·1673	120·5520	120·9366	121·3210	121·7054	122·0897	122·4738	122·8578
60·00	123·2417	123·6254	124·0090	124·3925	124·7759	125·1592	125·5423	125·9253	126·3082	126·6910
70·00	127·0737	127·4563	127·8387	128·2211	128·6033	128·9854	129·3674	129·7493	130·1310	130·5127
80·00	130·8942	131·2756	131·6569	132·0380	132·4190	132·8000	133·1808	133·5615	133·9420	134·3225
90·00	134·7028	135·0831	135·4632	135·8432	136·2231	136·6029	136·9825	137·3621	137·7415	138·1296
100·00	138·5000	138·8790	139·2580	139·6368	140·0155	140·3941	140·7726	141·1509	141·5291	141·9073
110·00	142·2853	142·6632	143·0410	143·4187	143·7963	144·1738	144·5512	144·9284	145·3056	145·6826
120·00	146·0595	146·4362	146·8128	147·1893	147·5656	147·9419	148·3180	148·6940	149·0700	149·4457
130·00	149·8214	150·1970	150·5725	150·9479	151·3231	151·6983	152·0733	152·4482	152·8230	153·1977
140·00	153·5723	153·9467	154·3211	154·6953	155·0694	155·4434	155·8173	156·1910	156·5647	156·9382
150·00	157·3116	157·6849	158·0580	158·4311	158·8040	159·1763	159·5495	159·9221	160·2946	160·6669
160·00	161·0391	161·4113	161·7833	162·1553	162·5271	162·8988	163·2704	163·6418	164·0132	164·4344
170·00	164·7555	165·1265	165·4974	165·8681	166·2387	166·6092	166·9796	167·3499	167·7201	168·0901
180·00	168·4601	168·8299	169·1997	169·5693	169·9388	170·3082	170·6775	171·0467	171·4158	171·7847
190·00	172·1535	172·5222	172·8908	173·2592	173·6275	173·9958	174·3639	174·7318	175·0997	175·4875
200·00	175·8351	176·2027	176·5701	176·9375	177·3047	177·6718	178·0388	178·4057	178·7724	179·1391
210·00	179·5056	179·8720	180·2383	180·6045	180·9706	181·3365	181·7024	182·0681	182·4337	182·7992
220·00	183·1646	183·5299	183·8951	184·2602	184·6251	184·9900	185·3547	185·7193	186·0838	186·4482
230·00	186·8125	187·1766	187·5406	187·9045	188·2683	188·6319	188·9955	189·3589	189·7222	190·0854
240·00	190·4485	190·8115	191·1744	191·5372	191·8998	192·2624	192·6248	192·9871	193·3493	193·7114
250·00	194·0734	194·4353	194·7970	195·1586	195·5202	195·8816	196·2428	196·6040	196·9651	197·3260
260·00	197·6868	198·0475	198·4081	198·7686	199·1290	199·4892	199·8494	200·2094	200·5693	200·9291
270·00	201·2888	201·6484	202·0079	202·3672	202·7265	203·0856	203·3447	203·8036	204·1624	204·5210
280·00	204·8796	205·2380	205·5964	205·9546	206·3127	206·6707	207·0286	207·3863	207·7440	208·1015
290·00	208·4589	208·8162	209·1734	209·3305	209·8874	210·2443	210·6010	210·9576	211·3141	211·6705

380

°C	+0·0	+1·0	+2·4	+3·0	+4·0	+5·0	+6·0	+7·0	+8·0	+9·0
300·00	212·0268	212·3830	212·7390	213·0950	213·4508	213·8066	214·1622	214·5177	214·8731	215·2283
310·00	215·5835	215·9385	216·2935	216·6483	217·0030	217·3575	217·7120	218·0664	218·4206	218·7747
320·00	219·1287	219·4826	219·8364	220·1900	220·5436	220·8970	221·2503	221·6035	221·9566	222·3096
330·00	222·6625	223·0152	223·3678	223·7203	224·0727	224·4250	224·7772	225·1292	225·4812	225·8330
340·00	226·1847	226·5363	226·8879	227·2393	227·5905	227·9417	228·2928	228·6437	228·9945	229·3452
350·00	229·6958	230·0463	230·3967	230·7469	231·0971	231·4471	231·7970	232·1468	232·4964	232·8460
360·00	233·1954	233·5448	233·8940	234·2431	234·5921	234·9409	235·2897	235·6383	235·9869	236·3353
370·00	236·6836	237·0318	237·3798	237·7278	238·0756	238·4233	238·7709	239·1184	239·4638	239·8131
380·00	240·1602	240·5073	240·8542	241·2011	241·5478	241·8944	242·2409	242·5873	242·9335	243·2797
390·00	243·6257	243·9716	244·3174	244·6630	245·0086	245·3540	245·6993	246·0445	246·3896	246·7345
400·00	247·0794	247·4241	247·7688	248·1133	248·4578	248·8021	249·1463	249·4903	249·8343	250·1781
410·00	250·5219	250·8655	251·2089	251·5523	251·8955	252·2387	252·5817	252·9246	253·2673	253·6108
420·00	253·9526	254·2950	254·6373	254·9795	255·3216	255·6636	256·0054	256·3472	256·6888	257·0303
430·00	257·3717	257·7131	258·0543	258·3954	258·7363	259·0772	259·4179	259·7586	260·0991	260·4395
440·00	260·7798	261·1199	261·4599	261·7998	262·1396	262·4793	262·8189	263·1583	263·4977	263·8369
450·00	264·1760	264·5150	264·8538	265·1926	265·5312	265·8698	266·2082	266·5465	266·8547	267·2227
460·00	267·5607	267·8985	268·2362	268·5738	268·9112	269·2486	269·5858	269·9229	270·2599	270·5968
470·00	270·9336	271·2702	271·6068	271·9432	272·2795	272·6157	272·9518	273·2878	273·6236	273·9593
480·00	274·2950	274·6305	274·9659	275·3011	275·6363	275·9714	276·3063	276·6411	276·9758	277·3104
490·00	277·6449	277·9792	278·3134	278·6475	278·9815	279·3154	279·6491	279·9827	280·3163	280·6497
500·00	280·9830	281·3161	281·6491	281·9820	282·3148	282·6475	282·9801	283·3126	283·6449	283·9771
510·00	284·3092	284·6412	284·9731	285·3049	285·6365	285·9681	286·2995	286·6308	286·9620	287·2930
520·00	287·6240	287·9548	288·2855	288·6161	288·9466	289·2769	289·6072	289·9373	290·2673	290·5972
530·00	290·9270	291·2566	291·5861	291·9155	292·2446	292·5740	292·9031	293·2320	293·5608	293·8895
540·00	294·2181	294·5466	294·8749	295·2031	295·5313	295·8592	296·1871	296·5149	296·8425	297·1701
550·00	297·4975	297·8247	298·1519	298·4789	298·8058	299·1325	299·4592	299·7858	300·1122	300·4385
560·00	300·7647	301·0908	301·4168	301·7426	302·0684	302·3940	302·7195	303·0449	303·3702	303·6954
570·00	304·0205	304·3453	304·6701	304·9948	305·3193	305·6437	305·9680	306·2922	306·6163	306·9402

°C	+0·0	+1·0	+2·0	+3·0	+4·0	+5·0	+6·0	+7·0	+8·0	+9·0
580·00	307·2641	307·5878	307·9114	308·2348	308·5582	308·8814	309·2046	309·5276	309·8505	310·1733
590·00	310·4959	310·8184	311·1408	311·4631	311·7852	312·1072	312·4291	312·7509	313·0726	313·3942
600·00	313·7157	314·0369	314·3581	314·6792	315·0001	315·3209	315·6416	315·9622	316·2827	316·6030
610·00	316·9233	317·2434	317·5633	317·8832	318·2029	318·5225	318·8420	319·1614	319·4806	319·7998
620·00	320·1188	320·4377	320·7565	321·0752	321·3937	321·7121	322·0305	322·3487	322·6668	322·9847
630·00	323·3026	323·6200	323·9372	324·2543	324·5714	324·8883	325·2051	325·5217	325·8383	326·1548
640·00	326·4711	326·7876	327·1041	327·4204	327·7366	328·0526	328·3686	328·6845	329·0002	329·3158
650·00	329·6313	329·9464	330·2614	330·5762	330·8910	331·2056	331·5201	331·8345	332·1488	332·4629
660·00	332·7770	333·0909	333·4047	333·7185	334·0321	334·3455	334·6589	334·9721	335·2853	335·5983
670·00	335·9112	336·2240	336·5367	336·8493	337·1617	337·4741	337·7863	338·0984	338·4104	338·7223
680·00	339·0340	339·3460	339·6578	339·9696	340·2812	340·5927	340·9041	341·2154	341·5265	341·8376
690·00	342·1485	342·4590	342·7694	343·0797	343·3899	343·6999	344·0099	344·3197	344·6294	344·9390
700·00	345·2485	345·5578	345·8671	346·1762	246·4853	346·7942	347·1030	347·4116	347·7202	348·0287
710·00	348·3370	348·6452	348·9533	349·2613	349·5692	349·8770	350·1846	350·4922	350·7996	351·1069
720·00	351·4141	351·7215	352·0288	352·3359	352·6430	352·9499	353·2567	353·5634	353·8700	354·1765
730·00	354·4829	354·7888	355·0946	355·4003	355·7059	356·0114	356·3168	356·6221	356·9272	357·2322
740·00	357·5371	357·8419	358·1466	358·4512	358·7556	359·0600	359·3642	359·6683	359·9723	360·2762
750·00	360·5800	360·8836	361·1872	361·4906	361·7939	362·0971	362·4002	362·7031	363·0060	363·3087
760·00	363·6114	363·9142	364·2169	364·5199	364·8219	365·1243	365·4265	365·7287	366·0307	366·3326
770·00	366·6344	366·9357	367·2370	367·5382	367·8392	368·1401	368·4409	368·7416	369·0421	369·3426
780·00	369·6429	369·9432	370·2433	370·5433	370·8432	371·1429	371·4426	371·7421	372·0416	372·3409
790·00	372·6401	372·9392	373·2381	373·5370	373·8357	374·1343	374·4329	374·7313	375·0295	375·3277
800·00	375·6258	—								—

Table 5.3　COEFFICIENTS OF LINEAR EXPANSION OF SOLIDS

Extracted from *Tables of physical and chemical constants* by Kaye and Laby (Longmans). The values given are per degree Celsius, and, except where some temperature is specified, for a range about 20° C.

Substance	$\alpha \times 10^{-6}$	Substance	$\alpha \times 10^{-6}$
		Gunmetal (admiralty)	18·1
Aluminium	25·5	Nickel Steel, 10% Ni	13·0
Graphite	0·63	,,　　　20% Ni	19·5
Copper	16·7	,,　　　30% Ni	12·0
Gold	13·9	,,　　　36% Ni	
Iron (cast)	10·2	(Invar)	−0·3 to +2·5
Iron (wrought)	11·9	,,　　　40% Ni	6·0
Steel	10·5 to 11·6	,,　　　50% Ni	9·7
Lead	29·1	,,　　　80% Ni	12·5
Nickel	12·8	Phosphor bronze, 97·6 Cu,	
Platinum	8·9	2 Sn, 0·2 P	16·8
Silver	18·8	Solder, 2 Pb, 1 Sn	25
Tin	21·4	Brick (Egyptian)	9·5
Tungsten, 27°	4·44	Cement and concrete	10–14
Tungsten, 2027°	7·26	Glass, flint	7·8
Zinc	25·8 to 26·3	Glass, soda	8·5
Brass (ordy) c. 66 Cu, 34		Glass, Pyrex	3
Zn	18·9	Silica (fused) −80° to 0°	0·22
Bronze, 32 Cu, 2 Zn, 5 Sn	17·7	Silica (fused) 0° to 30°	0·42
Constantan (Eureka), 60		Silica (fused) 0° to 100°	0·50
Cu, 40 Ni	17·0	Silica (fused) 0° to 1000°	0·54
Duralumin	22·6		

Table 5.4　COLD JUNCTION CORRECTIONS FOR PRECIOUS METAL THERMO-COUPLES

Temperature of cold junction, °C	Correction to be added to e.m.f. millivolt	Temperature of cold junction, °C	Correction to be added to e.m.f. millivolt
0	·000	13	·071
1	·005	14	·077
2	·011	15	·083
3	·016	16	·089
4	·022	17	·095
5	·027	18	·100
6	·032	19	·106
7	·038	20	·112
8	·043	21	·118
9	·049	22	·124
10	·054	23	·130
11	·060	24	·136
12	·066	25	·142

These corrections apply for either the 10 or 13 per cent rhodium-platinum couple.

Table 5.5 REFERENCE TABLES FOR PRECIOUS METAL THERMO-COUPLES
(Based on tables published by the National Physical Laboratory)*

E.M.F. values are given in millivolts (absolute) and temperature in degrees C (Int. 1948). Cold junction at 0° C.

To assist in interpolating between two printed values, the difference between successive e.m.f. readings is given in italics.

The final digit may differ in a few cases from that given in B.S. 1826:1952 and are slightly higher than those given by American National Bureau of Standards.

Platinum: 13 per cent, Rhodium-Platinum

	0° C	100° C	200° C	300° C	400° C	500° C	600° C	700° C	800° C
0°	0	0·644	1·463	2·392	3·397	4·460	5·571	6·735	7·952
	·054	*·075*	*·089*	*·097*	*·104*	*·109*	*·114*	*·120*	*·124*
10°	0·054	0·719	1·552	2·489	3·501	4·569	5·685	6·855	8·076
	·057	*·077*	*·089*	*·098*	*·104*	*·110*	*·114*	*·120*	*·124*
20°	0·111	0·796	1·641	2·587	3·605	4·679	5·799	6·975	8·200
	·059	*·079*	*·091*	*·099*	*·105*	*·110*	*·115*	*·120*	*·125*
30°	0·170	0·875	1·732	2·686	3·710	4·789	5·914	7·095	8·325
	·061	*·080*	*·091*	*·100*	*·105*	*·110*	*·116*	*·121*	*·125*
40°	0·231	0·955	1·823	2·786	3·815	4·899	6·030	7·216	8·450
	·064	*·081*	*·093*	*·100*	*·106*	*·111*	*·116*	*·122*	*·126*
50°	0·295	1·036	1·916	2·886	3·921	5·010	6·146	7·338	8·576
	·066	*·083*	*·094*	*·101*	*·107*	*·111*	*·117*	*·122*	*·126*
60°	0·361	1·119	2·010	2·987	4·028	5·121	6·263	7·460	8·702
	·068	*·084*	*·094*	*·102*	*·107*	*·112*	*·117*	*·122*	*·126*
70°	0·429	1·203	2·104	3·089	4·135	5·233	6·380	7·582	8·828
	·070	*·086*	*·095*	*·102*	*·108*	*·112*	*·118*	*·123*	*·127*
80°	0·499	1·289	2·199	3·191	4·243	5·345	6·498	7·705	8·955
	·072	*·086*	*·096*	*·103*	*·108*	*·113*	*·118*	*·123*	*·127*
90°	0·571	1·375	2·295	3·294	4·351	5·458	6·616	7·828	9·082
	·073	*·088*	*·097*	*·103*	*·109*	*·113*	*·119*	*·124*	*·127*

	900° C	1000° C	1100° C	1200° C	1300° C	1400° C	1500° C	1600° C	1700° C
0°	9·209	10·510	11·850	13·222	14·617	16·039	17·463	18·855	20·202
	·128	*·133*	*·136*	*·139*	*·141*	*·143*	*·141*	*·137*	*·132*
10°	9·337	10·643	11·986	13·361	14·758	16·182	17·604	18·992	20·334
	·128	*·133*	*·136*	*·139*	*·141*	*·143*	*·141*	*·136*	*·132*
20°	9·465	10·776	12·122	13·500	14·899	16·325	17·745	19·128	20·466
	·129	*·133*	*·136*	*·139*	*·142*	*·143*	*·140*	*·136*	*·132*
30°	9·594	10·909	12·258	13·639	15·041	16·468	17·885	19·264	20·598
	·129	*·133*	*·137*	*·139*	*·142*	*·142*	*·140*	*·135*	*·131*
40°	9·723	11·042	12·395	13·778	15·183	16·610	18·025	19·399	20·729
	·130	*·134*	*·137*	*·139*	*·142*	*·143*	*·140*	*·135*	*·131*
50°	9·853	11·176	12·532	13·917	15·325	16·753	18·165	19·534	20·860
	·130	*·134*	*·137*	*·140*	*·143*	*·142*	*·139*	*·134*	*·130*
60°	9·983	11·310	12·669	14·057	15·468	16·895	18·304	19·688	20·990
	·131	*·135*	*·138*	*·139*	*·142*	*·143*	*·138*	*·134*	
70°	10·114	11·445	12·807	14·196	15·610	17·038	18·442	19·802	
	·132	*·135*	*·138*	*·140*	*·143*	*·142*	*·138*	*·134*	
80°	10·246	11·580	12·945	14·336	15·753	17·180	18·580	19·936	
	·132	*·135*	*·139*	*·140*	*·143*	*·142*	*·138*	*·133*	
90°	10·378	11·715	13·084	14·476	15·896	17·322	18·718	20·069	
	·132	*·135*	*·138*	*·141*	*·143*	*·141*	*·137*	*·133*	

* C. R. Barber. The E.M.F.-Temperature Calibration of Platinum, 10% Rhodium-Platinum and Platinum, 13% Rhodium-Platinum Thermo-couples over the range 0°—1760°C. Proc. of Phys. Soc., Sect. B, Vol. 63, p. 492. July 1950.

Table 5.6   REFERENCE TABLES FOR PRECIOUS METAL THERMO-COUPLES
(*Based on tables published by the National Physical Laboratory*)*

E.M.F. values are given in millivolts (absolute) and temperature in degrees C (Int. 1948). Cold junction at 0° C.

To assist in interpolating between two printed values, the difference between successive e.m.f. readings is given in italics.

The final digit may differ in a few cases from that given in B.S. 1826:1952.

Platinum: 10 per cent. Rhodium-Platinum

	0°C.	100°C.	200°C.	300°C.	400°C.	500°C.	600°C.	700°C.	800°C.
0°	0	0·642	1·435	2·314	3·249	4·220	5·226	6·265	7·339
	·055	*·074*	*·085*	*·092*	*·095*	*·099*	*·103*	*·106*	*·109*
10°	0·055	0·716	1·520	2·406	3·344	4·319	5·329	6·371	7·448
	·057	*·075*	*·085*	*·092*	*·096*	*·100*	*·103*	*·106*	*·110*
20°	0·112	0·791	1·605	2·498	3·440	4·419	5·432	6·477	7·558
	·060	*·076*	*·086*	*·092*	*·097*	*·100*	*·103*	*·107*	*·110*
30°	0·172	0·867	1·691	2·590	3·537	4·519	5·535	6·584	7·668
	·061	*·078*	*·087*	*·093*	*·096*	*·100*	*·103*	*·107*	*·110*
40°	0·233	0·945	1·778	2·683	3·633	4·619	5·638	6·691	7·778
	·064	*·079*	*·088*	*·094*	*·097*	*·100*	*·104*	*·107*	*·110*
50°	0·297	1·024	1·866	2·777	3·730	4·719	5·742	6·798	7·888
	·066	*·080*	*·088*	*·094*	*·097*	*·101*	*·104*	*·108*	*·111*
60°	0·363	1·104	1·954	2·871	3·827	4·820	5·846	6·906	7·999
	·067	*·081*	*·089*	*·094*	*·098*	*·101*	*·104*	*·108*	*·111*
70°	0·430	1·185	2·043	2·965	3·925	4·921	5·950	7·014	8·110
	·069	*·083*	*·090*	*·094*	*·098*	*·101*	*·105*	*·108*	*·112*
80°	0·499	1·269	2·133	3·059	4·023	5·022	6·055	7·122	8·222
	·071	*·083*	*·090*	*·095*	*·098*	*·102*	*·105*	*·108*	*·112*
90°	0·570	1·351	2·223	3·154	4·121	5·124	6·160	7·230	8·334
	·072	*·084*	*·091*	*·095*	*·099*	*·102*	*·105*	*·109*	*·112*

	900°C.	1,000°C.	1,100°C.	1,200°C.	1,300°C.	1,400°C.	1,500°C.	1,600°C.	1,700°C.
0°	8·446	9·591	10·757	11·946	13·155	14·371	15·580	16·770	17·922
	·113	*·116*	*·118*	*·120*	*·122*	*·121*	*·120*	*·117*	*·113*
10°	8·559	9·707	10·875	12·066	13·277	14·492	15·700	16·887	18·035
	·113	*·116*	*·118*	*·121*	*·122*	*·121*	*·120*	*·117*	*·112*
20°	8·672	9·823	10·993	12·187	13·399	14·613	15·820	17·004	18·147
	·114	*·116*	*·118*	*·120*	*·121*	*·121*	*·120*	*·116*	*·112*
30°	8·786	9·939	11·111	12·307	13·520	14·734	15·940	17·120	18·259
	·114	*·116*	*·118*	*·121*	*·122*	*·121*	*·120*	*·116*	*·112*
40°	8·900	10·055	11·229	12·428	13·642	14·855	16·060	17·236	18·371
	·114	*·117*	*·119*	*·121*	*·122*	*·121*	*·119*	*·115*	*·111*
50°	9·014	10·172	11·348	12·549	13·764	14·976	16·179	17·351	18·482
	·115	*·117*	*·119*	*·121*	*·122*	*·121*	*·119*	*·115*	*·110*
60°	9·129	10·289	11·467	12·670	13·886	15·097	16·298	17·466	18·592
	·115	*·117*	*·119*	*·121*	*·121*	*·121*	*·118*	*·115*	
70°	9·244	10·406	11·586	12·791	14·007	15·218	16·416	17·581	
	·115	*·117*	*·120*	*·121*	*·121*	*·121*	*·118*	*·114*	
80°	9·359	10·523	11·706	12·912	14·128	15·339	16·534	17·695	
	·116	*·177*	*·120*	*·122*	*·122*	*·120*	*·118*	*·144*	
90°	9·475	10·640	11·826	13·034	14·250	15·459	16·652	17·809	
	·116	*·177*	*·120*	*·121*	*·121*	*·121*	*·118*	*·113*	

*See footnote on previous page

Table 5.7 REFERENCE TABLE FOR IRON V CONSTANTAN THERMO-COUPLES

E.M.F. values are given in millivolts (absolute) and temperature in degrees C (Int. 1948). Cold junction at 0° C.

Based on N.B.S. Circular 508 published by the American National Bureau of Standards, and conforming to Instrument Society of America Tentative Recommended Practice RP1.6. The values agree with those published in B.S. 1829:1962, although the B.S. gives the value to two decimal places only.

°C	−100	−0	+0	100	200	300	400	500	600	700	800
						MILLIVOLTS					
0	−4·628	0·000	0·000	5·267	10·781	16·325	21·849	27·389	33·114	39·153	45·532
2	−4·709	−0·101	0·100	5·376	10·892	16·436	21·960	27·502	33·231	39·278	45·661
4	−4·790	−0·202	0·200	5·484	11·003	16·546	22·070	27·614	33·348	39·403	45·789
6	−4·870	−0·303	0·301	5·593	11·115	16·657	22·181	27·727	33·465	39·529	45·918
8	−4·951	−0·404	0·402	5·702	11·225	16·767	22·291	27·839	33·583	39·654	46·046
10	−5·032	−0·505	0·504	5·811	11·337	16·877	22·401	27·952	33·700	39·780	46·175
12	−5·110	−0·604	0·606	5·920	11·449	16·987	22·511	28·065	33·818	39·906	46·303
14	−5·188	−0·703	0·709	6·029	11·559	17·098	22·621	28·177	33·936	40·032	46·431
16	−5·267	−0·803	0·811	6·138	11·671	17·209	22·732	28·290	34·053	40·158	46·560
18	−5·345	−0·902	0·914	6·247	11·782	17·319	22·842	28·403	34·171	40·285	46·688
20	−5·423	−1·001	1·017	6·356	11·893	17·430	22·953	28·516	34·289	40·412	46·817
22	−5·498	−1·098	1·120	6·466	12·005	17·540	23·063	28·629	34·408	40·538	46·945
24	−5·574	−1·195	1·224	6·576	12·116	17·651	23·173	28·742	34·526	40·665	47·073
26	−5·649	−1·291	1·327	6·685	12·227	17·763	23·283	28·856	34·644	40·792	47·201
28	−5·725	−1·388	1·431	6·794	12·338	17·873	23·393	28·969	34·763	40·919	47·329
30	−5·800	−1·485	1·535	6·904	12·449	17·983	23·504	29·082	34·882	41·046	47·457
32	−5·872	−1·579	1·639	7·013	12·561	18·094	23·614	29·196	35·001	41·173	47·585
34	−5·943	−1·673	1·744	7·123	12·672	18·204	23·724	29·309	35·121	41·300	47·712
36	−6·016	−1·861	1·848	7·233	12·783	18·315	23·835	29·423	35·239	41·427	47·840
38	−6·086	−1·955	1·953	7·343	12·894	18·425	23·945	29·537	35·359	41·555	47·967
40	−6·158	−2·003	2·057	7·453	13·006	18·536	24·053	29·051	35·479	41·682	48·094
42	−6·226	−2·051	2·162	7·564	13·117	18·647	24·166	29·765	35·598	41·810	48·221
44	−6·295	−2·147	2·267	7·674	13·228	18·758	24·276	29·879	35·718	41·938	48·348
46	−6·363	−2·243	2·373	7·784	13·339	18·868	24·386	29·993	35·838	42·066	48·475
48	−6·432	−2·339	2·478	7·894	13·449	18·979	24·497	30·107	35·959	42·139	48·602
50	−6·500	−2·435	2·584	8·004	13·561	19·089	24·607	30·222	36·079	42·321	48·728
52	−6·564	−2·526	2·690	8·115	13·673	19·200	24·718	30·336	36·200	42·449	48·854
54	−6·628	−2·617	2·796	8·225	13·783	19·310	24·828	30·450	36·321	42·577	48·980
56	−6·692	−2·708	2·902	8·336	13·894	19·421	24·939	30·566	36·442	42·705	49·105
58	−6·756	−2·799	3·008	8·447	14·005	19·532	25·050	30·680	36·564	42·833	49·231
60	−6·820	−2·890	3·114	8·558	14·116·	19·643	25·161	30·795	36·686	42·961	49·356
62	−6·880	−2·980	3·220	8·669	14·227	19·753	25·272	30·910	36·808	43·090	49·481
64	−6·940	−3·070	3·327	8·780	14·337	19·864	25·383	31·025	36·930	43·218	49·606
66	−7·000	−3·160	3·434	8·891	14·448	19·975	25·494	31·141	37·052	43·347	49·730
68	−7·060	−3·250	3·540	9·001	14·558	20·085	25·604	31·256	37·175	43·475	49·854
70	−7·120	−3·340	3·648	9·112	14·669	20·196	25·715	31·371	37·297	43·604	49·973
72	−7·176	−3·428	3·756	9·224	14·779	20·306	25·827	31·486	37·419	43·732	50·101
74	−7·232	−3·516	3·863	9·334	14·890	20·416	25·938	31·602	37·543	43·861	
76	−7·288	−3·604	3·970	9·445	15·000	20·526	26·049	31·718	37·665	43·989	
78	−7·344	−3·692	4·078	9·556	15·110	20·637	26·160	31·833	37·788	44·117	
80	−7·400	−3·780	4·185	9·667	15·221	20·747	26·271	31·949	37·912	44·246	
82	−7·452	−3·866	4·293	9·779	15·332	20·858	26·383	32·065	38·035	44·375	
84	−7·504	−3·952	4·401	9·890	15·442	20·968	26·494	32·181	38·159	44·503	
86	−7·556	−3·038	4·509	10·001	15·553	21·078	26·606	32·298	38·283	44·633	
88	−7·608	−3·124	4·617	10·112	15·663	21·188	26·717	32·414	38·407	44·761	
90	−7·660	−4·210	4·725	10·224	15·774	21·298	26·829	32·531	38·531	44·890	
92	−7·708	−4·294	4·834	10·335	15·884	21·409	26·941	32·647	38·655	45·018	
94	−7·756	−4·377	4·942	10·446	15·995	21·519	27·053	32·764	38·780	45·147	
96	−7·804	−4·461	5·050	10·558	16·105	21·629	27·165	32·880	38·904	45·276	
98	−7·852	−4·544	5·159	10·669	16·215	21·739	27·277	32·997	39·029	45·404	
100	−7·900	−4·628	5·267	10·781	16·325	21·849	27·389	33·114	39·153	45·532	

Table 5.8   REFERENCE TABLE FOR NICKEL CHROMIUM V. NICKEL ALUMINIUM
THERMO COUPLES*

E.M.F. values are given in millivolts (absolute) and temperatures in degrees C. (Int. 1948).
Cold junction 0°C. These values conform to the internationally accepted standards published
in American Bureau of Standards Circular 508 and British Standards Institution B.S. 1827:1952.

°C	−100	−0	0	100	200	300	400	500
0	−3·49	−0·00	0·00	4·10	8·13	12·21	16·40	20·65
2	−3·55	−0·08	0·08	4·18	8·21	12·29	16·48	20·73
4	−3·61	−0·16	0·16	4·26	8·29	12·38	16·57	20·82
6	−3·66	−0·23	0·24	4·35	8·37	12·46	16·65	20·90
8	−3·72	−0·31	0·32	4·43	8·46	12·54	16·74	20·99
10	−3·78	−0·39	0·40	4·51	8·54	12·63	16·82	21·07
12	−3·84	−0·46	0·48	4·60	8·62	12·71	16·91	21·16
14	−3·89	−0·54	0·56	4·68	8·70	12·79	16·99	21·24
16	−3·95	−0·62	0·64	4·76	8·78	12·88	17·07	21·32
18	−4·00	−0·69	0·72	4·84	8·86	12·96	17·16	21·41
20	−4·06	−0·77	0·80	4·92	8·94	13·04	17·24	21·50
22	−4·11	−0·84	0·88	5·01	9·02	13·12	17·33	21·58
24	−4·16	−0·92	0·96	5·09	9·10	13·21	17·41	21·67
26	−4·22	−0·99	1·04	5·17	9·18	13·29	17·50	21·75
28	−4·27	−1·06	1·12	5·25	9·26	13·37	17·58	21·84
30	−4·32	−1·14	1·20	5·33	9·34	13·46	17·67	21·92
32	−4·37	−1·21	1·28	5.41	9.42	13·54	17·75	22·01
34	−4·42	−1·28	1·36	5·49	9.50	13·62	17·84	22·09
36	−4·48	−1·36	1·44	5·57	9·59	13·71	17·92	22·18
38	−4·52	−1·43	1·53	5·65	9·67	13·79	18·01	22·26
40	−4·58	−1·50	1·61	5·73	9·75	13·88	18·09	22·35
42	−4.62	−1·57	1·69	5·81	9·83	13·96	18·17	22·43
44	−4·67	−1·64	1·77	5·89	9·91	14·04	18·26	22·52
46	−4·72	−1·72	1·85	5·97	9·99	14·13	18·34	22·61
48	−4·77	−1·79	1·94	6·05	10·07	14·21	18·43	22·69
50	−4·81	−1·86	2·02	6·13	10·16	14·29	18·51	22·78
52	−4·86	−1·93	2·10	6·21	10·24	14·38	18·60	22·86
54	−4·90	−2·00	2·18	6·29	10·32	14·46	18·68	22·95
56	−4·95	−2·07	2·27	6·37	10·40	14·55	18·77	23·03
58	−4·99	−2·13	2·35	6·45	10·48	14·63	18·85	23·12
60	−5·03	−2·20	2·43	6·53	10·57	14·71	18·94	23·20
62	−5·08	−2·27	2·51	6·61	10·65	14·80	19·02	23·29
64	−5·12	−2·34	2·60	6·69	10·73	14·88	19·11	23·38
66	−5·16	−2·41	2·68	6·77	10·81	14·97	19·19	23·46
68	−5·20	−2·47	2·76	6·85	10·89	15·05	19·28	23·54
70	−5·24	−2·54	2·85	6·93	10·98	15·13	19·36	23·63
72	−5·28	−2·61	2·93	7·01	11·06	15·22	19·45	23·72
74	−5·32	−2·67	3·01	7·09	11·14	15·30	19·54	23·80
76	−5·35	−2·74	3·10	7·17	11·22	15·39	19·62	23·89
78	−5·39	−2·80	3·18	7·25	11·30	15·47	19·71	23·97
80	−5·43	−2·87	3·26	7·33	11·39	15·55	19·79	24·06
82	−5·46	−2·93	3·35	7·41	11·47	15·64	19·88	24·14
84	−5·50	−3·00	3·43	7·49	11·55	15·72	19·96	24·23
86	−5·53	−3·06	3·51	7·57	11·63	15·81	20·05	24·31
88	−5·57	−3·12	3·60	7·65	11·72	15·89	20·13	24·40
90	−5·60	−3·19	3·68	7·73	11·80	15·98	20·22	24·49
92	−5·63	−3·25	3·76	7·81	11·88	16·06	20·31	24·57
94	−5·67	−3·31	3·85	7·89	11·96	16·14	20·39	24·65
96	−5·70	−3·37	3·93	7·97	12·05	16·23	20·48	24·74
98	−5·73	−3·43	4·01	8·05	12·13	16·31	20·56	24·83
100	−5·75	−3·49	4·10	8·13	12·21	16·40	20·65	24·91

*Also known as Chromel v Alumel when manufactured by Hoskins Manufacturing Co. Detroit, U.S.A.
or T$_1$ alloy v T$_2$ alloy when manufactured by Messrs. British Driver Harris, Ltd,

Table 5.8—*continued*

C.	600	700	800	900	1000	1100	1200	1300
0	24·91	29·14	33·30	37·36	41·31	45·16	48·89	52·46
2	25·00	29·22	33·38	37·44	41·39	45·24	48·96	52·53
4	25·08	29·30	33·46	37·52	41·47	45·31	49·03	52·60
6	25·17	29·39	33·54	37·60	41·55	45·39	49·11	52·67
8	25·25	29·47	33·63	37·68	41·63	45·46	49·18	52·74
10	25·34	29·56	33·71	37·76	41·70	45·54	49·25	52·81
12	25·42	29·64	33·79	37·84	41·78	45·62	49·32	52·88
14	25·51	29·72	33·87	37·92	41·86	45·69	49·40	52·95
16	25·59	29·81	33·95	38·00	41·94	45·77	49·47	53·02
18	25·68	29·89	34·04	38·08	42·02	45·84	49·54	53·09
20	25·76	29·97	34·12	38·16	42·09	45·92	49·62	53·16
22	25·85	30·06	34·20	38·24	42·17	45·99	49·69	53·23
24	25·93	30·14	34·28	38·32	42·25	46·07	49·76	53·30
26	26·02	30·23	34·36	38·40	42·33	46·14	49·83	53·37
28	26·10	30·31	34·44	38·48	42·40	46·22	49·90	53·44
30	26·19	30·39	34·53	38·56	42·48	46·29	49·98	53·51
32	26·27	30·48	34·61	38·64	42·56	46·37	50·05	53·58
34	26·36	30·56	34·69	38·72	42·63	46·44	50·12	53·65
36	26·44	30·65	34·77	38·80	42·71	46·52	50·19	53·72
38	26·53	30·73	34·85	38·88	42·79	46·59	50·26	53·79
40	26·61	30·81	34·93	38·95	42·87	46·67	50·34	53·85
42	26·70	30·90	35·02	39·03	42·94	46·74	50·41	53·92
44	26·78	30·98	35·10	39·11	43·02	46·82	50·48	53·99
46	26·86	31·06	35·18	39·19	43·10	46·89	50·55	54·06
48	26·95	31·15	35·26	39·27	43·17	46·97	50·62	54·13
50	27·03	31·23	35·34	39·35	43·25	47·04	50·69	54·20
52	27·12	31·31	35·42	39·43	43·33	47·12	50·77	54·27
54	27·20	31·40	35·50	39·51	43·41	47·19	50·84	54·34
56	27·28	31·48	35·58	39·59	43·48	47·26	50·91	54·40
58	27·37	31·56	35·67	39·67	43·56	47·34	50·98	54·47
60	27·45	31·65	35·75	39·75	43·63	47·41	51·05	54·54
62	27·54	31·73	35·83	39·83	43·71	47·49	51·12	54·61
64	27·62	31·81	35·91	39·90	43·79	47·56	51·19	54·68
66	27·71	31·90	35·99	39·98	43·87	47·63	51·27	54·74
68	27·79	31·98	36·07	40·06	43·94	47·71	51·34	54·81
70	27·87	32·06	36·15	40·14	44·02	47·78	51·41	54·88
72	27·96	32·15	36·23	40·22	44·10	47·86	51·48	
74	28·04	32·23	36·31	40·30	44·17	47·93	51·55	
76	28·13	32·31	36·39	40·38	44·25	48·00	51·62	
78	28·21	32·39	36·47	40·45	44·33	48·08	51·69	
80	28·29	32·48	36·55	40·53	44·40	48·15	51·76	
82	28·38	32·56	36·63	40·61	44·48	48·23	51·83	
84	28·46	32·64	36·72	40·69	44·55	48·30	51·90	
86	28·55	32·72	36·80	40·77	44·63	48·37	51·97	
88	28·63	32·81	36·88	40·85	44·71	48·45	52·04	
90	28·72	32·89	36·96	40·92	44·78	48·52	52·11	
92	28·80	32·97	37·04	41·00	44·86	48·59	52·18	
94	28·88	33·05	37·12	41·08	44·93	48·67	52·25	
96	28·97	33·13	37·20	41·16	45·01	48·74	52·32	
98	29·05	33·22	37·28	41·24	45·09	48·81	52·39	
100	29·14	33·30	37·36	41·31	45·16	48·89	52·46	

Table 5.9REFERENCE TABLE FOR COPPER V CONSTANTAN THERMO-COUPLES

E.M.F. values are given in millivolts (absolute) and the temperature in degrees C (Int. 1948). Cold junction 0° C.

These values conform to the data published in B.S. 1828:1961.

°C	−100	−0	+0	100	200	300
0	−3·341	−0·000	0·000	4·239	9·178	14·666
2	−3·397	−0·076	0·077	4·331	9·293	14·781
4	−3·452	−0·152	0·154	4·424	9·388	14·896
6	−3·508	−0·227	0·232	4·517	9·493	15·011
8	−3·562	−0·303	0·309	4·610	9·599	15·126
10	−3·617	−0·378	0·388	4·704	9·704	15·241
12	−3·670	−0·452	0·466	4·798	9·810	15·356
14	−2·724	−0·526	0·545	4·892	9·916	15·472
16	−3·777	−0·600	0·624	4·986	10·023	15·588
18	−3·829	−0·673	0·703	5·081	10·129	15·703
20	−3·881	−0·746	0·783	5·176	10·236	15·819
22	−3·933	−0·819	0·863	5·271	10·343	15·936
24	−3·984	−0·891	0·944	5·366	10·450	16·052
26	−4·034	−0·963	1·025	5·462	10·558	16·169
28	−4·084	−1·035	1·106	5·558	10·665	16·285
30	−4·134	−1·106	1·188	5·654	10·773	16·402
32	−4·183	−1·177	1·269	5·751	10·881	16·519
34	−4·232	−1·247	1·352	5·848	10·989	16·637
36	−4·280	−1·317	1·434	5·945	11·097	16·754
38	−4·327	−1·386	1·517	6·042	11·206	16·871
40	−4·374	−1·456	1·600	6·140	11·315	16·989
42	−4·421	−1·525	1·684	6·237	11·424	17·107
44	−4·467	−1·593	1·767	6·335	11·533	17·225
46	−4·513	−1·661	1·851	6·434	11·642	17·343
48	−4·558	−1·729	1·936	6·532	11·752	17·462
50	−4·603	−1·796	2·021	6·631	11·862	17·580

Table 5.9—*continued*

52	−4·647	−1·863	2·106	6·730	11·972	17·699
54	−4·691	−1·929	2·191	6·829	12·082	17·818
56	−4·734	−1·995	2·277	6·929	12·192	17·936
58	−4·776	−2·061	2·363	7·029	12·303	18·056
60	−4·818	−2·126	2·449	7·129	12·413	18·175
62	−4·860	−2·191	2·536	7·229	12·524	18·294
64	−4·901	−2·255	2·623	7·330	12·635	18·414
66	−4·942	−2·319	2·710	7·430	12·747	18·534
68	−4·982	−2·382	2·798	7·531	12·858	18·654
70	−5·021	−2·446	2·885	7·632	12·970	18·774
72	−5·060	−2·508	2·974	7·734	13·082	18·894
74	−5·099	−2·571	3·062	7·836	13·194	19·014
76	−5·136	−2·632	3·151	7·937	13·306	19·135
78	−5·174	−2·694	3·240	8·040	13·418	19·255
80	−5·211	−2·755	3·329	8·142	13·531	19·376
82	−5·247	−2·815	3·419	8·245	13·644	19·497
84	−5·283	−2·875	3·509	8·347	13·757	19·618
86	−5·138	−2·935	3·599	8·450	13·870	19·739
88	−5·353	−2·994	3·689	8·554	13·983	19·861
90	−5·387	−3·053	3·780	8·657	14·096	19·982
92		−3·112	3·871	8·761	14·210	20·104
94		−3·169	3·963	8·865	14·324	20·226
96		−3·227	4·054	8·969	14·438	20·348
98		−3·284	4·146	9·073	14·552	20·470
100		−3·341	4·239	9·178	14·666	20·592

Table 5.10   SPECTRAL EMISSIVITY OF MATERIALS, SURFACE UNOXIDIZED

The values given are for radiation of 0·65 microns wavelength, and are extracted from the tables given in *Temperature, Its Measurement and Control in Science and Industry* (American Institute of Physics, 1941) Rheinhold Publishing Co. The values relate to normal emission from plane polished or plane unoxidized liquid metal surfaces.

Material	Solid state	Liquid state	Material	Solid state	Liquid state
Carbon	0·80–0·93	—	Tungsten	0·43	—
Chromium	0·34	0·39	Steel	0·35	0·37
Copper	0·10	0·15	Cast Iron	0·37	0·40
Iron	0·35	0·37	Constantan	0·35	—
Manganese	0·59	0·59	Monel	0·37	—
Nickel	0·36	0·37	Chromel P (90Ni—10Cr)	0·35	—
Palladium	0·33	0·37	Chromel (80Ni—20Cr)	0·35	—
Platinum	0·30	0·38	,, (60Ni—24Fe—16Cr)	0·36	—
Rhodium	0·24	0·30	Alumel (95Ni—Bal. Al, Mn, Si)	0·37	—
Silver	0·07	0·07	90Pt—10Rh	0·27	—

Table 5.11   SPECTRAL EMISSIVITY OF OXIDES

The emissivity of oxides and oxidized metals depends to a large extent upon the roughness of the surface. In general, higher values of emissivity are obtained on the rough surfaces. The values given are for radiation of 0·65 microns wavelength, and are extracted from the tables given in *Temperature Its Measurement and Control in Science and Industry* (American Institute of Physics, 1941) Rheinhold Publishing Co.

Material	Range of observed values	Probable value for the oxide formed on smooth metal
Aluminium oxide	0·22 to 0·44	0·30
Beryllium oxide	0·07 to 0·37	0·35
Cerium oxide	0·58 to 0·80	—
Chromium oxide	0·60 to 0·80	0·70
Cobalt oxide	—	0·75
Columbium oxide	0·55 to 0·71	0·70
Copper oxide	0·60 to 0·80	0·70
Iron oxide	0·63 to 0·98	0·70
Magnesium oxide	0·10 to 0·43	0·20
Nickel oxide	0·85 to 0·96	0·90
Thorium oxide	0·20 to 0·57	0·50
Tin oxide	0·32 to 0·60	—
Titanium oxide	—	0·50
Uranium oxide	—	0·30
Vanadium oxide	—	0·70
Yttrium oxide	—	0·60
Zirconium oxide	0·18 to 0·43	0·40
Alumel (oxidized)	—	0·87
Cast Iron (oxidized)	—	0·70
Chromel P (90Ni—10Cr) (oxidized)	—	0·87
80Ni—20Cr (oxidized)	—	0·90
60Ni—24Fe—16Cr (oxidized)	—	0·83
55Fe—37·5Cr—7·5Al (oxidized)	—	0·78
70Fe—23Cr—5Al—2Co (oxidized)	—	0·75
Constantan (55Cu—45Ni) (oxidized)	—	0·84
Carbon Steel (oxidized)	—	0·80
Stainless Steel (18–8) (oxidized)	—	0·85
Porcelain	0·25 to 0·50	—

Table 5.12   TOTAL EMISSIVITY OF METALS (εt) SURFACE UNOXIDIZED

The values relate to normal emission from plane polished or plane unoxidized liquid metal surfaces. Reproduced from *Temperature, Its Measurement and Control in Science and Industry* (American Institute of Physics, 1941) Rheinhold Publishing Co.

Material	25°C.	100°C.	500°C.	1,000°C.	1,500°C.	2,000°C.
Aluminium	0·022	0·028	0·060	—	—	—
Bismuth	0·048	0·061	—	—	—	—
Carbon	0·081	0·081	0·079	—	—	—
Chromium	—	0·08	—	—	—	—
Cobalt	—	—	0·13	0·23	—	—
Columbium	—	—	—	—	0·19	0·24
Copper	—	0·02	—	(Liquid 0·15)	—	—
Gold	—	0·02	0·03	—	—	—
Iron	—	0·05	—	—	—	—
Lead	—	0·05	—	—	—	—
Mercury	0·10	0·12	—	—	—	—
Molybdenum	—	—	—	0·13	0·19	0·24
Nickel	0·045	0·06	0·12	0·19	—	—
Platinum	0·037	0·047	0·096	0·152	0·191	—
Silver	—	0·02	0·035	—	—	—
Tantalum	—	—	—	—	0·21	0·26
Tin	0·043	0·05	—	—	—	—
Tungsten	0·024	0·032	0·071	0·15	0·23	0·28
Zinc	(0·05 at 300°C.)					
Brass	0·035	0·035	—	—	—	—
Cast Iron	—	0·21	—	(Liquid 0·29)	—	—
Steel	—	0·08	—	(Liquid 0·28)	—	—

Table 5.13   TOTAL EMISSIVITY OF MISCELLANEOUS MATERIALS

(Most values are uncertain by 10% to 30%. In many cases value depends on particle size.) Reproduced from *Temperature, Its Measurement and Control in Science and Industry* (American Institute of Physics, 1941) Rheinhold Publishing Co.

Material	Temp. (°C.)	ε_t	Material	Temp. (°C.)	ε_t
Aluminium (oxidized)	200	0·11	Lead (oxidized)	200	0·63
	600	0·19	Monel (oxidized)	200	0·43
Brass (oxidized)	200	0·61		600	0·43
	600	0·59	Nickel (oxidized)	200	0·37
Calorized copper	100	0·26		1,200	0·85
	500	0·26	Silica brick	1,000	0·80
Calorized copper (oxidized)	200	0·18		1,100	0·85
	600	0·19	Steel (oxidized)	25	0·80
Calorized steel (oxidized)	200	0·52		200	0·79
	600	0·57		600	0·79
Cast iron (strongly oxidized)	40	0·95	Steel plate (rough)	40	0·94
	250	0·95		400	0·97
Cast iron (oxidized)	200	0·64		25	0·94
	600	0·78	Wrought iron (dull oxidized)	350	0·94
Copper (oxidized)	200	0·60	20Ni—25Cr—55Fe (oxidized)	200	0·90
	1,000	0·60		500	0·97
Fire Brick	1,000	0·75	60Ni—12Cr—28Fe (oxidized)	270	0·89
Gold enamel	100	0·37		560	0·82
Iron (oxidized)	100	0·74	80Ni—20Cr (oxidized)	100	0·87
	500	0·84		600	0·87
	1,200	0·89		1,300	0·89
Iron (rusted)	25	0·65			

Table 5.14

Correction to be added to the reading on an optical pyrometer when sighted on a surface of spectral emissivity e_λ. $\lambda = 0.65\mu$, $c_2 = 1.438$ cm deg.

The spectral emissivities of common substances are given in Tables 5.10 and 5.11, but where a high degree of accuracy is required the emissivity of the sighted surface should be determined. Reproduced from *Metals Reference Book* (1949) Butterworths Scientific Publications.

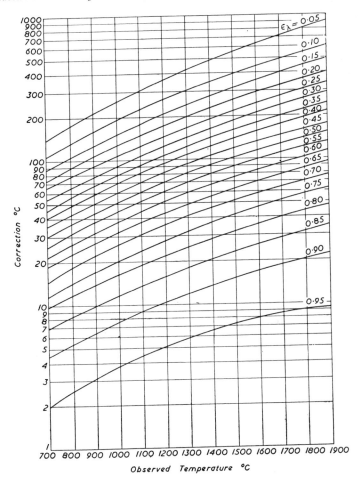

Table 5.15

Correction to be added to the reading on a total radiation pyrometer when sighted on a surface of total emissivity e_t.

The total emissivities of common substances are given in Tables 5.13 and 5.14, but where a high degree of accuracy is required the total emissivity of the sighted surface should be determined. Reproduced from *Metals Reference Book* (1949) Butterworths Scientific Publications.

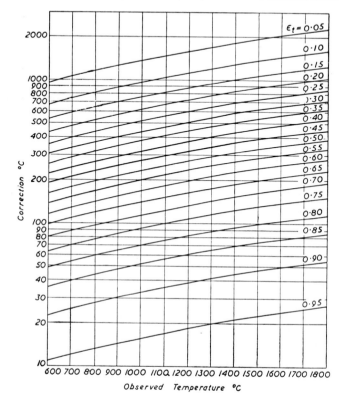

INDEX